D1600158

CLEANROOMS

Facilities and Practices

M. Kozicki

Arizona State University

with

S. Hoenig

University of Arizona

and

P. Robinson

Motorola, Inc., SPS

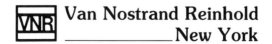 Van Nostrand Reinhold
New York

Copyright © 1991 by Van Nostrand Reinhold

Library of Congress Catalog Card Number 90-35627
ISBN 0-442-31950-9

Van Nostrand Reinhold
115 Fifth Avenue
New York, New York 10003

Chapman and Hall
2-6 Boundary Row
London, SE1 8HN, England

Thomas Nelson Australia
102 Dodds Street
South Melbourne 3205
Victoria, Australia

Nelson Canada
1120 Birchmount Road
Scarborough, Ontario M1K 5G4, Canada

16 15 14 13 12 11 10 9 8 7 6 5 4 3 2 1

Library of Congress Cataloging in Publication Data

Kozicki, M. N. (Michael N.), 1958-
    Cleanrooms, facilities and practices/by M.N. Kozicki, S.A.
Hoenig, and P.J. Robinson.
        p. cm.
    ISBN 0-442-31950-9
    1. Clean rooms. I. Hoenig, Stuart A.   II. Robinson, P.J.
(Patrick J.) III. Title. IV. Title: Clean rooms, facilities and
practices.
TH7694.K69 1990
696 — dc20   90-35627
                                                          CIP

Contents

5. Cleanroom Layout   80

6. Preconditioning, Control, and Static   111

Preface

In writing this book, our goal was to produce a much needed teaching and reference text with a fresh approach to cleanroom technology. The most obvious technological reason for bringing this book into being is that cleanrooms have become vital to the manufacture and development of high-technology products in both the commercial and military sectors, and therefore people have to develop an understanding of them. Examples of cleanroom applications include the manufacture of integrated circuits and other electronic components, precision mechanical assemblies, computer disks and drives, compact disks, optical components, medical implants and prostheses, pharmaceuticals and biochemicals, and so on.

The book is written for anyone who is currently involved, or intends to become involved, with cleanrooms. We intend it to be used by a wide range of professional groups including process engineers, production engineers, plant mechanical and electrical engineers, research engineers and scientists, managers, and so on. In addition, we believe it will be beneficial to those who design, build, service, and supply cleanrooms, and may be used as a training aid for students who intend to pursue a career involving controlled environments and others such as cleanroom operators and maintenance staff. We have attempted to steer clear of complex theory, which may be pursued in many other specialist texts, and keep the book as understandable and applicable as possible.

The subject matter of the book is broad. A multiplicity of topics is necessary as cleanrooms are extremely complex systems, and as such have a number of vital parts which make up the whole. We attempt to show the reader just exactly why cleanrooms are necessary in a manufacturing or research sense and how the problems of microcontamination may be lessened by the construction and utilization of appropriate controlled environments and by the regulated behavior of the personnel who work in and around these environments. Apart from the discussions of the many technological areas, for example, filtration and ultrapure water production, there are also treatments of subjects such as work force safety and legal issues. This text is unique because it covers the apparently unconnected areas of contamination control and hazard management. However, we contend that these areas are actually often highly interdependent and that knowledge of these diverse fields is necessary to enable one to operate in today's high technology industries in a fully effective manner.

Cleanrooms is principally about contamination; where it comes from, what it does in an industrial (or research) sense, and how we may construct facilities and utilize specialized practices to control it. Many of the concepts introduced are just good common sense; others have been the result of years of experience and research in the field. Above all, the book is intended to be practical, avoiding unnecessary theory and mathematics. Although we primarily use examples from the semiconductor industry, due to the direct experience of the authors in this technology sector, the information on facilities and practices is not specific to this area alone.

This book is designed to promote a better understanding of the basics of cleanroom technology. By doing this we hope to amplify the impact of this technology in existing and future industries and in research programs.

We would like to acknowledge the help of a number of people at Arizona State University, including Don Gerber for assistance with the cleanroom models, Ed Bawolek for the 209D worksheet, and Chris Ricciuti and Nagesh Sreedhara for assistance with the manuscript. We would also like to say "thank you" to our long-suffering wives (particularly Rhona K., who got the worst of it) for putting up with the entire process.

1

Introduction to Cleanroom Technology

THE NEED FOR CLEANROOMS

The Contaminated World

We live in a contaminated world. Contamination is material, usually in small quantities, which we do not want. There are a multitude of materials and processes, both biological and industrial, that are sensitive to contamination. The contamination arises from a number of sources, in natural and industrial categories. It may be obvious to us when it is in the form of dust and lint on a work surface, or it may be in the less distinguishable form of vapor in the air.

Manufacturing Yield and Product Reliability

For a component or system to be successful in a marketplace, it must be capable of being produced in sufficient quantities at costs which are competitive. Also, the component or system must be capable of performing its function throughout its intended lifetime. To produce units that meet these criteria, a knowledge of why high costs and unreliable devices occur is necessary. This is the realm of yield and reliability. *Yield* is the proportion of units which are functional at the end of the manufacturing process, expressed as a percentage of the total possible working units. *Reliability* is the quality of the unit to survive beyond its infancy and continue to operate for some time after this.

There are essentially three reasons why yield may not be 100 percent.

Poor product and process design can result in lower yields, for example, inappropriate specified materials or dimensions or tolerances that are too tight. Problems during processing, such as operator or equipment error, can also put product performance out of specification. However, the yield-reducing mechanism of interest to us in this text is disruption by contamination. This factor is discussed in detail in Chapter 2. We will further examine yield and reliability, specifically as applied to the semiconductor industry, in the next section of this chapter.

Even though we are emphasizing yield and reliability with regard to production processes, we should not lose sight of the fact that contamination can also have a significant impact on research work. Here we are not so concerned about economics or large numbers of working components, but we must be certain of the integrity of the results. If contamination is not taken into account here, we may end up basing large programs on erroneous information.

How do we deal with the problems of contamination in manufacturing and research? The answer to this question is not particularly easy (otherwise this book would be considerably shorter), but the key word is *control*. We must control contamination by the use of a *controlled environment*. The word *cleanroom* is often used for such an environment, although it can be somewhat misleading as control is not only applied to particulate and gaseous contamination but also to other physical parameters such as vibration and static electricity. To further complicate matters, we also have to control other factors, particularly work force behavior and operational practices.

AN OVERVIEW OF SEMICONDUCTOR TECHNOLOGY

The semiconductor industry is one of the largest users of cleanrooms. In this section we discuss the main aspects of semiconductor technology and many of the definitions used.

History

Originally, all electronic devices were discrete units existing as separate entities within their own packages (vacuum tubes, transistors, and so forth). The work of Kilby and then Noyce in 1958 gave rise to the technology which made it possible to create many devices arranged as an electrical circuit on a piece, or chip, of semiconducting material. As the technology matured, increasing numbers of devices could be integrated to form increasingly complex circuits. The levels of integration (with approximate chronology) are frequently given as:

Late 1950s—SSI (Small Scale Integration), a few tens of components per circuit

Late 1960s—MSI (Medium Scale Integration), several hundred components per circuit

Late 1970s—LSI (Large Scale Integration), a few thousand components per circuit

1980s—VLSI (Very Large Scale Integration), several tens of thousands to over one million components per circuit

1990s—ULSI (Ultra Large Scale Integration), several million components per circuit

and

VHSIC (Very High Speed Integrated Circuits), a military program to create extremely fast operating VLSI/ULSI circuits)

Larger scale integration produces circuits which are smaller, faster, more reliable, and cheaper. How do we achieve higher levels of integration? We make the devices smaller and pack them more densely.

Integration

Devices are essentially the component parts which make up a circuit. Discrete devices are individual components which may be brought together and interconnected to form a circuit. Each discrete device may be taken out and replaced if desired. The distinguishing feature of a device is that its behavior may be described in terms of voltage-current relations at its terminals. An electrical circuit is a collection of devices, interconnected by conductors (wires) in a specific configuration to perform a predetermined function; that is, produce a specific output for a particular input.

Printed wiring boards (PWBs), or printed circuit boards (PCBs), are copper-clad "boards" (for example, fiberglass) which hold and connect the components. The copper is patterned and etched to provide wires to interconnect components which are soldered directly to copper tracks. The patterning process is by lithography and etching; that is, the copper is coated with a photosensitive material which is exposed through a mask which has an image of the desired pattern of interconnections; the board is developed and the copper is left covered only by the photosensitive, chemically resistant coating on the areas where the tracks have to be formed. The board is then immersed in ferric chloride solution which etches away the uncovered copper but leaves the covered copper intact. The remaining photosensitive material is then removed by an appropriate solvent, and the board is drilled to accept the *leads* or connections of the discrete devices. The components (resistors, capacitors, transistors, and so forth) are inserted and soldered in place. The solder holds the components and also creates the electrical connection between component and copper track.

The PWB has been used since the days of vacuum tubes. Its main drawbacks are:

- For a complex circuit with many components the board can become unacceptably large—even splitting the circuit into many smaller boards involves great expense in inter-board connections,
- Reliability decreases drastically with circuit complexity, as the more soldered joints or inter-board connections there are, the greater the likelihood of a bad connection which will affect circuit performance,
- Mass production of boards is labor-intensive and requires expensive, complex equipment for component insertion and so forth.

To eliminate or reduce the problems with PWBs, we fabricate and interconnect all the devices on a semiconductor substrate. Devices are no longer discrete but part of an integrated circuit.

Planar technology allows the fabrication of complex electrical circuits on a semiconductor substrate. With the exception of high inductance components, all electronic circuit elements can be fabricated in semiconductor form. All devices are formed by performing operations or processes on one (flat) surface of a semiconductor substrate (hence "planar"). All devices are built up simultaneously. Since devices do not have to be individually packaged and interconnected on a PWB, circuit size is very much smaller and the circuits are considerably more reliable. Since interconnection lengths are much smaller, the resistance and capacitance associated with interconnects is reduced; this reduces the time delay for electrical signals and the circuits can therefore be faster. The technique lends itself well to mass production; hundreds of circuits may be fabricated simultaneously, therefore cost is reduced.

A measure of the technology is the *linewidth*—the smallest defined dimension within the circuit, that is, the defined width of a conductor or element within a device, usually measured in microns (1 micron = one millionth of a meter). In the original discrete devices, linewidth could be thousands of microns, SSI several hundred microns, MSI several tens of microns, LSI 3 to 10 microns, VLSI 1 to 3 microns, ULSI less than 1 micron. In essence, higher levels of integration demand smaller linewidths. Smaller linewidths may be achieved through a greater understanding and control of materials and fabrication technology. Higher levels of integration can allow circuits to be produced which would be impossible with any other technology due to cost and reliability considerations. However, the fabrication of integrated circuits would not be cost-effective if the yield was low. Yield is a function of the degree of control over the manufacturing facility and production practices.

The semiconductor substrates are actually flat, single crystal wafers or slices cut from a larger cylindrical crystal. The wafer diameter may be 1/2 inch to 12 inches, although 6 inches is currently common in U.S. production plants. Circuits are fabricated on one side of the wafer only. Since a typical integrated circuit is less than 1 square centimeter in area, many hundreds of ICs may be formed simultaneously on a single wafer. Wafers are processed in batches of 25 (typically). It is not uncommon to have 10,000 circuits being fabricated simultaneously. Once complete, the individual circuits on the wafer are separated from each other and packaged to form a usable IC.

Yield and Reliability in Semiconductor Processing

In practice, the yield in semiconductor processes may be close to 100 percent (all devices working) or close to zero. As with many production processes, there are essentially three categories of causes for yields to be less than 100 percent:

1. Parametric processing problems
2. Circuit design problems
3. Random point defects

Processing techniques are not ideal, and variation of process parameters is not uncommon. Process variations may lead to the nonfunctioning of circuits if they occur in critical areas. Process variations include:

- Variations in thickness of grown or deposited layers (gate oxide, poly-Si, and so forth)
- Variations in the resistance of diffused or implanted layers (source-drains, threshold adjust, and so forth)
- Variations in the width of lithographically defined lines and features
- Variations in the registration of a photomask with respect to previously defined layers

All of these variations could render a circuit inoperative if they fall outside tolerance limits, for example, if a gate oxide is too thin, then the device characteristics will be severely altered and the device may even break down when a voltage is applied to the gate. Variation in a process parameter can be uneven across a wafer. It is possible for one half of a wafer to contain perfect devices while none of the devices on the other half works.

Poor circuit design can lead to sensitivities of the circuit to process parameters which are within the tolerance of the process. This can happen if too little thought has been given to the correlation between variations in different process parameters, for example, the circuit may fail due to the worst-case combination of process variations.

Point defects are small (much less than the size of the IC) regions of imperfection. These may be caused by slight imperfections in the crystal substrate formed when the crystal is grown or cut, or created when the substrate is heated or mishandled. A further source of point defects is particulates. Tiny particles from the air (even in a cleanroom), from the materials used, and even from the processing equipment will create defects. These defects may form during the creation of patterns (photolithography) when the particle acts as a mask, the particles may be burned-in during a furnace process and chemically contaminate the wafer, or they may disrupt the continuity of a thin deposited layer causing pinholes. We discuss this more in Chapter 2.

A process problem need not render a circuit inoperative immediately. It may take hours or years before any effect is noticed. There are also other processes which can cause a circuit to fail in operation. The number of failures against time is usually represented by a bathtub (u-shaped) curve. Infant mortality, or early breakdown, is high due to processing and assembly problems. Old age victims are mainly due to wear-out. Time-dependent failure mechanisms include:

- Contamination (especially mobile alkalai metal ions) diffusing into gate oxides
- Electromigration (the thinning of metal tracks under current flow) causing conductor failure
- Corrosion of interconnects and bonds by ingress of chemicals into leaky packages
- Growth of unwanted compounds at contact points between metals

All of these processes are accelerated by increasing the temperature. Other mechanisms are:

- Electrostatic discharge (static) which will "blow" the gate oxide
- Mechanical shock (vibration) which can disrupt bonds
- Radiation damage which can create transient errors in MOS memories

OTHER CLEANROOM APPLICATIONS

Probably the first serious user of the cleanroom concept was the aerospace industry. Miniaturized components for satellites needed to be assembled in an atmosphere that would assure that no contamination could cause failure when the vehicle was placed into earth orbit. Cleanrooms are still extensively used for the testing of space vehicles, the assembly of instrumentation components of space vehicles, and other fields such as research in medical biophysics. Cleanrooms are employed in research and educational institutions for medical and pharmaceutical research. They are also used in the fields of biology, chemistry, electronics, and physics. Many modern universities have cleanrooms on campus. The pharmaceutical industry employs large cleanrooms for the production of contaminant-free pharmaceuticals and biological materials. The radioactive materials industry uses cleanrooms for the segregation of materials to eliminate radioactive contamination in research and industry. They also use them for the production of fuel rods and related materials and assembly of components for radioactive devices. Other users of cleanroom technology include the food industry, plastics industry, and photographic industry.

THE IMPACT OF CLEANROOM TECHNOLOGY

The concept of the cleanroom has had an impact upon many existing industries and has given birth to new ones. It is safe to say that many of these industries would not exist in their present form if it had not been for advances in the technology of controlled environments.

Due to the contamination-related problems inherent in semiconductor processing, without cleanrooms we would never have seen the types of products which are now commonplace, for example, the personal computer would still be a dream in a designer's head, unable to be manufactured at reasonable cost. Medical implants and many ultrapure, bacteria-free preparations would also be less common. The aerospace industry and military contractors know only too well how cleanroom technology has changed their existence. Failure in those businesses can be disastrous, not to mention expensive.

2

Microcontamination

TYPES OF CONTAMINATION

Organic and Inorganic Substances

We mentioned in Chapter 1 that contamination is, effectively, material which we do not want. It is worthwhile pointing out at this stage that what is contamination in some processes is actually benign or even beneficial in others. We may divide contamination into a number of categories, but perhaps the most obvious division is between organic and inorganic materials.

Inorganic materials can be natural or man-made solids, liquids, or gases. They can be elemental (consisting of one element of the periodic table), compounds (chemical combinations of two or more elements, such as sodium chloride), or mixtures such as metal alloys (for example, brass). Organic substances frequently have natural sources, although they may also be synthesized industrially. They are always compounds of carbon and may also be solids (for example, plastics), liquids (for example, alcohols), or gases (for example, methane). The elements in the compounds besides carbon are usually hydrogen, oxygen, and halogens such as chlorine, although other elements can also be included (for example, metals).

Complex chemical structures are typical of organic contaminants if they are of biological origin. They may be as basic as low molecular weight hydrocarbon chains or as complex as groups of whole cells. The latter form is particularly interesting, as cells, whether of plant or animal origin, are loaded with a large variety of compounds which in turn are composed of a number of different elements. Some organic contamination may even be alive and growing if it is in the form of bacteria.

To further illustrate the diversity of contamination, we may put contaminants into an alternative three categories:

1. Particulate materials. These entities might consist of metal fragments, silica from damaged semiconductor wafers, droplets of solidified organics

(for example, photoresist materials), lint from clothing, skin flakes, and human hair. Clearly every effort must be made to reduce the number of these particles to an absolute minimum.

2. Bacteria, fungi, or other materials generated by biological processes. In this area we must recognize that bacteria and fungi can grow on a wide variety of materials in light and darkness. The residue of these growth processes can be highly toxic or sufficiently alkaline to etch materials. In any case, they will contain oils or chemicals that can interfere with sensitive manufacturing processes. There is no way of eliminating bacteria or fungi, in that they are everywhere, but we can limit their rate of growth to the point where they do not present a severe problem.

3. Gases and vapors. Condensed organics can serve as growth media for bacteria or fungi or a variety of materials can evaporate and then condense as films or particles. The magnitude of this problem is only now beginning to be appreciated.

Airborne Particulates

As may be seen from the previous section, all materials in any form may potentially be regarded as contamination. Since contamination is usually in small quantities compared to whatever is being contaminated, solid contamination typically will take the form of small particles in a range of sizes. Visible particles are in the order of 50 microns (note that a human hair is around 100 microns in diameter). Bacteria range from 0.3 microns to a few tens of microns, face powder is in the range 0.5 micron to 10 micron, and smoke ranges from 0.01 micron to 1 micron. Depending on the location and form of the source, particles may be transported to our sensitive site by air, water, process chemicals, and gases, or on the surfaces of process solids, packaging, clothing, skin, and so forth. The particles will become contamination in a material when they land on that material and are somehow incorporated. We will investigate various transport mechanisms in later chapters, but here we will look at airborne particles.

Once particles are generated and released from their source materials, they may be carried by air currents generated by meteorological conditions (wind), movement of people or equipment, or convection (rising hot air). Not all particles can be transported this way, as the drag forces exerted on the particle surfaces by the air must be greater than the force on the particle due to gravity, the latter being dependent on the mass of the particle (gravitational force = mass × acceleration due to gravity, g). The drag forces are not so easily represented and will depend on a number of factors such as the density and velocity of the air and the nature of the particle, for example, roughness, surface to volume ratio.

A simple rule of thumb for particles in still air is that solid, near-spherical particles greater than 1 micron in diameter will settle quite quickly under the force of gravity. This is rather important, as this effect will result in the removal of around 99.9 percent of these larger particles from still air; and this should be taken into account when testing for airborne particulate contamination. Particles smaller than 1 micron but larger than 0.1 micron in diameter will settle very slowly and will be carried readily by even slight air movements.

An interesting departure from our simple particle movement theory occurs for particles less than 0.1 micron in diameter. At this size, even in totally still air, these particles will tend to remain suspended almost indefinitely, even though there are no moving air drag forces, as such, acting. The mechanism here involves the motion of the gas molecules. In air at room temperature, the gas molecules are moving with an average velocity (dependent on temperature) in random directions. It is this movement, or more specifically the collisions of these moving molecules with surfaces, which produces pressure. If the surface in question is a small particle, which has a large surface to volume or weight ratio, these collisions will push the particle and prevent it from settling to the ground. This effect can actually be seen with smoke, brightly lit so that the particles scatter light and may thus be viewed through an optical microscope. The particles "dance around" in the still air as they are hit from all sides by the air molecules; the effect is known as Brownian Motion.

Gases and Vapors

We have concentrated so far on particulate contamination, but it must be stressed that low-level contamination need not be localized into a small mass such as a particle. It may also be dispersed in the form of a gas or vapor, the latter originating from a liquid surface. As with particles, the gaseous contamination may also be transported in moving air by mixing, or in still air by diffusion. Transport due to mixing will depend on the velocity and turbulence or the airflow, and transport due to diffusion depends on the diffusion rate of the gas in air, which in turn depends on the type of gas and the concentration gradient.

In order for gaseous substances to become contaminants in a material, they have to make contact and become incorporated in that material. In the case of true gases, the reaction has to occur from the gas phase, for example, the oxidation of a surface, which is the chemical reaction of oxygen with a material, can be considered to be contamination of a material from the gas phase. Vapors may react from the vapor phase, but they may also condense as a thin layer of liquid or semisolid on the surface, for example, oil vapors may recondense on a cool surface to create a thin oil film.

Contamination in Liquids

Contamination in various forms may be transported by liquids as well as by air. We will discuss the contaminants in water in Chapter 9, as this is a special case, but it is worthwhile to introduce contamination in liquids here.

If particles are created or fall into a liquid, they may be readily transported by that liquid in much the same way as air transports them. However, in this case the density and viscosity are considerably higher and hence the drag forces are larger. Therefore, liquids may transport larger particles than air at the same velocity. A further problem with particles in liquids is that the particle material may be soluble to some extent in the liquid, for example, many minerals dissolve in water and some organic substances in alcohols. This disperses the contamination and makes its removal very difficult. Gases and vapors may also become dissolved in liquids, and, of course, unwanted liquid materials may also contaminate liquids (for example, alcohols in water). Liquid contamination need not mix with the water to be transported (for instance, most oils do not mix with water but will be carried with it).

SOURCES OF CONTAMINATION

External Natural Sources

As one may expect, the natural environment is a huge source of particulate and gaseous contamination. The thousands of years of freeze–thaw cycles which create the deserts of the world also provide billions of tons of fine surface contaminants of various sizes which may become airborne in even light winds to give contamination control personnel major problems. The particulates may be made up of a vast range of minerals. The most abundant will be various forms of silica (oxides of silicon), as silicon is the most abundant element in the earth's crust. There will also be various other oxides, sulfides and sulfates, carbonates, nitrates, and so forth. In a general sense, the types of contaminants will vary with the region, for example, desert or beach areas will have mainly silica and little else, whereas farm areas will have larger proportions of natural nitrates from the soil. Although there are not too many on the surface of the earth, active volcanos will contribute considerable amounts of airborne sulfur-containing compounds. Be aware that global weather conditions can provide us with a few surprises; there is a documented case of a light red dust appearing in London in large quantities which was analyzed and found to have come directly from the Sahara Desert in northern Africa.

Since most of the earth's surface is covered with salt water, we must not underestimate its contaminative power. Breaking waves will produce a fine saltwater spray. If the facility of interest is near to an oceanside region, this spray will constitute particulate (although not solid) contamination. The spray may also ultimately form small, solid, salt particles when the water content is reduced by evaporation.

As many allergy sufferers will tell you, plants can also contribute to environmental contamination in a big way. The main element here is pollen, which is produced by virtually all plants in vast quantities as part of the reproductive process. Once again, the material is readily carried on the breeze and consequently may be found just about anywhere on earth. Dead plant and animal cells may also be introduced into the air if the material dries and is ground up or abraded. Other significant biological materials that are prevalent are spores, molds, and bacteria.

Natural sources can also produce inorganic and organic gas and vapor contaminants. Volcanic areas, including hot springs, and coal and other mineral mines will provide significant local levels of gases such as hydrogen sulfide and sulfur dioxide. Marsh areas by lakes and rivers will tend to produce gases such as methane by the action of bacteria on decaying organic matter. Whether these gases are a significant contamination source will depend very much on the local concentrations, the proximity of the manufacturing plant, and of course the sensitivity of the process. Many people, on reading this section, will immediately deny that there is any problem with naturally produced gases in a contamination sense. If you are one of the unconvinced, a further example may be necessary. Think of the outcry over the accumulation of the radioactive gas radon in homes, offices, and enclosed factories. This particular gas is a health hazard to humans but, as we will see on page 23, such poisonous gases can also be a threat to products.

External Industrial Sources

The industry of the world is also a large generator of external contamination. Once again, we may partition the contamination into particulate and gaseous forms. One of the largest types of industrial particulate contamination is carbon-based fly ash and carbon particles (soot) produced by the combustion of fossil fuels such as coal and oil. Other large-scale industrial processes can produce particulate contamination; for example, the copper-smelting process can produce some exotic looking particles composed of copper compounds. Parking your car near a steel foundry is a good demonstration of local high-level particulate contamination; as you will probably return to a vehicle coated with a fine layer of iron oxide dust.

The burning of fossil fuels in facilities such as power plants also produces vast amounts of gases such as carbon dioxide and sulfur dioxide. Both gases are produced in sufficient amounts to increase global or continental levels leading to an apparent greenhouse effect, or global warming, in the case of carbon dioxide and acid rain in the case of sulfur dioxide. Another gaseous contaminant of global concern is the class of materials called fluorinated hydrocarbons, for example, Freon 14 (CF_4). These gases are released from industrial, domestic, and automobile refrigerant systems and are blamed for the destruction of the earth's ozone layer. On a scale that is closer to home, the levels of carbon monoxide and ozone in cities are increased to unhealthful levels by gasoline-burning car engines. In many cases, high levels of gaseous contaminants can be localized in a geographical sense. This localization is caused by the local intensity of the source (for example, cars jamming city streets), the local geography (for example, cities like Phoenix and parts of

Los Angeles are surrounded by mountains and hence gases are held in a "bowl"), or by a combination of factors such as geography and prevailing winds. A good example of highly localized gaseous contamination was the case of a semiconductor plant which was plagued by strange smells within the cleanroom. The cause was a camping gas (propane) bottle storage yard 100 yards away which was releasing gas in legal amounts, and this was then being carried by the prevailing winds into the plant's air-conditioning system.

As with external natural sources, the actual concentrations of particulate and gaseous contamination in the outside air depend very much on the geographical location. Maximum particle sizes are generally in the region of a few hundreds of microns (mainly plant material in windy areas) or a few tens of microns (organic and inorganic materials in calm areas) maximum. It is not uncommon to see several thousands of these larger particles per cubic foot of air. However, due to the ease of transport of smaller particles, the numbers of these smaller entities are inevitably higher, for example, several million per cubic foot of around 0.5 micron in size.

Personnel

One highly significant source of contamination internal to the production environment is the personnel who work in that environment. Personnel are critical for two very good reasons: (1) humans are animals and as such are biological contamination "factories," and (2) the operators are frequently the closest thing to the actual elements we wish to keep free from contamination. We will deal with many aspects of the latter factor throughout this text, but we will take a closer look at "dirty" people in this section.

One point we should make clear at the outset is that one should not take offense at being called dirty. Even the cleanest people produce huge amounts of contamination as part of the process of living. The most significant contamination produced by humans is skin flakes. The skin we are covered with is there to protect us from the environment we live in. It is actually a very severe environment in a great many respects (we have just discussed some of the contamination floating around out there), especially when we consider the fact that we are regularly bombarded by harmful ultraviolet rays which can actually destroy cells and tissue. The human body defends itself by sacrificing its outermost layer of skin and allowing it to become disposable. This may keep us alive, but it is a nightmare from a contamination control viewpoint. The dead cells flake off when the skin is gently abraded and will be carried off by the convection currents surrounding the body or channeled through the clothing. Since the cells are organic, they are complex in a chemical sense, and they also carry other substances which are naturally found on the skin, for example, salt microcrystals (mainly sodium chloride from perspiration) and skin oils.

The number of skin flake particles shed depends upon the state of the skin and the activity of the subject in question. If the skin is moistened, for example, by perspiration or by the application of a moisturizing skin cream, the skin flakes will tend to stick to the surface better. Unfortunately, body heat will result in the drying out of the skin and the subsequent release of the cells. Dry cells will also tend to fracture into many small particles. A good rule of thumb is that a motionless person will generate some 100,000 particles of approximately 0.3 to 0.5 microns diameter per minute. This figure changes dramatically when the subject begins an activity, as the increased abrasion of clothing against skin and limbs against body, as well as the constant expansion and contraction of the flexible skin, will result in the expulsion of a greater number of particles. A person sitting at a workbench moving just the arms and body will generate on average 1,000,000 particles per minute. If the whole body is in motion, for example, during walking, this number increases by at least a factor of five.

Skin flakes are not the only contamination that the human body produces. Small flakes of the protein keratin are released from the hair. Hair itself will also be released from all over the body. The act of breathing produces a great many contaminants. In addition to large quantities of carbon dioxide and water vapor, there will be droplets of water with dissolved salts and suspended biological material (for example, parts of cells from the inside of the mouth). This respiration-related contamination can be projected large distances from the body, carried by the exhaled air. A cough or sneeze can project the contaminants at high velocity onto critical surfaces. Even the eyes can cause problems, as the tear ducts regularly produce a saline solution to wet the eyes. When we blink, small droplets are splashed out of the eyes.

Unfortunately, we occasionally make the human contamination factor worse by adding to the materials which may be shed from the surface of the skin. For example, most cosmetics have a powder base which is mixed with an appropriate liquid carrier for ease of application. Once this carrier evaporates, we are left with a semi-dry material which will either flake off by itself or be carried on skin cell particles. For instance, mascara and other eye cosmetics contain large amounts of fine carbon particles which can be shed with time. The colorants in cosmetics are a veritable periodic table with such elements as zinc and iron regularly appearing. Although this may appear to be an attack on the more fashion conscious woman, men can be just as guilty, as they also wear powder-based deodorants and antiperspirants, the latter containing aluminum compounds. The use of aftershave is also problematic, as the alcohol content will dissolve the natural skin oils and thereby allow easier shedding. There is also a reasonably well-defined relationship between smoking and particle generation. We will return to this in Chapter 11.

Equipment and Facilities

A further internal source of contamination is the equipment used in the manufacturing process. Once again this is a particularly significant source, as equipment will be in close proximity to the sensitive product in many cases. The construction materials of the facility and the plant equipment (air-conditioning systems, compressors, water pumps, and so forth) will also contribute.

Many pieces of equipment contain moving components. Any moving part is a potential source of contamination, especially if friction is involved (and it usually is). Pistons and other components which make sliding contact between surfaces will generate particles by abrasion. This loss of material manifests itself as wear in the components over a long period. Lubrication or the use of reduced friction components helps the problem but will not make it go away altogether. In some cases, lubrication can make the contamination situation worse, as the lubricant can be expelled from the component as fine droplets. Gears are also a particle source, as the meshing of the teeth involves friction. Gears are also attached to rotating shafts which are held by some sort of bearing. Any rotating element in a bearing will produce contamination in much the same way as sliding components, even with rolling bearings. Another somewhat overlooked source of contamination in equipment is belts. Belts are not only used as mechanical drive connections but also to convey components. Since belts are merely loops of flexible materials, they are typically directed by rollers or pulley wheels. When the belt is stretched around a pulley, material can be ejected from the surface due to the deformation. For example, if a rubber material is used it may be treated to seal the surface. Unfortunately, the treatment can make the surface slightly less flexible and thus it will flake off when the material is stressed.

Much the same observation can be made regarding plant equipment, as this is typically large-scale rotating machinery with an assortment of shafts, pistons, gears, and belts. Although it may not be in close proximity to the product, it will still add contamination to the working environment. Another form of contamination which industrial plants add is not particulate or gaseous but vibration. One does not usually think of vibration as contamination but in many processes, especially in the semiconductor industry, vibration can be a great reducer of yield. We will deal with vibration more fully in Chapter 7.

Nonmoving materials in equipment can also be a contamination source. Coated materials such as painted steel can cause problems when the paint is dried out by heat or chemical action and begins to flake off. Most paints will outgas for many years after their application due to the retention of small amounts of their liquid component. This effect gets much worse if the surface is heated. Even "clean" metal surfaces are a contamination source, as the process of forming the components (casting, machining, and so forth) will leave residual particles. For example, a heat-treated "stainless" steel can still have microparticles of iron oxide (and other metal oxides) embedded in the surface. These can subsequently be released by mechanical shock. Since the equipment is typically stored in a warehouse prior to installation and transported in less than clean trucks, it will become dirty, and this dirt will be carried into our production environment unless a strict decontamination procedure is followed.

Since many pieces of equipment contain vacuum systems, we should not leave them out of this discussion. Most rotary pumps used to "rough" vacuum chambers and oil diffusion pumps are filled with oil. Small concentrations of this oil will diffuse out of the pump and into the process chamber unless a suitable oil trap is used. Oil mist will also be introduced into the work environment outside the equipment if the rotary pump outlet is not properly exhausted. Components inside the vacuum chamber will also outgas under vacuum. This outgassing is actually a release of the material adsorbed on the surface when the chamber was exposed to room air, mainly water vapor and condensed hydrocarbon vapors. Keep in mind that even though the air has been removed from the system, particles can still be transported by the action of static charge.

We will look closely at construction materials and techniques in Chapter 4, but it is important to realize that most common building materials are a major source of particulate and even gaseous contamination. Many materials, like brick, sheetrock, and particle board, are formed from particles, and it is relatively easy to re-release these back into the environment. Fibrous natural materials such as wood are bad, as they release cells and resins. As mentioned previously, painted surfaces may not be suitable due to outgassing. Sealants and adhesives used to join panels are also prime outgasers.

Production Materials and Processes

In many cases production materials and processes can be sources of contamination. Just getting the materials into the production environment can introduce contamination from external sources, for example, particles can be carried in on packaging which has become dirty in the warehouse or in transit. This problem is different from the equipment packaging problem, as materials used in the process will be brought into the facility regularly and hence the effect will persist as long as operations continue.

The materials themselves may be a source of contamination, as avoiding too much material or material in the wrong place is just what contamination control is about. For instance, production gases may leak during a cylinder change or acid or solvent vapors may escape from a fume hood into the air during an etching process. Hot components, such as silicon wafers which have newly emerged from a diffusion furnace, may emit vapors of dopant metals. Some processes are inherently dirty and create byproducts which are undesirable, for example, many plasma-etching processes which involve relatively harmless reactants can produce extremely corrosive by-products. The formation of polycrystalline-silicon layers by chemical vapor deposition can lead to particulate contamination from the silicon, which deposits on the process tube and is subsequently scraped off and powdered when the wafers are unloaded. Many other deposition processes are also sources of particulate contamination, for example, in the evaporation or sputtering of metals, layers build up on chamber walls until they flake off.

Improper Work Practices

No introduction to microcontamination would be complete without mentioning the contribution from improper work practices. This is a very large area in industrial contamination control and one which we will return to frequently in this text. Since we have established that people, equipment, and processes can be very dirty, personnel can play a large part in reducing or increasing the effects of this contamination on the product. The most important factor in work practices with contamination-sensitive materials is an awareness of the problem; this is a case where "ignorance is not bliss." That is why we place such a great emphasis on training in the chapters to come.

Some classic examples of improper work practices are: not cleaning equipment regularly, not cleaning up spills as they occur, handling product or equipment without proper contamination control equipment, poor equipment maintenance techniques (the list is, unfortunately, virtually endless).

EFFECTS OF CONTAMINATION
Circuit Yield and Reliability

Linewidths in integrated circuits are growing smaller; oxide thicknesses in MOS devices are changing from 20 or 30 nm to around 10 nm. A rule of thumb which has been established in the integrated circuit industry is that "killer" particles are those which are one tenth of the minimum linewidth in the circuit. Looking at Figures 2.1 and 2.2 it is clear that our concern about particulates in the tens of nm size range is well placed. Why is yield and reliability so dependent on contamination in integrated circuit manufacture? Contamination can affect the integrity of the materials and structures in a number of ways. We know quite a lot about the effects of particulate contamination, as our discussion will reveal. However, the effects of low-level vapor contamination are a bit more of a mystery due to the dispersion factor. One certain fact is that as dimensions shrink, we will see a greater emphasis placed on the control of this form of contamination.

One of the most significant yield-reducing mechanisms occurs during photolithography operations. If a particle lands on a wafer or photomask during lithography, it may disrupt the circuits being formed by two means:

1. Light shining through the mask will be blocked or diffracted by particles which are larger than the light wavelength, and thus they will form an image in the photoresist. The image in the resist will ultimately be trans-

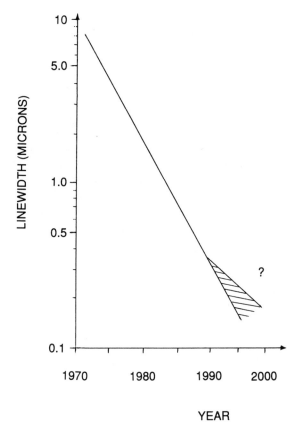

Figure 2-1   Minimum semiconductor integrated circuit linewidth as a function of year.

ferred to whatever layer is being patterned. As an example, if a metal layer is being patterned using a positive resist, then the image of the particle will appear as an extraneous metal area. This extra area could short-out two adjacent metal tracks. If negative resist is used in the same case, then holes or nicks will appear in the metal lines. Any narrowing of current-carrying lines will result in eventual failure due to *electromigration*, an effect which further narrows the metal at weak points. The higher current density at these points allows the electrons to sweep the metal away and hence further thin the area until it eventually fails. This failure could occur during testing or burn-in or later in the field. It should be noted that if the particle is on the mask, we can get a defect which repeats on many wafers. This can be a very expensive problem.

2. If a contact lithography process is used and a particle is trapped between the mask and wafer, as the patterns are aligned the particle will be ground

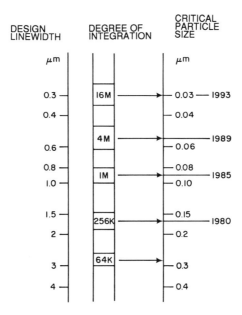

Figure 2-2 Correlation between degree of integration and the nominally critical particle.

in to both surfaces, and this will cause mechanical damage to both mask and wafer. Since the alignment process involves relative movement of the mask and wafer, a comparatively large area can be damaged.

Fortunately, contact processes are giving way to projection processes, and these problems may thus be reduced. The mechanical damage aspect is removed, and particles on the mask (or reticle in this case) may be defocussed by use of a transparent stand-off called a *pellicle*. However, we still have the problem of particles on the wafer. Our other form of contamination, vibration, is also a killer in lithographical operations. These operations require precise alignment of one layer to another, with tolerances of a fraction of a micron. Vibration in equipment will tend to result in relative movement of the mask and wafer, and hence images become blurred and registration is poor.

Contamination will also reduce yield in a number of other fabrication steps, including furnace, deposition, etching, and even ion-implantation processes. In hot furnace processes, organic particles, such as skin flakes, on the wafer surface will melt and decompose (hydrogen and oxygen are removed) to leave carbon and other nonvolatile elements on the surface, which may then be incorporated into a grown oxide or diffuse into the wafer. Elements from inorganic particles, for example, mineral salts, can diffuse into the wafer and remain there indefinitely. Any elements such as carbon will interfere with the bulk or interfacial structure of the semiconductor; these defects can alter device specifications and hence reduce yield. Mobile ions from mineral salts such as sodium will have a longer term effect on electrical characteristics, as they are able to diffuse through oxide layers under the influence of an electric field at room temperature. Therefore, device characteristics will change slowly until the device eventually fails.

Particles which land on a wafer prior to the deposition of a layer (for example, aluminum) will become trapped by the deposited layer. This may cause discontinuities or small holes which can cause circuit failure immediately or after some time. Alternatively, a fully trapped particle may outgas during subsequent heating and can blister or crack the overlying film. In the longer term, a trapped particle may react chemically with the circuit materials and corrosion can occur.

Particulate presence during etching can preferentially mask a region of a layer to leave an unetched area (this is true for both wet and plasma etching). Similarly, larger particles can mask ion implantation so that an area does not receive the correct number of ions. A high current beam will also melt organic materials onto the surface or the wafer. Apart from the direct effect that

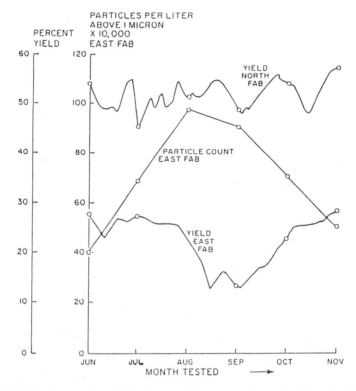

Figure 2-3 Yield vs. time and DI water particle count at two wafer fabrication facilities (from J. Gilliland, Hewlett Packard).

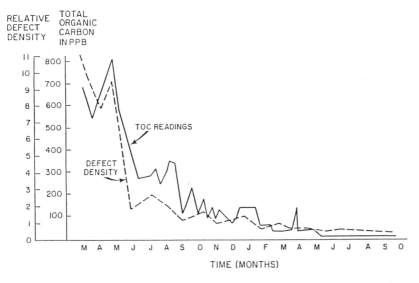

Figure 2-4 Correlation between total organic carbon in DI rinse water and the relative defect density in a VLSI manufacturing system vs. time (from *Proc. 5th Annual Semiconductor Pure Water Conference,* Jan. 1986).

contamination can have on wafers carrying the particles, the equipment can also become contaminated by this contamination and transfer it to subsequent wafers and batches. For instance, if particles are carried into an acid etch bath, all batches processed in that bath after that will become contaminated.

A big aspect of the contamination and yield problem is associated with the materials used for semiconductor manufacturing. If the chemicals, gases, and so forth, are not of the highest purity, yield will be low. Two examples of the effect of water contamination on yield are shown in Figures 2.3 and 2.4. There is a definite correlation between particle counts or total organic carbon (which we will discuss in more detail in Chapter 9) and yield or defect density. Finally, Figure 2.5 illustrates empirically the relationship between defects per square inch and probe yield for a typical product. The "real life" results reinforce our theories only too well.

Mechanical Breakdown

Contamination can also have a profound effect on the production and reliability of precision mechanical systems. Early experiments in rocketry showed how critical it was to avoid contamination in units such as the valves which control the cryogenic propellants. Tiny particles in the mechanism often led to catastrophic in-flight failures such as sticking valves (thereby demonstrating that what goes up can come down sooner than expected). As mentioned

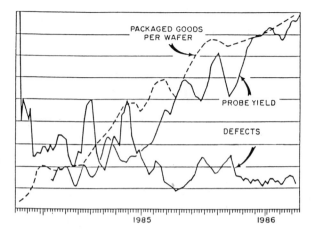

Figure 2-5 Defects per square inch, probe yield, and packaged goods shipped per wafer vs. time for a large area circuit. Four-week moving averages (from *Microcontamination*, April 1987).

previously, this is why the aerospace industry had such a large role in the development of cleanrooms.

An excellent example of how critical contamination can be in a mechanical sense is the magnetic disk industry. Particles on or in the surface of a disk can cause "head crashes," where the read/write head makes contact with the uneven disk surface. This usually results in a permanent loss of data and occasionally damage to the head. Vapor contamination can also be an extreme nuisance in industries which deposit metal coatings on components, as in the manufacture of compact disks (CDs) where a thin metal film is deposited on a plastic disk by evaporation or sputtering. If condensed oil vapors are present on the surface of the plastic before metal deposition, the surface film will reduce the metal adhesion and the metal will thus flake off, rendering the CD useless.

Biomedical and Pharmaceutical Problems

It should be readily apparent that dealing with biologically active agents in the laboratory atmosphere is a risky business to start with. The idea of handling virus or bacterial strains capable of causing degenerative disease processes at all is a prospect most would rather not face. In the laboratory, pure strains of agents capable of causing cancer and other deadly disease must be cultured for testing and research purposes. In this case the cleanroom assures that these disease-causing agents are effectively restricted to an area. Cross contamination could be disastrous to research projects and delay them for years.

In the pharmaceutical industry the same care for cross contamination is demonstrated as in the biomedical laboratory. We expect that when our physician prescribes a medication for us, the pharmacist will fill the prescription with the purest possible medication, certainly nothing deadly. The care in compounding pharmaceutical formulations and products requires the use of cleanrooms to assure that cross contamination does not occur between manufacturing areas.

Relating Product Health to Work Force Health

The driving need for contamination control for product safety or reliability dovetails nicely with the need to protect the worker from exposure to hazardous materials used in the cleanroom environment. Exposure to biological materials or pharmaceutical compounds can cause a biological response in the workers in the cleanroom. In the microelectronics industry where hazardous materials are handled in the cleanroom, safety regulations require the employer to limit exposures to this material in order to prevent injury to people. In humans, there are established legal limits of exposure to toxic or harmful agents. These limits of exposure are based upon either animal experiments or upon actual experience with accidental exposure to people.

What levels of contamination are significant with regard to the product being manufactured or assembled in the cleanroom? With regard to humans the limits are generally well specified. Arsenic, used as a doping material in the semiconductor industry, has an exposure limit of 0.2 milligrams per cubic meter of air. This is the exposure limit for humans, and it is expressed as a *time weighted average,* meaning that a worker may have excursions above the limit provided there is exposure below the limit as well to produce an eight-hour average exposure not in excess of that number. Safety professionals and industrial hygienists in the industry observe and measure exposures to arsenic in this setting to assure that employees are not overexposed. If time-weighted average exposures are significant, engineering controls are prescribed to control the exposure problem.

Suppose that this air-monitoring data were to be integrated with the yield data from the production operation. If we were to consolidate the exposure-monitoring data with variations in product quality, we might well find that there is a direct relationship between quality problems in the workplace and contaminant measurements taken by the safety organization. Given that there is no exposure limit for silicon used in the microelectronics industry to build semiconductor devices, we have little chance to relate airborne contamination levels directly to the "health" of the product being produced. Since arsenic causes electrical changes in silicon, we can measure the result of exposure in terms of the quality of the electrical changes beyond the parameters expected by the process. Given that changes are occurring in the produced material, the same engineering controls used to protect the worker can frequently be employed to protect the product. In this manner, the safety and health organization becomes an adjunct of the quality control organization and can play a significant role in providing a measurement of product quality in situ.

SUMMARY

In summary, we may divide contamination into a number of different categories. When considering air, we are most concerned with small particles, gases, and vapors, as these may be transported easily from their sources to our contamination-sensitive areas. Liquids, such as the water used for cleaning, can transport more diverse forms of contamiantion, for instance, it may carry particles, other liquids (even if they do not mix with water), dissolved solids, and dissolved gases. We may also place sources of contamination into external and internal, natural and industrial categories. In many cases, the most significant internal source of contamination is the people who work in our sensitive environment. Finally, contamination can have a devastating effect in sensitive processes in electrical and mechanical engineering and in the medical industries.

3

Controlled Environment Concepts

In the last chapter we examined types, sources, and effects of contamination. We are now ready to begin to discuss how we may keep this contamination away from our sensitive materials, structures, and processes. It is perhaps worthwhile at this point to introduce the idea of the *contamination pathway*. This pathway consists of the following elements: source, transport, contact (with a sensitive area), and retention. We can break the pathway at any stage. We will see how we can do this in the next three chapters. This chapter is an introduction to the concept of controlled environments in general.

PERFORMANCE CONSIDERATIONS

Cleanroom Parameters

In setting up a controlled environment, the first question we must ask is, "Exactly what do we want to control?" If our controlled environment is a so-called cleanroom, it is obvious that our first priority will be the control of contamination, both particles and vapors. Therefore we must have some control over the intake and generation of contamination in our controlled area. We must also be able to measure the results of our control action, for example, in order to judge how effective our control of particles is, we have to be able to measure particle densities. There is one point we must note and keep in mind at all times. We usually have a reason for desiring this control, for example, we may wish to keep substrates free of particulate contamination to maximize yield. The main object of our attention must therefore be that substrate. It is easy to get into the kind of situation where much effort is expended in the room without much benefit at the substrate, for example, we know of instances where vast sums of money were spent

25

on replacement air filters, when a new type of operator glove would have produced a greater improvement in yield. A recurring reminder in this book is to find out what will give you the biggest improvement at the wafer and do that first.

In many cleanroom environments, we are interested in controlling other parameters. One such parameter is temperature. It is an extremely good idea to control temperature in any production area for two reasons: the staff will not work efficiently in freezing or sweltering conditions, and so temperature must be adjusted for comfort, and many systems and processes are temperature-sensitive. This latter factor has two aspects: some equipment will overheat if the ambient temperature is too high, leading to a destructive condition known as thermal runaway in electronic systems, and temperature changes of more than one degree can result in expansion or contraction of mechanical components, leading to offsets in position-sensitive equipment such as projection aligners. Relative humidity (rh) is also a controlled parameter, for both operator comfort (dry air is very unpleasant to breathe for long periods as it dries out the mouth and throat, and humid work areas are uncomfortable) and for control of static electricity (more humidity means less static build-up as the charge can leak away more easily). Some chemical processes, such as certain photoresist reactions, even need a certain level of humidity in order to proceed.

Room air pressure and velocity are also parameters which must be controlled in many cases. Keeping a room at positive pressure is an important weapon against the ingress of contamination, as air is forced out of openings such as doors and unsealed joints, rather than being sucked in, thereby making it difficult for contamination to flow or diffuse in. This task may not be as simple as it sounds, as pressurization may have to be performed in a dynamic fashion, that is, if large doors are opened periodically to bring in raw materials, the flow of air into the room may have to be adjusted to maintain positive pressure. Dynamic air balancing can be tricky: there is a possibility that ceilings may be blown out by control systems which have not responded quickly enough to changing conditions. Air velocity is also critically important, as we will see in Chapter 4. Under certain circumstances, airflow which is too slow or too fast can result in a lack of control over contaminant densities in our controlled volume.

Ideal vs. The Real World

What are the problems in setting up and maintaining a controlled environment? Unfortunately, the problems are many. We will take a look at some of them here.

First, we have to establish a volume which we will (attempt to) control. We can do this by enclosing an area with hard walls or flexible soft sheets supported on an appropriate frame. We also must have a floor and a ceiling. Problem number one arises with building materials and the techniques we use to erect the structure. Will the material act as a source of particulate or gaseous contamination from the surface or cut edges? Will material contamination problems become worse with time? If the material is suitable with regard to contamination, does it still meet building and fire codes? If we can find a clean material which we can legally use, we now have to supply air to our room. Since we know that ambient air is highly contaminated, we must clean it before putting it into the room. To take out particles, we can use a filter. However, there is no such thing as a 100 percent efficient filter because we have the dichotomy of having to create a physical barrier which traps particles but still allows air to pass freely. Therefore, a compromise is in order.

The situation becomes considerably worse when we put personnel, equipment, and processes into our controlled volume. Since people are tremendous sources of contamination, we either have to keep them out or put them in protective clothing. Equipment should also be covered, but we still have to get the raw materials in and the processed items out. Solutions are often difficult.

PERFORMANCE STANDARDS

Particulate Standards

The task of developing any standard is usually a difficult one, requiring co-operation among, input from, and agreement among a number of different groups who have a vested interest in the topic. In the case of cleanrooms, the Federal Government in the United States has produced appropriate standards. The relevant standard is FED-STD-209, *Clean Room and Work Station Requirements, Controlled Environments,* and at the time this book was being written, we were on revision D. A copy of this standard is reproduced in Appendix I at the end of this text. (As testimony to the difficulty of producing standards, revision B lasted for many years, whereas revision C only lasted 8 months.) The standard is designed to be understandable in its own right, and so we will not go into too much detail here, other than to emphasize the main points.

The first and most obvious aspect of 209D is that it applies only to particulate contamination and not to gases and vapors. This latter class of materials is discussed in the next section. The second point to note is that it spends a great deal of time discussing how and where to measure particle concentrations, emphasizing particle-counting techniques and statistical methods. This is vitally important, as 209B had nothing to say in this respect and hence two different people could come up with two completely different determinations of "cleanliness." (This allowed unscrupulous cleanroom contractors to bend the measurement data to whatever they wanted it to say by making measurements only in the cleanest areas and not where the work was actually going to be done.)

The two most important features of 209D as far as we are concerned are Table 1 and Figure 1, reproduced here as Table 3.1 and Figure 3.1. These are effectively definitions of cleanroom *class*. For instance, we can see that in the definition of class 10, we must count the numbers of 0.5, 0.3, 0.2, and 0.1 micron particles, and these numbers cannot exceed 10, 30, 75, and 350 per cubic foot of sampled air respectively. However, for class 10,000, we are concerned only with 5.0 and 0.5 micron particles, and these should not be present in numbers greater than 70 and 10,000 per cubic foot respectively. The common particle size in all classes is 0.5 micron, and it is the numbers of these particles which give the class its numerical value, that is, class 100 has no more than 100 particles of 0.5 micron per cubic foot of air, class 1 has no more than 1 particle of 0.5 micron per cubic foot, and so forth. Note that whenever we talk about particle size, we are really referring to the maximum linear dimension, for example, if the particle was a cuboid, the size in this context would be the length of the longest side. The reason the classes are based on 0.5 micron particles is essentially historic, as this was the size that commercially available particle counters could readily count (in large numbers). Now we can readily count 0.1 micron particles, and the latest standard reflects this. Looking at Figure 3.1, we can see that if measurements at the specified particle sizes fall below a class line, then the cleanroom is attaining that particular class. Other classes, such as class 50, may also be defined in much the same way by drawing a line through 0.5 microns at 50 particles per cubic foot parallel to the other lines on the graph. From 209D, if the new class is greater than class 1,000, the particle sizes to be measured are 5.0 and 0.5 micron; classes between 1,000 and 10 should use 0.5, 0.3,

Table 3.1   Class Limits in Particles per Cubic Foot of Size Equal to or Greater than Particle Sizes Shown (micrometers)

Class	Measured Particle Size (Micrometers)				
	0.1	0.2	0.3	0.5	5.0
1	35	7.5	3	1	NA.
10	350	75	30	10	NA.
100	NA.	750	300	100	NA.
1,000	NA.	NA.	NA.	1,000	7
10,000	NA.	NA.	NA.	10,000	70
100,000	NA.	NA.	NA.	100,000	700

[From *Fed. Std.* 209D.]

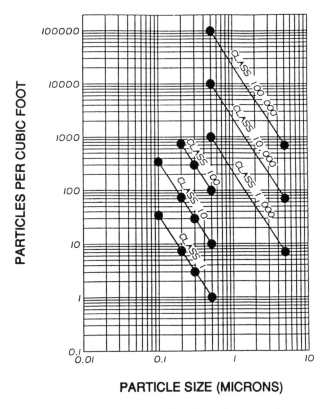

Figure 3-1   Class limits in particles per cubic foot of size equal to or greater than particle sizes shown (from *Fed. Std. 209D*).

and 0.2 micron, and classes less than 10 should use 0.5, 0.3, 0.2, and 0.1 micron.

The number of samples to be taken is also defined in 209D and depends on the physical situation. If the flow is laminar or unidirectional (both of these mean nonturbulent; Chapter 4), the minimum number of sample locations is determined by the area of the entrance plane (in square feet) divided by 25 or divided by the square root of the class designation, whichever turns out to be the smallest number. For example, if the smoothly flowing laminar air enters a work space volume, nominally defined as class 100, with a cross-sectional area of 100 square feet (upstream of the work area), we should make a minimum of four measurements (100 divided by 25). If the volume is class 10,000, we need only make one measurement per 100 square feet (100 divided by the square root of 10,000, which is smaller than dividing by 25). For non-unidirectional airflow (turbulent), the minimum number of sample locations is the square footage of the floor area divided by the square root of the class designation. These multiple measurements are then used in the statistical formulae given in the standard to produce a meaningful error-corrected determination of cleanroom class. For an example of how to use these formulae, see Appendix C of 209D, in Appendix I.

We may use the information and techniques given in the standard to certify the class of our cleanroom with respect to particulate contamination. A full certification procedure would be used to determine whether the cleanroom meets the intended class after construction or retrofitting work. However, this may prove too time-consuming for routine monitoring, especially in large high cleanliness rooms, hence the standard allows for a watered-down version of the procedure for more frequent use. A simplified version of the procedure is given in Appendix II in the form of a worksheet to aid in the monitoring/certification process.

Gas and Vapor Standards

When using hazardous production materials (HPM), it is an unfortunate but nevertheless inescapable fact that exposure of the work force is inevitable, especially in the case of gases and vapors. No system is foolproof (or leak-proof) enough to prevent this. There will always be some quantity of HPM in the work environment. We have to ensure that the levels present cause no harm to the work force or the product. Traditionally, Federal Standard 209 does not fully address the question of gaseous contamination, and this leaves us with a question, that is, "How much *is* too much?"

Fortunately, we do not have to generate "how much is too much" data as it is done for us from an industrial toxicology rather than a contamination control viewpoint. U.S. Government organizations such as the National Institute for Occupational Safety and Health (NIOSH), part of the U.S. Department of Health and Human Services (which also looks after the Public Health Service and the Centers for Disease Control), are set up to perform the necessary laboratory, literature, or incident research work to assess material hazards. From reported data, the American Conference of Governmental Industrial Hygienists (ACGIH) produces its Threshold Limit Values (TLVs) for various substances. The TLV really tries to tell us how much of a particular substance is safe in a work environment. Since data regarding human exposure is generally scarce (the little data of this type which is available comes from accidents, warfare, and the like, and thus the experimental conditions are not well controlled), laboratory animal studies and biochemical modelling are frequently used. The damage thresholds are determined for specific organs using monkeys, guinea pigs, and so forth, and the dose which would cause the same effect in humans is determined by extrapolation. This is an imprecise technique, therefore TLVs have a large scaling safety margin built in. The limits set for many substances are still a point of contention in some circles.

There are actually a number of forms of the TLV. TLV-TWA (time weighted average) is a time weighted average for exposures over 8-hour days in a 40-hour work week. TLV-STEL (short term exposure limit) considers a 15-minute TWA with 60 minutes between exposures and no more than four exposure periods per day. The STEL is therefore higher than the TWA, as the exposure period is much shorter, but it should not be used independently of the TWA. Both deal with a time weighted average. This is not a license to grossly overexpose the workforce for a short time and then remove them from the area so that the average exposure is below the TLV, but it does allow for fluctuations in levels which naturally occur during the work day. The third TLV is the TLV-C (ceiling) which is the concentration which should not be exceeded during any part of the working exposure. The TLV-C is usually applied to very hazardous substances which can cause an immediate detrimental effect. Updated TLVs are published by the ACGIH in an annual booklet which also contains other information on work hazards such as cold stress, vibration, lasers, and ionizing radiation. It should be noted that the TLVs are scientifically determined *recommended* limits and are not legal limits. TLVs for gases and vapors (a vapor originates from a liquid surface) are generally given in parts per million (ppm) in air, whereas for dusts (solid particles), fumes (solid particles which condense from a gas), and mists (liquid droplets) the measure is mg/m^3 of air. The TLVs are given for pure or single substances as very little is known or understood about the effects of mixtures due to the increased number of experimental variables they present.

The legal limits are the PELs (permissible exposure limits) which are set by the Federal and State Occupational Safety and Health Administrations (OSHA). OSHA has two main functions with regard to industry; regulatory/enforcement and advisory. The TLVs are generally adopted as the PELs with some exceptions where OSHA adopts a lower value when there is some doubt regarding the accuracy of the interpretation of data. OSHA enforces the PELs in industry. (Federal OSHA has little to do in states which have a strong State OSHA.) It should be noted that compliance with OSHA regulations is an absolute minimum a company should do. Noncompliance is seen as negligence and is now resulting in prison sentences for company executives in fatal accident cases. OSHA will always advise when possible on how to meet compliance.

We should say at this juncture that the limits set for industrial hygiene purposes have been superb in protecting sensitive products. This is because human beings are very sensitive mechanisms, and contamination levels must be low enough to prevent any detrimental effects on personnel. However, as we move into the realms of quantum structures, for instan , superlattice devices, in which we throw away the ruler and start counting the atoms instead, the levels set for humans will probably be too high to protect the product. A rethink of the technology is therefore probably in order!

PARAMETER MEASUREMENT

Microscopic Techniques

The most basic tool for the detection of particulate contamination is the human eye. The eye is surprisingly sensitive to particles on smooth surfaces, and anything over 50 microns in diameter can usually be resolved in strong illumination by a person with good eyesight. The oldest technique used in contamination control is the well known bright light inspection system where the surface to be inspected is illuminated at an oblique angle and a human inspector looks for scattered light from topographical features (that is, particles sticking up from a flat surface). If the system is set up properly and the inspector is experienced, bright light inspection can be quite satisfactory. Of course, we do not get any quantitative data about particle size, but we can get a good idea of the numbers of "big" particles.

The sensitivity of this technique can be increased for organic materials, such as tiny pieces of lint, by using low intensity ultraviolet light. In this case, organic particles will tend to fluoresce, that is, give off light in the visible range, and therefore will become more obvious to the eye (we will return to the finer points of fluorescence later). If a clean flat plate, such as a silicon wafer or polished glass slide, is left on a surface in the region of interest, it will collect particles which are large enough to settle. The number of particles collected by this witness plate will be equal to the product of the particle fallout rate and the dwell time of the plate in the region. The plates are typically marked with appropriate grids so that the particles may be counted per unit area. The fallout rate per unit area can then be determined quite easily by dividing the count by the dwell time. This is an important measurement as it provides an indication of what is actually landing on a surface as opposed to what doesn't land. In many cases what doesn't land is not of interest.

To improve the witness plate technique, we need to increase the range of the human eye, and the simplest way to do this is to employ some method of magnification. An optical microscope is the usual tool, as 1-micron particles may be readily resolved. With a good quality microscope we may also begin to see features on the particles and measure them so that contamination characterization and identification become possible. The importance of this will be discussed later in this chapter. Federal Standard 209D describes the use of optical microscopy, particularly with regard to counting 5-micron particles trapped in sampling filters. In this technique, air is drawn through a membrane filter and the particles become trapped on the smooth surface of the filter medium. This medium is then placed under a binocular microscope, typically $250 \times$ so that 5-micron particles can be resolved. Appendix A of Federal Standard 209D (See Appendix I) gives more complete details of the measurement techniques required by this standard. Note that this is also an excellent method for determining gross particulate contamination from garments. The area to be tested is vacuumed and any particles which are drawn off become trapped in an appropriate filter (for example, a 0.4 micron pore size membrane filter).

If optical microscopy lacks the desired level of resolution, one may turn to scanning electron microscopy. In this technique, the area of interest is illuminated with a scanned electron beam. An appropriate detector is employed which gathers the electrons scattered from the surface and converts them to a signal which may then be used to create a meaningful (topographical) image on a monitor. Since the effective wavelength of electrons is much shorter than light, a far greater resolution is possible. The disadvantages of this technique from a practical viewpoint are that a higher degree of skill and training is required to operate the instrument, and the sample has to be placed in a high-vacuum chamber. The latter factor reduces the ease of use of the technique. Also, certain organic materials, such as skin flakes, will be damaged (fried) by the electron beam if high voltages are used. Higher energy electrons will also damage sensitive materials (for example, gate oxides in MOS devices), and so the technique may be somewhat too destructive for in-line inspection of certain products.

All of the above techniques involve the collection of particles on a witness plate of some type and the subsequent manual counting of these. Whereas this can be very useful in many respects, it is painfully slow and prone to error by weary/inexperienced operators or accidental contamination from other regions during moving and counting. A more instantaneous method, using some automation technique, is considerably more desirable.

Light-scattering Techniques

When we, as observers, view particles on a surface, our eyes are receiving scattered light from these topographical features. There is a great deal of physics and mathematical theory associated with this effect, and we may use some of it to allow us to create automated particle counting and, to some extent, characterization systems.

In the semiconductor industry, optical scattering systems using lasers are often used. The units sweep the laser beam across the surface of a wafer and collect the scattered light. The information on scattered light intensity as a function of scattering angle is sent through a computer program that relates the data to what is called *scattering theory*. This in turn yields information on particle size and location on the wafer. Readers who use the laser scattering data should be aware that scattering theory involves a number of assumptions, the chief of which is that the particle is a smooth sphere with known electromagnetic properties. In the actual application of these instruments, the above assumptions are seldom fulfilled. The particles are not spheres, they may not be smooth, and they may vary widely in color and electromagnetic properties. This suggests that laser scanning data be used to compare various levels of contamination before and after a process but care should be exercised when considering absolute measurements of particles on the surface. This is particularly the case with small (submicron) particles on a wafer that may have a layer of silicon dioxide. The theory predicts very small signal to noise ratios for particles less than 0.5 microns in diameter. If there is a layer of silicon dioxide on the substrate, the particle light scattering may be greatly distorted by scattering in the silicon dioxide.

Many of the same comments apply to the use of laser or bright light monitors for the detection of airborne (not surface) contamination. These systems take in a measured volume of air and pass it through a chamber in which laser light or bright white light is scattered off any particles into a detector (Figure 3.2). It is thus a relatively simple task to count large numbers of particles automatically. Scattering theory is again used to provide an indication of the particle size. In this case there is no substrate to interfere, but the particle shape, color, and so forth, is still an unknown. Here again the laser system tends to have a very low signal to noise ratio for the particles below 0.5 micron. Large particles and fibers do not go through the laser system at all. Just because you don't see them with an optical counter doesn't mean they are not in the environment. The volume in which the actual sensing takes place is chosen so that coincidence is minimized, that is, the probability of more than one particle being at the detection point at any one time is less than 5 percent. The best commercially available systems are usually reasonably good down to 0.1 microns (as long as their limitations are kept in mind). One way of increasing the range of light-scattering systems is to make the smallest particles bigger. We do this by condensing an organic vapor around them. Under the right conditions, sub-tenth micron particles will act as nuclei for the condensation process in air saturated with an appropriate organic material. Such systems are therefore known as condensation nuclei counters (CNC).

Other problems with light-scattering monitors for measurements of airborne submicron particles involve the need to use very low airflows to allow the particles a long time in the scattering region. If you are measuring less than ten 0.5 micron particles per cubic foot, you have to run for a long time to

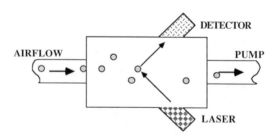

Figure 3-2 Light-scattering technique for the detection/counting of airborne particles. Light from the source is scattered by particles in the beam into the detector.

get statistically valid data. The problem here is that conditions in cleanrooms change rapidly so that long testing periods are not really practical. It therefore seems that there is a need for other particle detection technologies.

Other Detection Methods

In the discussion above we noted the need for new evaluation technologies that were not operator sensitive and could be used by untrained personnel. The more astute reader will have noticed that we also limited our discussion to particles and have said little about condensed vapors. This was quite deliberate, as the technology for routine evaluation of this type of contamination is not nearly as mature as particle detection (some would say the technology is nonexistent). In this section we present some developing technologies for the evaluation of surface contamination.

The use of ultraviolet (uv) light induced emission of electrons as a detection system for organics on surfaces was first reported by NASA. The principle of the technique is simple. When the relatively energetic ultraviolet light strikes a material, electrons may be emitted, depending on the frequency of the uv light and the target material, that is, the energy of the photons (equal to Planck's constant multiplied by the photon frequency) has to be greater than the *work function* (a physical constant) of the target. This is known as the

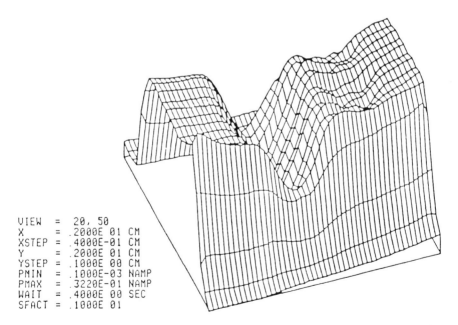

```
VIEW  =  20, 50
X     = .2000E 01  CM
XSTEP = .4000E-01  CM
Y     = .2000E 01  CM
YSTEP = .1000E 00  CM
PMIN  = .1000E-03  NAMP
PMAX  = .3220E-01  NAMP
WAIT  = .4000E 00  SEC
SFACT = .1000E 01
```

Figure 3-3 Photoelectric scan data on a floppy disc with an oil streak (from PAT, Inc., Newbury Park, CA).

photoelectric effect. Commercially available systems make use of this effect by irradiating the sample with an ultraviolet light source and collecting the emitted electrons by means of an electrode near the surface. An example of the data obtained with such a system is shown in Figure 3.3. A unit of this type might well be used for qualitative surface evaluation before and after cleaning. Unfortunately, there is a problem with the present generation of detection systems based on the ultraviolet/electron emission technology. The difficulty involves the measurement of the very small dc electrical currents emitted from the target, in that slight temperature changes can result in significant baseline drift. This is not a severe problem in the environmentally controlled laboratory where trained technicians are available. However, it almost precludes the use of the system in the factory where skilled technicians and controlled conditions may not be present. An alternative method is to use ac chopping and phase-locking techniques for the measurement of the photoelectric currents. This approach is currently under development. ac chopping and phase-locking systems have been shown to be essentially drift free and are therefore well suited to this application. Figure 3.4 shows some test results with such a system. It is apparent from these results that it may also be possible to identify the types of surface contamination to some degree, as different surface materials affect the way electrons are emitted.

One system that has been investigated in the United States and Japan involves the use of a corona discharge. The corona discharge is known to pulse as dust particles pass through the discharge region as the presence of the particle alters conduction in the discharge path. A corona system of this

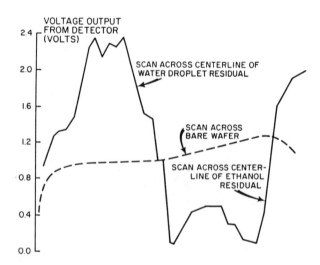

Figure 3-4   Change in photoelectric current, in terms of voltage, as a function of surface contamination (A.C. System).

type can detect nanometer-sized particles and thus goes beyond any light-scattering techniques. Another alternative laser technique has been pioneered in Japan. In this case, the scattered laser light is imaged by a vidicon to present a real time representation of the dust particles on the surface. With a unit of this type it is possible to observe the dust particles as they fall on the surface and to monitor the dust removal process. This system might well be adapted for evaluation of a variety of surfaces before and after cleaning.

Contamination Identification

In evaluating or improving a cleanroom, the first question should always be "What contaminants are present?" This is a very big question indeed, for if we know what the contamination is, we can attempt to reduce its presence. As in many disciplines, prevention is always better than cure, thus eliminating the sources of contamination is the ultimate form of contamination control. Unfortunately, this is not always possible, but identifying the contamination source can also give us a clue to how it is transported to our controlled area. Knowing the route, we can break the pathway accordingly.

Our experience is that very few users have any idea of what contaminants are actually present in the cleanroom. In some cases particle counts are done, but the users do not appreciate that the counter will only see small (under 10 micron) spherical particles. If there are 100-micron lint fibers, they will never get through the dust counter. We have done experiments that involve putting clean microscope slides at various locations in the cleanroom, letting them sit for an hour or two, and then looking at them with a microscope. The results are that the major contaminant is lint from personnel clothing and overalls. In another example, we have frequently found dried photoresist in many areas in the cleanroom and traced it to the spin-bake system. Spin-bake systems are supposed to be vented, but the appearance of dried photoresist in the cleanroom is a good sign that ventilation can be inadequate. Optical microscopy can be a powerful technique in contamination identification (as long as the contamination is big enough to be seen). The best people at this job are actually forensic scientists, who tend to be more active in solving crimes than solving cleanroom problems. However, using the techniques developed by the McCrone organization (which specializes in the art of forensic identification and classification), most people can become quite proficient at solving the "cleanroom crime."

Another thing the microscope will tell you is the state of cleanroom discipline. If you find a lot of human hair, there is no question that head coverings are not being properly used. If you find blue lint from personnel clothing, it is a sign that the bunny suits are not being worn properly. If you find metal shavings or lint from bunny suits, it is a clear indication that there is a problem, and you don't have to guess about the cause. The point here is that a little microscopy can go a long way in helping you to clean up the facility. If bunny suits are the problem, putting money into better air filters in the ceiling will not be very effective.

Analysis of cleanroom particulates is not limited to just looking at them with a microscope. Microchemical analysis can tell you what elements or compounds are involved. At one time the type of equipment required for analysis of particulate contamination on surfaces was either too complex or too expensive for routine on-line use. Now, many facilities have at least one scanning electron microscope for high resolution inspection. An SEM is extremely useful for topographical analysis, but it cannot give any chemical information. However, analysis systems can be added to many SEMs. An excellent addition is an energy dispersive x-ray analysis (EDXA) system. The energies of the x rays generated by the interaction of the electron beam and the target material depend on the elements present. Thus, under computer setup and control, a relatively inexperienced operator can obtain an elemental spectrum of a contaminant sample. EDXA tends to work best for the heavier elements and the sensitivity is reasonably good (around 0.1 atomic percent). If lighter elements are involved, wavelength dispersive x-ray analysis would be more suitable. Alternatively, if greater sensitivity is required, other techniques such as auger electron spectroscopy or x-ray photoelectron spectroscopy would be required. These latter techniques are not as user friendly and

Table 3.2 Particle Type Analysis by Area

Particle Type	Number of Particles per 100 cu ft Areas		
	Gowning	Inspection	Processing
Organic	1038	267	59
Alumino-Silicate	159	23	27
SiO_2	53	26	36
Al_{2O3}	40	3	13
KCl	35	23	23
Talc	28	6	6
$CaCO_3$	26	7	9
Gypsum	23	2	——
Brass	14	8	——
Dolomite	——	2	——
NaCl	——	——	1
Rubber	23	1	——
PVC	——	——	1
Cr_{2O3}	——	1	1
FeOx	——	——	6
$CuSO_4$	——	2	1
Other	25	5	34
Total	1472	354	231

From *Proc. 1987 Microcontamination Control Meeting* by T. B. Vander Wood and J. M. Rebstock, p. 127.

tend to require support personnel for operation and interpretation. In Table 3.2 we present some data from a cleanroom facility to show the power of chemical analysis. This information can be of critical importance in the improvement of a facility. For example, it is evident from this table that body salts, including potassium chloride (KCl), are not being effectively controlled (by the garments), as they appear in near-uniform quantities in each of the three test areas (gowning, inspection, and processing). It is also evident that cosmetic products are adding considerable amounts of particulates to the gowning area (talc and aluminum compounds).

Fluorescence can also be used in an analytical sense. An example of the type of data that can be obtained with fluorescence is shown in Figure 3.5.

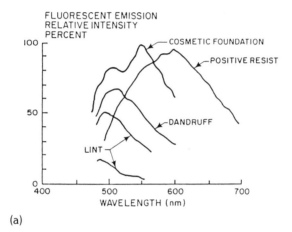

(a)

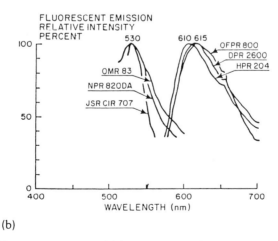

(b)

Figure 3-5 Fluorescent spectra of (a) organic particulate contaminants and (b) photoresist residues (from T. Hattori, Sony Corp.).

As may be seen, the fluorescence signature of each material is quite distinct. This enables us to identify many organic substances and some inorganic materials. For practical use, these signatures could be kept on file (contamination "mugshots") and unknown fluorescence plots compared against them in the identification process. Information of this type can be of great value in deciding what the problem is. Once these results are available, the solution is usually rather simple.

We might note at this point that there is a tendency in the industry to look for quick solutions to problems. This is sometimes called the "instant happiness in a spray can" syndrome, and those who live by it are the prey of the first salesman who comes in with the magic gadget. Long experience has convinced us that there is no substitute for understanding the problem and finding an intelligent solution. Looking for the quick way out is generally a total failure. Here the point to appreciate is that the first step in improving the cleanroom is to find out what contaminants are present. When those data are available, you begin with the worst case first. It may be easy to say "Buy new air filters," but that won't help if the problem is employees' wearing cheap gloves that shed particles of latex.

Monitoring

The modern cleanroom is monitored extensively for safety. The basic reason for this is that there are typically toxic and hazardous materials handled in the cleanroom that require instant response if something goes wrong. For instance, hydrogen gas may be used. Hydrogen is odorless and tasteless, so it has poor warning properties. If there is a hydrogen gas leak, the first indication is likely to be an explosion or fire, neither of which is acceptable. Many chemicals and gases used in the cleanroom fit this same description, that is, they pose a hazard but offer little or no warning of trouble.

There are three classes of monitoring performed in the cleanroom:

1. Continuous
2. Routine
3. Episodic

Continuous monitoring is intended to address the issue raised above. That is, a monitor is put into place to continuously scan the area of concern for a particular gas or vapor. Monitors come with a variety of options to assist in reacting to whatever is monitored. They can simply detect and alarm at some set point. Or they can be made to detect, alarm at multiple set points of increasing risk, and record levels detected while automatically taking some preprogrammed action steps.

Hydrogen Monitors

It is usual to monitor for hydrogen in the cleanroom using an appropriate hydrogen meter. These are relatively simple wheatstone bridge systems which contain hydrogen-sensitive elements and hence can determine the quantity of hydrogen in air. The unit usually uses either an analog or digital meter which reads 0 to 100 percent of the lower explosive limit of hydrogen in air. The lower explosive limit of hydrogen in air is 4 percent. All that is necessary is to select a level for the equipment to announce an alarm. Most locations use 20 percent of the lower explosive limit of hydrogen as the alarm level. This translates to 0.8 percent hydrogen in air.

Once the setpoint for alarm is decided, the next decision involves what action steps will be taken if and when that level is seen by the monitor. Most of the semiconductor industry takes a lead from the alarm circuit and runs it to a relay panel so that they can automatically shut off the hydrogen and turn on an inert purge gas. We also allow the monitor to ring a local alarm at this point to notify the equipment operator that something is wrong and that action has been taken automatically.

Hydrogen monitors are generally reliable and require little maintenance. They are frequently placed within the equipment which uses hydrogen to monitor for potential leaks as close to the source of the leak as possible.

Smoke Detectors

Smoke detectors are required to be placed in the air return duct or plenum of the cleanroom by the fire codes. These come as either photocell units or have a very small ionizing source. In either case, they "see" smoke or particles of combustion, either by these particles blocking out light to the photocell or becoming charged and altering the characteristics of an electrical circuit, and send an alarm signal at a preset density.

As with hydrogen detectors, it is possible to take a signal from the alarm circuit and route it to a relay panel to accomplish other tasks in addition to the pure alarm function. Having detected smoke in the air return, we might want to shut down recirculation fans, cycle vents to a 100 percent make-up (fresh air) air cycle, and allow the local exhaust ventilation system to clear smoke from the area. Signals from these units can also be transmitted to a central station on the property. This would provide notice to the security guard, for instance, that at 3:00 A.M. while the area is empty, something has gone wrong.

Smoke alarms are generally set up to respond in pairs. That is, if one detector sees smoke it goes into alarm, but the transmitted signal is sent as "trouble" until a second signal is received by another unit. The reason for this is to avoid dust-generated alarms from someone working nearby. In alarm technicians' language, this is called *cross zoning*.

Velometers

A velometer is a meter which will read the velocity of air past a given point. It is extensively used by air conditioning technicians and industrial hygienists to monitor air flow. The air conditioning department sets air flow into rooms from the air conditioning system using a velometer at the ceiling register. The industrial hygienist uses the instrument to measure air flow into an exhaust hood to assure that the equipment is operating to safety requirements.

Velometers are frequently mounted permanently into some equipment where local exhaust ventilation is considered to be critical. The analog gauge is usually a simple red/green display with a needle pointing to green when the exhaust is sufficient. These units, too, can be modified by a simple contactor in the red zone. In this manner, when something goes wrong with the local exhaust ventilation, we can ring an alarm or take other action steps through a relay panel again.

A variant of the velometer is a differential pressure switch. This unit measures exhaust by a quarter-inch tube usually tapped into the exhaust duct near the equipment. Vacuum caused by the exhaust holds an electric contact open as long as the exhaust is working properly. If the exhaust fails for any reason, the contactor closes, an alarm is sounded, and power is shut down to heaters, and so forth.

Vapor Monitors

Vapor monitors are a variation of the type of monitor used to detect and measure hydrogen. That is, they read resistance changes across an electrical bridge and can be calibrated to a variety of vapors. One can monitor for xylene, acetone, alcohol, and a host of other organic solvent vapors via this equipment. Vapor monitors are particularly useful in the cleanroom for monitoring chemical dispensing areas, underfloor areas where chemical drains run, storage rooms, and so forth. Like the other monitoring systems, they can be used not only to look for a particular substance in vapor form, but having found the substance, they can alarm at some preset level and take certain preprogrammed actions.

Chlorine Monitors

With reactive ion etch systems in use in the semiconductor industry, chlorine was moved from the water treatment facility to the cleanroom. This material is not only toxic, but also extremely corrosive, so it is fairly standard to monitor for it.

Chlorine monitors range from the relatively old wet cell technology to modern solid state units which will read chlorine directly in parts per million ranges with some ease. As a matter of general monitoring philosophy it is prudent to monitor as close to the operation of concern as is practical. The user normally has the option of setting alarm levels, so it will be necessary for these to be rationalized in advance.

Toxic Gas Detectors

Toxic gas detectors are probably the most critical alarms in the cleanroom from a safety point of view. None of the hydride gases offers good warning properties, so it is given that exposures will occur above exposure limits if something goes wrong.

There are a variety of technologies for detecting and measuring the concentrations of toxic gases. All of them are intended to continuously monitor the cleanroom air and advise the users when something has started to leak. Paper tape systems employ a chemical-impregnated paper tape on a system of reels something like a cassette. The chemical is sensitive to a particular gas and reacts to produce a discoloration on the tape. This discoloration is then read optically and is directly proportional in terms of concentration to the intensity of the stain. Flame photometry is used on another type of unit. Gas chromatography coupled to a mass spectrophotometer is the basis for a third type of unit, and photoionization is used in a fourth. They all share the ability to monitor more than one location, given a number of sampling lines and points. They all, likewise, are able to generate printed reports of gas levels throughout the area. Finally, they are all capable of more than one alarm level, so they can *notify* at one level by simply sounding an alarm and can take *action* at a second level by program or through a relay panel.

Multiple point monitors are the usual scheme because of the cost of the detection units contained within these sophisticated systems. Eight to sixteen points monitored by a single system is not uncommon. The points are brought to the central monitor via tubing and an analysis is made. The equipment cycles through the monitored points, one point at a time. The speed with which this monitoring cycle is completed is dependent upon the levels of detection demanded of the unit.

All of these units can monitor at sub Threshold Limit Values or Permissible Exposure Limit, but this is usually at the expense of cycle time. The user must make the decision based upon what is most important to him. An option being used more frequently is to have the unit monitor all points simultaneously. Upon detecting anything, it goes into a search cycle where it monitors each point individually until it finds the leak. It can then quantify the leak, print the concentration, and ring alarms or take preprogrammed action. The use of microprocessors has been instrumental in allowing for this modification of these systems.

All of these monitors more or less continuously analyze the cleanroom. They look for the leak, spill, or system failure that would allow hazardous materials into the cleanroom air stream. As a generic term, they are *area monitors,* that is, they are responsible to monitor a discrete area for a substance or series of substances and record data, announce alarms, or take response actions as preprogrammed. By inference we then generally monitor the people within that area, but only very indirectly. Standard practice is to mount a gas detector, for instance, within the gas cabinet, close to the source of the potential leak. Unless a gas cabinet is empty, there is no way people could be inside. So measurement of concentration of a gas in a gas cabinet is only very poorly reflective of exposures of people. This is not to say that cabinet monitoring is unnecessary or invalid or unimportant. It is needed. However, it is intended to monitor for the upset condition and react in such a way as to prevent exposure, or at least limit exposure to people.

Routine monitoring is designed to measure the exposure of people on a more direct basis. It is the business of the industrial hygienist to devise people-monitoring programs. The industrial hygienist does a premonitoring survey of the cleanroom and divides it into exposure zones. One area may use hydrofluoric, nitric, and acetic acid. Another area may use photo resist, strippers, and developers, and a third area may use hydride gases. Having zoned the area logically into exposure potential zones, the industrial hygienist's next activity is to estimate the number of potentially exposed people. This will allow him or her to establish a statistically valid representative sampling scheme. For instance, it may require that twelve samples over three shifts are necessary to obtain valid exposure data in the photolithography area.

Individual sampling is generally done with specialized equipment. Sampling pumps draw air through charcoal tubes for organics, through filter paper for solid materials such as arsenic, phosphorus, and so forth, or through a bubbler for acid vapors. Diffusion badge collectors, which operate on room temperature diffusion without pumps, are also frequently used.

The industrial hygienist samples employee exposure for the best part of the workday. She generally observes the work in progress and makes note of a number of control or upset factors during the sample. She will measure exhaust velocity in a hood. She will note whether or not the employee uses protective clothing and whether or not during the monitoring period there are spills or failures. These observations help to validate the monitoring data in terms of confidence that the exposure is being controlled. The samples, once taken, are analyzed and gross data is reported to the industrial hygienist. She will then calculate a time weighted average exposure using the gross analytical data and sample time. This is then compared to both previous readings for the same operation and established exposure limits. The industrial hygienist will often take a surface wipe sample during the monitoring cycle. This is to measure cleanliness procedures in the area and is a good indicator of how well the people in the area pay attention to clean-up operations.

Routine monitoring is done to monitor employee exposure or potential exposure. It is a valid measure of the adequacy of both engineering and administrative or procedural controls put in place to control exposure. Its limitation is that it cannot be done for all employees on a continuous basis. It does, however, characterize exposures in terms of risk. It shows those operations which are well controlled and those which are not. It allows the emphasis to be placed where the need is located.

Episodic monitoring is the most difficult monitoring to perform, for it requires either emergency response capability or good planning with a great deal of coordination. Episodic monitoring deals with analysis of exposures during either upset condition or nonroutine operations. Upset condition is represented, for example, by a container of chemicals spilled onto the floor of the cleanroom. It also covers the response to an alarm in a gas cabinet. The nonroutine operation is generally characterized by maintenance activity or equipment modification or removal.

In the upset situation, the first thing that usually happens is that people leave the area for a place of safety. An emergency crew puts on the necessary protective clothing and reenters to deal with the situation. They clean up the spill, or neutralize, or shut off or shut down as required, and restore the area or equipment to pre-upset status.

When is it safe to allow people to reenter and return to work? This is where episodic monitoring enters the picture. Portability of monitoring equipment and direct reading of results are two important keys to episodic monitoring. At this point the question, "Is the area safe for people?", must be answered, usually under some pressure. Operations are interrupted and things have to be unloaded, loaded, moved, and so forth, to maintain operations without damage to product.

Break-tip tube detectors are particularly useful for this type of monitoring. They are available for a wide variety of materials and supply data on concentrations of material in air almost immediately. They are not accurate (usually about 20–25 percent error is assumed), but they are fast and a good indicator of the presence or absence of a material. Portable hydrogen detectors, flammable gas and vapor monitors, and oxygen monitors are also useful field instruments for making decisions regarding reentry. Microelectronic applications have produced portable gas chromatographs that could and have been used in this process. It is insufficient to put a nose into the area and sniff to announce the "all clear." Neither is it correct to assume that enough time will make it safe enough. The old concept of the dead canary does not belong in the modern cleanroom.

The other episodic measurement requirement deals with maintenance of equipment or the removal or modification of equipment that is potentially contaminated by hazardous materials. The first thing a maintenance technician does to a piece of equipment is to remove panels and covers. These same panels and covers provide enclosure for containment and exhaust operation. Unless the equipment is clean (and it never is), the technician is potentially exposed to whatever hazardous material is used. The same thing is true during tear-down and removal of equipment from the area. Panels are removed and exhaust lines are severed. All engineering exposure control devices are rendered useless.

If the industrial hygienist can learn in advance that equipment removal or major maintenance operations are to be conducted, he can arrange to do some specialized sampling. The routine industrial hygiene monitoring is to deal with 8-hour time-weighted exposures. Equipment removal or maintenance exposures cannot be time weighted, as one has to expect excursions well above exposure limits to occur as operations proceed. Fortunately, most of these are for very short durations. High-volume sampling is usually done during these types of jobs to characterize the exposure patterns and assure that protections and precautions are adequate. A high-volume sample may last only a few minutes but collects enough sample material in that time to allow for peak exposures to be identified by positive analysis in the laboratory. Routine sampling and high-volume sampling are frequently done together on those jobs such as tear-down and removal which last all day or several days. This sampling has the advantage of producing both average exposure data and peak information.

This discussion of monitoring is by no means complete. It is intended to convey the idea that monitoring and its strategies must be a well-thought-out process. No single monitoring approach will suffice for the cleanroom environment. The cleanroom is becoming as sensitive to injury by contamination as man is sensitive in terms of illness. Regular, careful monitoring satisfies both of these sensitivities.

Other Parameters

In the first secton of this chapter we discussed briefly the main parameters we have to deal with in our controlled environment, particularly particle densities, vapor concentrations, temperature, relative humidity, room pressure, and air velocity. We have dealt with the contamination factors already in this section, and now it is time to turn to the measurement of the other parameters. We will not discuss the measurement of purely process variables, such as gas flows to a piece of apparatus or oxide thicknesses in a semiconductor device, but restrict our attention to environmental control examples.

If you were told to measure room temperature you would undoubtedly use an alcohol or mercury thermometer in which the temperature dependent (linear) expansion of the liquid in a thin tube, measured against a calibrated scale, is used to determine temperature. This is a good measurement method if you are the temperature controller and will personally reduce the heat input if the reading is too high. This would be a rather tedious job since it would have to be carried out whenever control was required, which could be 24 hours a day. Therefore, some type of automatic control system is desirable. Such control systems are discussed in Chapter 6, but we have to be able to measure temperature in a manner which is compatible with the system (for example, computers don't like to read thermometers). Typical temperature-sensing elements are listed below:

1. Bimetal element. Consists of dissimilar metals, with different thermal coefficients of expansion, fused together in a straight, U-shaped, or spiral configuration. Due to the differential expansion or contraction of the metals, the strip deflects as temperature changes. This type of element is usually used in an "on-off" control, for example, the deflection of the strip is used to activate a switch which turns the heating/cooling system on or off. An alternative but similar scheme is a rod and tube element in which a high expansion metal rod is attached at one end inside a low expansion tube. Once again the differential expansion may be used to activate a switch or a fluid valve (steam, chilled water, and so forth).
2. Sealed bellows. Consists of sealed bellows which are filled with a liquid, gas, or vapor which expands or contracts according to temperature. The ensuing movement of the bellows can be used to open or close a valve or change the value of a (mechanically) variable electrical resistor. A remote bulb can be used which is connected to the bellows by a capillary tube. This reduces the size of the point of measurement element.
3. Resistance element. Consists of a pattern of a resistive material on a suitable substrate. The electrical resistivity changes with temperature, and hence this scheme has no moving parts and may readily be interfaced to electronic control systems. Metals tend to have a positive temperature coefficient of resistance or TCR (resistance increases with temperature) whereas semiconductors have a negative TCR.

The manual measurement of relative humidity involves two thermometers, the bulb of one being covered by an absorbent material which is soaked with water. The rate of evaporation of the water from the wet bulb depends on two factors; air temperature (which is registered by the dry bulb) and the amount of moisture in the air (dry air allows a higher rate of evaporation). The evaporation process cools the wet bulb; a lower temperature on this thermometer therefore means drier air and vice versa. This wet-and-dry bulb arrangement has been used for many years and is still used to verify the action of automatic control systems. However, as with thermometers for temperature measurement, the wet-and-dry bulb thermometer is an unsuitable measurement instrument for automated control. In mechanical humidity control systems, a hygroscopic element, such as an organic fiber (hair), is typically used. This fiber expands or contracts according to the level of humidity and thus may be used to alter a valve or variable resistor. A non-mechanical approach involves the use of a sensing element comprised of a film, the resistance of which depends on the amount of water absorbed, on a supporting substrate. Two interleaved electrodes are deposited on the surface of the measurement material, and the resistance between these is directly related to the air humidity.

Air pressure and velocity may also be measured by a number of different means. A popular visual indicator is the water manometer, which is essentially a U-shaped tube, one end of which is connected to the pressure to be measured (for example, in a duct or room volume) and the other to a reference pressure. A pressure higher than the reference pressure will push the water against its weight to support a column of water in the downstream side of the U-tube. A lower pressure will pull a column of water in the upstream side. Hence the pressure can be measured on a scale in inches of water supported. A similar, more accurate system uses mercury instead of water and a vacuum as the reference pressure. The corresponding unit is millimeters of mercury. A better measurement element as far as automated systems are concerned is the one-sided bellows. A known reference pressure is contained within the sealed bellows; a pressure higher than this will deflect the bellows surface inward; and a lower pressure will allow them to expand outward. The mechanical movement may then be used to alter a variable resistor or mechanical valve. There is a variety of types of anemometers which may be used for air velocity measurement. The simplest are the vane anemometers which either deflect or rotate with a flow of air into them. The deflection or speed of rotation may be used as an indication of air velocity. Elements for automated systems are generally of the thermal anemometer type, which tend to rely on the measurement of the change in resistance of a heated element when it is cooled by an airflow. A reference sensor is used to compensate for ambient temperature effects.

SUMMARY

In this chapter we have considered what we typically require of a controlled environment and how we may judge the performance of a cleanroom by the use of standards. The current U.S. standard (Appendix I of this book) should be reviewed by the reader. We have also seen how we may use measurement techniques to determine not only how clean our environment is but also how we may eliminate some of the sources of contamination. Light-scattering techniques are excellent for regular counting of large populations of particles but witness plates combined with microscopy (sometimes with EDXA) are required to tell what the contamination is.

4

Creating Clean Areas

We should have an idea at this point about what a cleanroom is and what it is supposed to do. We will now consider the basics of how to create a clean area. We begin by discussing the materials which go into making our cleanroom.

CONSTRUCTION MATERIALS AND TECHNIQUES

Basic Material Properties

The first question we have in the consideration of how to create a clean zone is what do we use to delineate our clean space? It should be obvious from the discussions in the first three chapters that no ordinary construction materials will do, but exactly what do we require of our walls, floor, and ceiling?

We might sum it up by saying that we want materials that don't outgas, don't generate particles, are mechanically and chemically stable, can be cleaned, and will withstand the abuses of alterations, maintenance, and everyday life. The walls, floors, and ceiling should release no particles, bacteria, vapors, or other deleterious matter that will hurt the product. There is nothing on this planet which does not contaminate the environment to some degree, no matter how small the release of contamination. However, in our choice of clean materials, we should strive to select those which contaminate little, or at least choose to achieve a successful trade-off between performance and cost.

Unfortunately this is the point where most discussions on this matter stop, giving those who do not know any better the impression that data of this nature on materials is available from the manufacturers. Let us settle this point in no uncertain terms: in many cases the information does not exist or is at best inadequate. The scientific testing of materials in a comprehensive manner is an expensive, complex, and time-consuming business, beyond the abilities and resources of many suppliers. It is easy to put a particle counter next to the material and see how many particles of a particular size are released, but this is hardly what we can call a comprehensive scientific test. It tells us nothing of how the characteristics of the material change over a long period of time or under duress, for example, from vibration, trauma/ shock, chemical vapor attack, liquid spills, and so forth. This information is 'important if we are to predict how the materials are going to perform under real-life use conditions. This kind of testing tends to be done by experts in big chemical companies or in independent research laboratories, such as some of government labs. In summary, we may think some materials are better than others, but without definitive comparative data it is difficult to make a truly informed choice.

Another trap we can fall into when choosing cleanroom materials is to make the choice based on what others have done before. This method is dangerous, as the choice of others could have been made using different criteria from your own. A good analogy is that you would never take medicine prescribed to someone else if all you knew about that person was that he was ill. What, then, can you do to help make a decision? The best procedure is to run what tests you can on the materials in your own facility under test conditions as near to your operating situation as possible. Testing requires time, resources, and ingenuity, but is well worth it. Self-testing may not yield the most comprehensive results but can frequently tell you what you want to know.

One thing left to mention is frequently forgotten, yet it is extremely important. During construction and after the job is finished, it will be necessary for holes to be cut in the wall and floor. These holes have to be sealed after formation; this is not a trivial matter, as the sealants should not release particles or outgas to any large degree. The National Aeronautics and Space Administration (NASA) has the same problems with holes, and they can't easily fix some of their devices after launch. In two volumes, which are available at cost from NASA, they give the properties of a host of solid, liquid, semiliquid, and gaseous materials. Most cleanroom people don't know about this information, but it is there for the taking, all the necessary research having been done or collected by NASA. There is none of this business of "company A" versus "company B"; they give names and model numbers.

Wall Materials

Which materials have been examined to some degree, been passed as being suitable for cleanroom use, and are finding their way into existing rooms? We will begin with a general discussion of wall materials.

Starting at the lowest cost end of the spectrum, we have standard construction materials which have been covered with a low-contamination layer. This is inexpensive, as we can take a material like sheetrock and bond a thin layer of vinyl to it using an appropriate cement. Why use an actual vinyl sheet rather than vinyl paint, for instance? Paints are not too bad on surfaces which are already smooth but they will tend to coat particles on relatively loose surfaces such as sheetrock or wood. These vinyl-coated particles can become detached when the surface is rubbed or knocked hard. Paints also tend to be bad outgassers, but this improves with age. However, old paint which has lost most of its volatile content has a tendency to flake. This is unacceptable in a cleanroom environment.

Vinyl will outgas to some degree, but high quality vinyls with low added volatile materials will not cause too much of a problem in this respect, and they do not spontaneously generate a great deal of particulate matter. How much particulate matter? The quantity is not precisely determined, but this material has proven to be suitable for class 100 cleanrooms for some time now; therefore it apparently does not introduce enough particles into the environment to cause problems at this level of cleanliness. The vinyl may be heat or chemically sealed at any joints. A somewhat more expensive option is to use a laminated structure with a wood (plywood, for instance) core and a hard plastic laminate (for example, Formica) on top. These laminates are generally hard wearing and easy to clean. Polystyrene foam may also be used for the core to cut down on weight (and cost), but the resulting panel is not as strong and may present a fire hazard. Hard plastic laminate is superior to the soft vinyl coating as it is more robust and can be cleaned more readily. An alternative core is honeycombed aluminum, which leads to a strong and safe structure.

Plastic sheets are rarely used alone for walls as most plastics do not have sufficient structural integrity; the more flexible materials warp and seals fail, and the more rigid ones crack if struck. However, flexible clear vinyl sheets are used for soft-walled cleanrooms. Soft walls are frequently used around special areas or equipment to help isolate them from the rest of the room. This is an excellent way of creating a mini-environment at low cost, but one should remember that these walls or curtains require frequent cleaning as they will pick up contamination from personnel brushing against them.

The problem with the vinyl-on-board or laminate approaches is that when holes are deliberately cut or when the surface is accidently torn or cracked, we expose the incredibly dirty underlying material. We must reseal any breaches in the surface as soon as possible. Which sealants can we use? Fortunately we actually have some data on this from NASA. When there is any relative movement in the wall surface or in the pipes and so on that pass through the wall, due to vibration or thermal expansion for instance, it is best to use an elastic sealant. If rigid sealants are used, the displacement can crack them and the seal will become broken. A non-setting gel sealant could also be used, but there is a danger that this will accidently become rubbed off at some later time if the repair area is exposed to personnel. Because most non-heat treated elastic materials and gel sealants remain extremely flexible, they stick together well and particle counts are low (almost negligible, unless the seals are around fast-rotating shafts and the like, which can cause the ejection of droplets). However, since all contain volatile components, we have to be concerned with outgassing properties. This is where the NASA test results prove useful.

Figures 4.1 and 4.2 show the two sides of the outgassing story. Figure 4.1 shows the percentage mass loss by the outgassing of volatile components under standard accelerated test conditions. As we can see, materials such as silicone rubber sealant, an extremely inert elastic material, and silicone grease fare very well in this test. Most paints, perhaps with the exception of silicone based material, seem to be poor. However, when we look at Figure 4.2, which shows the percentage mass of recondensed material on a 25° C surface, we see a different picture. Here the silicone rubber sealant is worse than the silicone grease (a reversal from the outgas test), being not much better than some paints. However, in our minds it is better not to have too much material in the air in the first place, which makes the silicone rubber the winner in this competition. There is one warning. The silicone rubber sealant comes with an acetic acid based solvent to keep it from setting before you want it to. This must be persuaded to outgas as quickly as possible, perhaps by the application of gentle heating, so that it does not cause a contamination problem later.

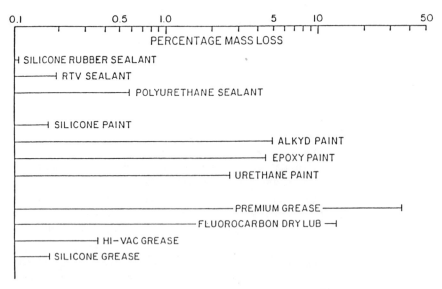

Figure 4-1 Experimental results showing percentage of original mass lost after 24 hours at 125° C per ASTM test E595-77. Note results for a particular type of material (for example, silicone grease) may vary with manufacturer, compounding, aging, and thermal cycle for curing. Data is to be considered as representative. For details refer to original publication (from NASA ref. publication 1124, June 1984, *Outgassing Data for Selecting Spacecraft Materials* by W. Campbell, J. J. Park, R. S. Marriott).

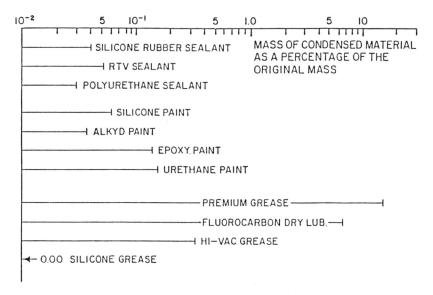

Figure 4-2 Experimental results showing condensed organics collected on 25° C surface adjacent to heating materials as listed. Data is presented as a percentage of the original mass heated as per ASTM text E595-77. Note results for a particular type of material (for example, silicone grease) may vary with manufacturer, compounding, aging, and thermal cycle for curing. Data is to be considered as representative. For details refer to original publication (from NASA ref. publication 1124, June 1984, *Outgassing Data for Selecting Spacecraft Materials* by W. Campbell, J. J. Park, R. S. Marriott).

At the more expensive end of the cost spectrum for walls are the metallic materials, which are strong and do not suffer from the problems of the materials described above. Aluminum is produced in sheet form for wall panels. It is a strong, light material but is actually quite reactive. There are usually few reactions under normal conditions because aluminum quickly forms a thin protective oxide film in air. However, if acid vapors are present, some degradation of the material due to uncontrolled corrosion reactions will occur. To prevent this the aluminum surface is usually covered with a hard, chemically stable coating. For instance, if we deliberately increase the surface oxide coating thickness by an electrochemical process known as anodization, the aluminum sheet becomes extremely inert. The big advantage of anodization is that the hard oxide coating and the metal are chemically bonded and therefore can withstand a lot of wear and tear without the coating becoming detached. An alternative coating for aluminum is enamel. Enamel is sprayed on and subsequently fired to create a hard, scratch-resistant coating. The process is a little easier to perform than anodization and therefore the resulting panels are less expensive. There are also wall materials that use thin anodized or enamelled aluminum sheets on a lightweight synthetic core. These panels are very strong but light as we do not have the same amount of relatively heavy metal as in a solid panel. However, some types are banned by fire authorities in certain applications as the synthetic core can cause problems in the event of a fire.

Another metallic wall material is sheet steel, which is less expensive than aluminum. Unfortunately, the properties of untreated steel are undesirable for cleanroom use as it corrodes easily and therefore has to be coated for protection. Coatings used are enamel and spray vinyl, which is applied in much the same way as paint, but the resulting layer is more plastic than traditional vinyl paints. The problem with this scheme is that any breach of the coating can result in a corrosion "scar," and the chemical products of this corrosion could flake off as particles. We cannot anodize steel as such, as iron oxide is not as smooth and mechanically stable as aluminum oxide. We can, however, galvanize the steel. Galvanizing is an electrochemical process in which a more inert metal, for example, zinc, is coated onto the steel. Unfortunately, this may not be the best solution if strong chemical vapors are present, as the surface coating can start to corrode also. We also have the problem of resealing the surface whenever we cut the panel, although this is not as problematic as in the case of laminated materials. A better alternative, and a more expensive one, is to use stainless steel. Stainless steel has had other elements added to make it corrosion resistant. High quality stainless steel usually does not require any coatings. However, it is important to remember that stainless steel can still generate particles—not a great many, but some nevertheless. This phenomenon is related to the way the steel is formed. Since high temperatures are used, precipitates form inside the steel. The precipitates typically consist of oxides of impurities and additives, for example, manganese oxide. Precipitates have very different mechanical properties from the surrounding steel and if they appear at the surface, can detach. The way to eliminate the problem is to chemically etch the surface to remove surface precipitates, but this is not economically viable for large sheet sections. All sheet metal panels can be joined fairly easily by using a custom seal system. The panels to be joined slot into both sides of an I-shaped seal unit. Flexible rubber elements line the insides of the slots to ensure a good seal.

Finally, a good low-contamination material which can be used in the cleanroom walls is glass. Glass allows people to see in, including visitors who should be discouraged from entering the room to add to microcontamination problems, and cleanroom employees to see out (to help relieve the monotony of the inside sterile environment). The glass should be tempered to reduce flying glass hazards in the event of an explosion, and coated with an appropriate ultraviolet blocking material.

Floors

The cleanroom floor is a very problematic area as it has to take a great deal of abuse. Pedestrian traffic, regular cleaning, heavy wheeled carts, chemical spills, equipment moves, and room alterations all have a negative effect on the floor. Any sliding action, even just shuffling feet, has the potential to destroy the floor surface and generate particles in the process. Heavy particles, which naturally fall to the floor under the influence of gravity, help to abrade the surface when pressure is applied. Very few flooring materials can stand up to this heavy usage. Kinds of coverings to avoid are linoleum type materials, which have a hard, crackable surface over a coarse mat base, and any form of tiled floor. In the latter case, the joins will harbor dirt, allowing it to escape at a later time. We need a continuous, non-cracking material which is mechanically and chemically robust.

The lowest cost alternatives are vinyl floorcoverings which are laid on top of a concrete structural floor. Various grades of vinyl are available. What is required is a low volatile component mix so that the material is not so soft that it will tear when equipment is moved over it and so that outgassing will be minimized. A brittle material is not practical because it will break up over time and create particles. Different available vinyl flooring materials have vastly different properties. Figure 4.3 shows the results of an accelerated test performed on two typical controlled environment floor materials. This test simulates floor wear by abrasion and illustrates how the properties of a floorcovering can change with time, as well as how two materials can have very different contamination properties. Remember, when you begin to see floor wear, it means that the floor is actually in an advanced stage of breaking up, that is, the wear you see is actually due to the degradation of the surface finish. Once this is damaged, the particle generation rate goes up dramatically. Floor wax will reseal the surface, but its use is not advocated as it will dry and flake off, especially when wax layers build up.

As we will see in the following chapter, many room designs call for a perforated floor which allows air to pass through with as little resistance as possible. The original cleanroom perforated floors were little more than computer room floor systems. They consisted of drilled laminated sections, supported at each corner by adjustable jacks. These floors were not particularly clean due to the choice of material, but recently things have improved dramatically. Perforated floor sections in high standard cleanrooms now are more likely to be vinyl coated steel or aluminum over steel. This provides a good strong floor with low contamination properties. These floors are usually cast rather than drilled, as it is easy to make rounded edges on the holes or slots by casting, thus ensuring that the perforations do not harm cleanroom footwear to any great degree.

Ceiling Systems

The requirements of a cleanroom ceiling are very different from those for the floor. The ceiling has to contain the filter units, or allow the support and supply of such units under the ceiling level, as well as lighting, sprinkler heads, and so forth. It is important in many cleanrooms to have a ceiling which is

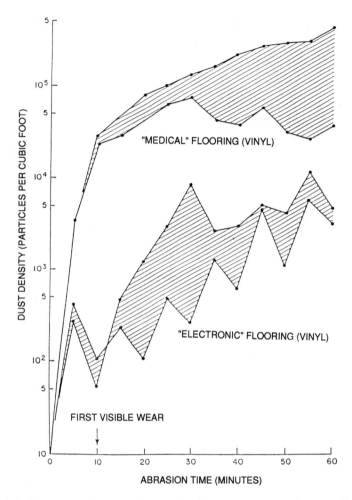

Figure 4-3 Experimental results showing dust generation during abrasion of various cleanroom flooring materials. Particle count covers 0.2 to 12 micrometer range. Test apparatus modified Taber type abrader.

flexible to the changing requirements of the room but at the same time is standardized. Hence, the most popular type of ceiling system used in cleanrooms is the suspended ceiling.

The suspended ceiling is based on a support grid which is, as the name implies, suspended from the building structure. The grid is a standard size, for example, 2-feet by 4-feet sections, to allow the interchange of standard components. The aluminum or coated steel frame elements which make up the grid have an inverted T cross section so that ceiling components may slot in from the top and thus be supported. The components may be as simple as blank panels made of the same materials as the walls, blanks with holes cut for pipework and the like, or they may be filter units supplying clean air to the room, or even fluorescent lighting units.

The marvelous feature of the suspended ceiling is that components may be placed wherever needed without having to embark on any major structural alterations. The bad aspect is that they have great potential for particle leakage from the space above, and great care must be taken in sealing the components to the frame. An elastic sealant is usually employed here. In addition, the components should be clamped down to prevent lifting due to sudden pressure changes in the room. The method of suspension must be strong enough to support whatever is added.

Building Techniques

Now that we have looked at the basic materials of a cleanroom, we come to the point of putting it all together. It should be borne in mind that if the cleanest materials in the world are not put together properly, the cleanroom is a lost cause.

The first thing that everyone involved in the construction of a cleanroom should understand is that it is no ordinary building project. A cleanroom becomes a special place as soon as groundbreaking begins and not just when the keys are handed over. All actions and procedures should be reviewed with regard to microcontamination aspects before they are carried out.

The materials which will form walls, floor, and ceiling should be treated with respect as soon as they arrive on site. They should never be stored outside in the construction yard as they will inevitably be treated like all the other non-cleanroom materials by unknowledgeable contractors. Packaging on the materials must never be removed (or torn, for that matter) until it is time to use the materials, and only under cover, out of the dirty outside air and damaging effects of sunlight. A general wipedown, using the techniques described later in this chapter, is recommended before and after installation. The pre-installation wipedown helps to prevent the trapping of contaminants in seals. After any cutting or drilling operation, the resealing operation should take place as soon as possible to prevent any long-term damage or deep contamination of the materials.

It is good practice once the flooring material is in, an operation which should be late in the construction schedule, to gown construction employees to some extent to help prevent damage to the floor and reduce ground-in contamination. Gowning means more than just overshoes. A Tyvek jumpsuit (see Chapter 11), booties, and a bouffant cap are optimal at this stage, as contamination will come off working clothes. Any spills, filings, and so forth should be cleaned up immediately as they tend to become dispersed with time and are more difficult to clean up. Once the room has become a room (with walls, ceiling, and floor), even if the clean air supply has not yet been turned on, all nonessential operations should be prohibited. Prohibited are work on noncritical materials which will go outside the room, eating, smoking, and so forth. This may sound like overkill, but employee attitude toward the room is every bit as important as the microcontamination science aspect.

Once the clean air supply is turned on to the room, it is critical for construction workers to be appropriately gowned to protect the cleanroom and associated systems. The minimum attire should be whatever the operating apparel will be, except in the case of lower-class rooms. For rooms which will ultimately be class 10,000 or worse, a jumpsuit is still suggested, even if lesser protecting smocks will be used in production. The room should be as near to perfect as possible before sign-off and certification, and if this means that the contractors have to go overboard, then so be it. At this stage, if possible, the room should be cleaned each day, including wiping the walls and ceiling to remove the day's construction dirt.

AIR FILTRATION

The completed room may be intrinsically clean, but the air put into it will most certainly be filthy unless we do something with it. What must be done is to clean it up by filtration and related techniques. We will assume that the raw air has been prefiltered to remove the bulk of the contamination (see Chapter 6), and we will therefore restrict our discussion in this section to precision filtering immediately before the air enters the room. This discussion will be as simple as possible in the hope that it will be as useful as possible.

Basic Air Filtration Theory

In this section we will go over some of the aspects of removing particulate contamination from the air. The problem here is that we have a fluid medium (the air) which has a definite viscosity, that is, it exhibits a resistance to flow, and it has solid matter in it, namely particles. The obvious solution is to place a solid object in the airflow with holes formed in it so that the air goes through, but the particles are too big to pass and therefore become trapped on the upstream side. This sounds reasonable but has a couple of major flaws. For one thing, the holes would have to be extremely small to catch the sizes of particles we are interested in (down to 0.1 micron). Small holes are quite easy to produce with polymer technology, but the catch is that the air would not pass very easily through these small holes due to its viscosity. We would either get a tiny trickle of air past the filter or would have to use immense pressures to push the air through, neither of which are acceptable for cleanroom air supply applications. The other problem with this approach is that once a hole is plugged with a particle, no air can pass through that particular hole. As more holes become plugged, less air can get through. In filter jargon, as the filter becomes loaded, the differential pressure (the pressure difference between the upstream and downstream sides) rises. Eventually, no air would flow at all.

Clearly we need a different solution to this problem. The solution actually came from two different areas, both of which had a great need to precision-filter large quantities of air. The chemical warfare and nuclear materials industries realized early in their existence that filtration of small dimensional solids was vital to their existence (and to the existence of the people living around their plants). Hence the CWS (Chemical Warfare Service) filter and other similar types were born. How they work is rather complex and interesting, but the basic theory is simple. First we will look at the filter structure.

Very fine glass fiber filaments, ranging from a fraction of a millimeter to less than a micron in diameter, are formed into a thin pad and held in place using a resin bonding agent. Both the fibers and the resin are inert and very low contamination generators. The pad is made to have an open structure such that the holes (or interstices) are several microns to several tens of microns wide so that the air can flow through with as low a pressure drop as possible (0.3 to 0.5 inches of water pressure drop is typical of commercially available filters). The cleanroom industry has perfected the technology of these filters and modified them to provide maximum particle removal efficiency. In fact, the name we use for them is HEPA (high efficiency particulate air) filters. Efficiencies of 99.99 percent to 99.9999 percent or more at 0.3 micron are possible with this type of filter. (Note that 99.9999 percent efficient means that one 0.3 micron particle out of 1,000,000 will penetrate). To achieve a large filter surface area (to allow as large an airflow as possible) in a small filter unit size, the glass fiber pad is folded around corrugated separators which allow the air to get to and from the filter surfaces (Figure 4.4). The frame and separators were originally wood and craft paper respectively, but these have been replaced in recent times by anodized aluminum. The latest filters rely on stiffening elements at strategic places across the pleats so that the corrugated separators are unnecessary.

Why are these filters so efficient? There are actually a number of mechanisms at work. Big particles (by which we mean above 5 microns) are caught by interference between the particles and the filter itself, that is, they become caught in the interstices as they are too large to pass through the filter structure. Very small particles (a few tenths of a micron and below) are caught by colliding with the fibers themselves as they pass through the mesh of the filter. If we assume that a typical opening in the filter mesh is a diameter D (around 5 microns), particles larger than D will be collected on or near the surface of the filter. Particles down to 0.1 D may be collected by collisions, whereby they have enough momentum to break free of the airstream and crash into the fibers. Still smaller particles (less than 0.1 D) will tend to follow the airflow lines and can only be collected by diffusion. This involves random movement whereby the particles undergo Brownian Motion, jostled by the air molecules,

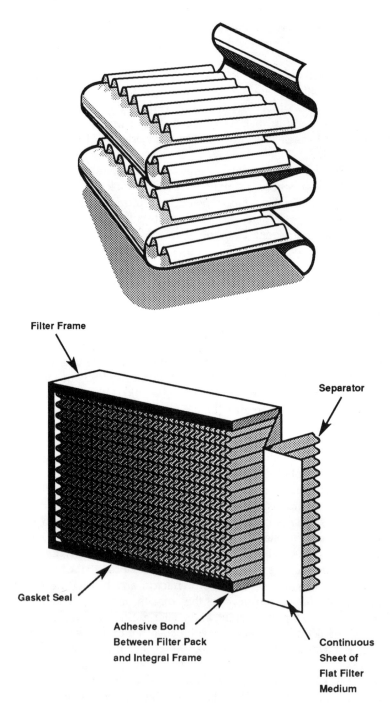

Figure 4-4   HEPA filter construction.

until they make contact with the filter fibers and stick to them. This latter process requires time and will not be effective unless the filter is thick or the air flow velocity is low. The smaller particles will easily stick to the surface of the fibers, the attractive force being supplied by surface chemistry, electrostatic charge, van der Waals force, and combinations of these. Once the particles adhere to the surface, it is very unlikely that they will become dislodged. However, if the particle-sticking coefficient is not unity, the particles may gradually diffuse through the filter and appear on the downstream side. So it is possible that filters may not be particularly efficient in trapping small particles in some cases. We will return to this thought later.

The more efficient, and more expensive alternative to HEPA filters is the ULPA (ultra low particulate/penetration air) filter. The most recent variety of this type of filter uses a chemically formed pad of expanded PTFE (polyte-trafluoroethylene). The expanded PTFE is naturally like a filter in its chemical structure, as it has stringy, fiber-like elements with ample space between them. The ratio of space to solid volume is much better than in the HEPA filter, and hence we can have a filter which passes air with reasonable ease but can trap extremely small particles. Typical ULPA filter efficiencies are in the region of 99.99999 percent (claims of even better efficiencies have been made).

Gas and Vapor Removal

The removal of gases and vapors has become an item of increasing interest in controlled environment technology. In the early days of cleanroom technology, the effects of particles were well understood but the detrimental effects of vapors, including water vapor, were not. Now we have a much better idea of how dispersed contamination can cause problems in sensitive processes. The problems with the removal of gases and vapors from a mixture of gases (air) are considerably greater than with the filtration of particles. We will discuss how we handle these problems in the following paragraphs.

The simplest method of vapor control is to reduce the concentration by dilution. This method assumes that you have a supply of clean air to dilute the contaminated air with. If the bulk of the gaseous contamination originates within the room, we may use outside air as the dilutant. However, if the outside air is contaminated, we have a problem. The other drawback of dilution is that we have to dispose of some air in order to add some clean material. This means that we may end up throwing away air we have gone to the trouble of preconditioning and filtering. We may do this anyway as the chemical extract systems within the room will remove copious quantities of air.

It is clear that it is better to remove rather than dilute the gaseous contamination, especially if we are concerned about the cleanliness of our dilutant/outside air. The way we remove water vapor, the largest gaseous contaminant, is discussed in Chapter 6. What else is left? As we discussed in Chapter 2, we have industrial gases in our outside air, such as carbon dioxide and sulfur dioxide, but the most significant external and internal gaseous contaminant classification is hydrocarbons or organic vapors. These vapors escape from solvent cleaning processes, outgas from materials (especially paints and plastics), and leak from vacuum pumps. Fortunately, these materials are quite easy to trap.

The best trapping agent for organic vapors is activated charcoal, a dry porous form of carbon. It has a very large net surface area, and the surface has a strong affinity for hydrocarbon molecules. If the airstream to be treated is passed over a bed of activated charcoal, most of the organic vapors will be removed. Usually several charcoal beds are arranged in a cascade fashion in a tower to allow maximal contact with the air. The charcoal may be regenerated by heating it to temperatures in excess of the desorption temperatures of the various absorbed materials. The only point to watch with these systems is that the charcoal beds can be a considerable source of particulate contamination. Therefore filtration should be an integral part of the absorption unit.

Other Contamination Reduction Techniques

The particle trapping mechanisms for small particles in HEPA filters tend to limit the effectiveness of these filters. Filter efficiency can be increased by the use of electrostatic technologies. If the filter fibers are charged, they will induce charges of the opposite sign on particles that pass nearby and collect them. This process is clearly most effective for the very small entities that are usually collected by the rather slow diffusion process.

Practical electrostatic filter enhancement systems involve the use of corona discharge systems ahead of the filter or electrostatic field gradients across the filter element itself. These techniques are most suitable for collection of the smaller particles that are not effectively removed by conventional systems. Figure 4.5 shows a practical electrostatic filter (ESF). The increased collection efficiency allows the use of a high permeability filter medium so that the pressure drop across the filter is low.

Electrets may also be used in particle reduction. The term electret refers to a class of materials that has been heated and then allowed to cool in a strong electrostatic field gradient. This process either injects electrons or aligns the molecules of the material, that is, it becomes permanently polarized, and therefore retains a strong, permanent electrostatic charge. Any insulating material (for example, wax, plastic, or ceramic) can be turned into an electret but, at present, the only electrets commercially available are made from plastic materials. Of this group, PTFE appears to be the best in terms of quantity of charge held and stability over time. Electrets can be used to collect floating

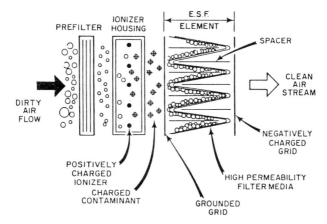

Figure 4-5 Anatomy of an ESF. Highly permeable, electrically charged microglass media permanently capture sub-micron particles (from Dollinger Corp., NY).

dust from the air. In Figures 4.6 and 4.7, we show the results of some calculations of the electrostatic forces exerted by a substrate, which we could liken to our electret, charged to a number of different potentials. In Figure 4.6 the 1-micron dust particles were uncharged, in Figure 4.7 charged. There

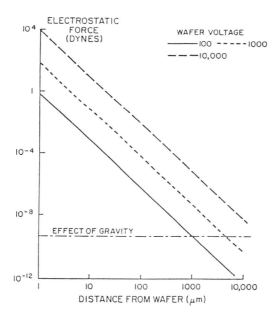

Figure 4-6 Electrostatic force on a neutral 1-μm particle (from Yost, M., A. Steinman, Electrostatic Attraction and Particle Control, *Microcontamination,* Vol. 4, No. 6, p. 18, 1986).

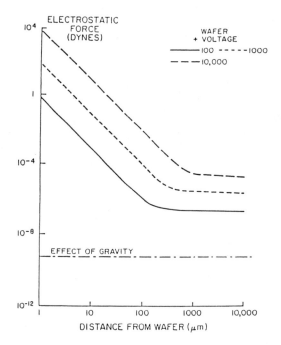

Figure 4-7   Electrostatic force on a charged 1-μm particle (n = 1000) (from Yost, M., A. Steinman, Electrostatic Attraction and Particle Control, *Microcontamination,* Vol. 4, No. 6, p. 18, 1986).

is no question that charged surfaces will reach out and collect free floating particles, and therefore electrets could be quite efficient at particle control. The other point we could take from these graphs (the actual purpose of the study which resulted in these graphs) is that charged wafers will also grab particles from the air. This necessitates charge reduction schemes near the wafers to keep them clean.

Practical Filtration Problems

Before we leave our discussion of filtration, we should mention some of the practical problems one is likely to encounter. The first problem lies with the choice of filter, or if you like, the choice of efficiency. It may be tempting to think that you should always use the most efficient filter available as this will allow the least amount of particles to pass into your room. This may be valid if you have vast sums of money to spend, but the hard fact is that higher efficiency filters not only mean a higher capital outlay but they are also more expensive to run. For instance, a good medium efficiency filter may cost around $300 while a high efficiency unit may be $500 or more. However, the pressure drop across the higher efficiency filter may be greater to begin with, for example, 0.5 inches of water as opposed to 0.3 inches of water for the less efficient variety, and therefore we have to use more energy to push the air through, which means higher electricity bills. In addition, as the filters become loaded with particles, it is more difficult for the air to flow through. Since we typically do not want to run the filters beyond a pressure drop of 1 inch water for reasons of power economy (not to mention fan stress), the higher efficiency filters will have a shorter working life and will therefore have to be replaced more often. The higher cost does not end with that point, as higher efficiency filters also tend to be more expensive to install and maintain.

Occasionally, we see particles coming through HEPA filters which have been in service for some time. This coincides with a high pressure drop across the filter and indicates that the filter is so full that high velocity air jets within the few remaining clear areas within the filter matrix are blasting particles through. Alternatively, the surfaces of the fibers may be so coated with particles (and perhaps other condensed matter) that their sticking coefficient is reduced. Unfortunately, there are continuing reports of progressive failures in HEPA filters not associated with an increase in pressure drop. The mechanisms here are by no means clear, but the suggestion of salt or corrosive vapor damage to the separators may be one factor.

Another aspect of the practicalities of HEPA filter usage is how we seal them into the ceiling system. Since we do not want to have a pathway for contamination between the area above the filters and the cleanroom, it is important to have a good seal between the filter and the ceiling frame. What complicates this problem is that the seal cannot be rigid as there will almost certainly be some relative movement, no matter how small, due to changes in room pressurization. This would crack any hard seal, and thus a flexible sealing material is required. What is typically used here is either a fluid seal, such as petroleum jelly in a trough to prevent it from escaping, or an elastic seal, such as a bead of silicone rubber. The fluid seal is much better in terms of flexibility, but it will tend to outgas more than the elastic variety unless a low vapor pressure fluid is used.

When discussing practical problems, we must include the effects of bacterial growth. A HEPA filter is dark, dirty, warm and around 45 percent RH (if we are conditioning our air properly)—providing almost ideal conditions for bacterial growth. As the dust builds up, the bacteria level may increase due to the increased supply of nutrients, and bacteria may eventually come through by becoming detached and entering the airstream. The fact that the cleanroom will acquire an odor after being shut off for a few days may be a sign that there is bacterial growth. Studies of bacteria, mold, and fungi in cleanrooms have indicated that these three categories are definitely present to some degree, particularly on any diffusers/protectors placed downstream of the HEPA filters. Some of the bacteria are typical of those shed by people. However, we have observed spore-forming species that are usually associated with outdoor environments. These results suggest that bacteria are being shed by personnel, carried up through the return path, and deposited in the HEPA filter where they grow. Other bacterial spores are drawn in with the make-up air, pass through the prefilters, and deposit in the HEPA filters. Once again, if the conditions are right the spores will germinate and grow.

One problem with bacterial/mold/fungi contamination is that it multiplies. Fungi are particularly effective at generating little stalks with spores on top. When the spores are ripe, they are released into the air and travel downwind to deposit on any available surface. In effect, the contamination can not only move with the airflow, it can multiply at exponential rates if the proper conditions are available. All we can say at the moment is that HEPA filters may be an ideal site for growth of bacteria, fungi, and mold, and we should all be aware of this. Over time, bacteria may even grow through filters to appear at the cleanroom side. In fact, all old HEPA filters should be treated as a biohazard, handled by gowned and masked personnel and bagged and burned.

Can we reduce the problem of bacteria in the HEPA filters? Once they are in it is difficult to do very much about it. Fumigating the filter with formaldehyde vapor may help to some degree, but this could damage the filter and will probably not kill all the bacteria. The process could not be performed during a working period as the vapors are harmful to us as well as the bacteria. Using ultraviolet lamps in the duct upstream from the filters will also help, as high intensity uv destroys bacteria fairly well. In addition, wet benches, spill/drip areas, damp wipes, and so forth may all be sites for growth of these contaminants. Keeping everything as clean as is humanly possible may be the only solution to this problem.

AIRFLOW

In considering how to develop a clean environment, it is not enough to merely clean up the air and put it into an intrinsically clean area. We also have to consider what happens to the air once it enters the room. We will discuss the possibilities in this section.

Turbulent Flow

Unconfined airflow, more often than not, will tend to be turbulent. This basically means that the air mixes and swirls, perhaps forming eddies and vortices. Vortices (the singular is "vortex") are spinning air masses which occur particularly when a moving airstream flows past a stationary region, causing elements near the boundary to rotate. Thus air, in this rotating form, may pass a single point a number of times, that is, it has a multiple pass characteristic. The air pressure in these regions is generally slightly reduced due to the movement. Vortices may also be formed when the air is made to flow around a object which is not a good aerodynamic shape, for example, a cylinder, in which case turbulence forms at the downstream side of the object. Vortices can also form when air is made to flow into a dead end corner of a room; the air is deflected back on itself. The essence of turbulent flow is its lack of unidirectionality. Hence we also call turbulent air non-unidirectional or multidirectional. The lines of flow, or streamlines, within turbulent air are generally non-parallel.

The main problem with turbulent air, as far as contamination control is concerned, is that it can reduce what we might call the cleaning efficiency of our controlled environment. To understand this, we must keep in mind that even the cleanest facilities will have a definite particle generation rate associated with the inside of the cleanroom. The incoming clean air must be able to sweep out this contamination. The presence of turbulence within the room will result in contamination being drawn into and held in the low pressure vortices for as long as each vortex exists. The vortices may be stationary or may move with a velocity in the general direction of the airflow, but in any case the removal of contamination is effectively slowed down or even stopped in some areas. The other problem with turbulent air is that fast moving vortices or eddies can move contamination from the floor, which will usually be the dirtiest region of the room as all the particles will tend to settle there under the influence of gravity. This contamination may be redeposited on work-surfaces or equipment.

We are not stating that any turbulent flow in a cleanroom is totally unacceptable, as no real room will be without its share of turbulent regions, especially when we put people inside. However, where turbulence exists, the concentration of contaminants will inevitably be higher. We must therefore ensure that our most contamination-sensitive areas are free of severe turbulence.

Laminar or Unidirectional Flow

The alternative to turbulent flow is laminar or unidirectional flow. In this case we do not have the vortices of turbulent flow. The streamlines remain parallel, and the air has a single pass characteristic. Unidirectional flow is obviously the best choice for clean areas. A pertinent question at this point would be, "How do we create laminar flow?" It is actually quite simple in practice: keeping the air laminar is another matter altogether.

In the absence of any disturbing influence such as a nonaerodynamic form or a rapidly moving object in the airflow, the air exiting an opening, such as our HEPA filter, will be laminar if we limit its velocity to the range 50 to 150 feet per minute (approximately). A slower airflow will lack the momentum to maintain parallel streamlines, and a faster airflow will tend to intermix. The velocity usually chosen for cleanrooms is around 90 feet per minute (with a 20 percent tolerance). This unidirectional air will effectively sweep out contamination from our clean volume.

Will our airflow remain unidirectional indefinitely? The answer to this question depends on what the physical situation is. For instance, if we confine the flow in the vertical direction using a vinyl curtain or if we have a total laminar flow area (see Chapter 5), in the absence of any turbulence-creating object in the flow the air will remain laminar. If we do not confine the airflow we have a completely different situation. Air exiting from a HEPA filter into an open room can be likened to an air jet. The mathematics of air jets is fairly well understood, but we will not tackle the calculations here. Instead, we will look at the general results of these calculations.

When the air leaves the HEPA filter, it will tend to move laterally as well as downward, that is, the lack of confinement allows a sideways motion of the air due to the push of the air immediately behind. The filter air instantly begins to mix with the room air; if the velocity is too high or too low, turbulence will occur within a few inches from the outlet. Even air with a velocity in the laminar flow range will begin to create turbulence at its edges as momentum is transferred from the moving air to the relatively still air of the room. Despite the shape or dimensions of our filter, and ignoring the turbulent regions at the edges and any influence of the room walls and floor, the moving air will spread out to form a near-circular cone with a spread angle around 25°. This means that the cross section of our airflow becomes progressively larger as we move away from the outlet. This is extremely important as the increasing cross section means a decreasing downward velocity, for example, 1 cubic foot of air passing through a 1-square-foot area at 100 feet per minute velocity will be reduced to 10 feet per minute when spread over a 10-square-foot area. In addition, the velocity of the air will not be constant across the cross section of the flow region. The highest velocity will exist where the air has spread out least, that is, in the center of the flow. The flow decreases from this point out in an approximately Gaussian fashion, as shown in Figure 4.8. The centerline velocity is a function of distance from the outlet, the outlet dimensions, and the initial air velocity, but for cleanroom dimensions it tends to be proportional to the inverse of the square root of the distance.

In practical terms, unconfined air exiting from a single HEPA filter will have slowed down sufficiently to have a significant non-unidirectional component at around 3 feet below the exit point. This means that we should either:

1. Place the filter about 3 feet above the area to be kept clean
2. Partially confine the airflow by vinyl curtains or a hard extension (for example, a sash) from the filter to bring the effective outlet point closer to the clean area
3. Use a bank of HEPA filters (that is, a larger area of HEPA filters) to extend the unidirectional zone

On this last point, since the reduction in velocity depends on the area of the outlet, increasing this area by adding more HEPA filters will result in a

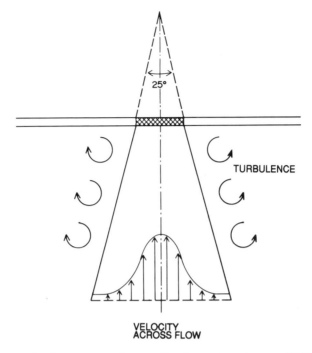

Figure 4-8 Approximate flow profile of air exiting from an HEPA filter.

higher velocity than the single HEPA filter case at the same distance away from the outlet. Hence the air will remain laminar further out. Doubling the HEPA filter area from one to two HEPA filters can provide up to an additional 2 to 3 feet of unidirectional air at the center of the area.

If a turbulence-creating object, for example, a cylindrical pipe, is placed in the path of the unidirectional air, the laminarity will be disturbed a distance of approximately three times the diameter of the object downstream (the "y = 3x" rule), at which point the laminarity will return. However, if the air is made to strike a solid bench or otherwise made to change direction suddenly, turbulence occurs and laminarity can be irretrievably lost. It is thus very important to ensure that work benches are designed in such a way as to smooth the airflow over them. One way of doing this is to taper the wall onto the bench to remove the angular corner and provide a gentle change in direction for the air. If this is not possible, placing the bench to create a gap between the bench top and the wall will help to reduce any turbulent effects in this region by allowing the air to spill behind the bench. Unfortunately, not all equipment we find in our controlled environment will be particularly well shaped in an aerodynamic sense, but as long as the loading points are in unidirectional flow, they should be cleaner than the rest of the system.

The main disruptor of unidirectional flow is the cleanroom personnel. Since

90 feet per minute is around 1 mile per hour, a person walking at a leisurely 3 mph can easily disrupt the flow. What is worse is that a fast moving person can create a low pressure region behind him or her which sucks in particles and carries them to wherever our walking/running individual stops. A little training in airflow and its significance would not be inappropriate here.

We will continue to discuss what happens to the air once it gets into the cleanroom in Chapter 5.

Pressurization

The final subject in this section relates to the pressurization of the room. Proper pressurization, as an important factor in keeping the environment clean, can act as a barrier to contamination.

In typical semiconductor and electronics cleanrooms, positive pressurization is used. The pressure inside the room is typically around 0.05 inches of water above the surrounding areas. This means that there will be a net outflow of air through any cracks in the room, including poor wall seals and deliberate openings such as open doors. The outpouring of air tends to push external contamination away from the openings to reduce ingress.

However, it should be kept in mind that positive pressurization will not completely remove contamination ingress. Personnel entering through an open door will create a region of slightly negative pressure behind them as they move, and this will create a mixing of outside and inside air. It is also possible for small particles to diffuse in air. If the flowrate through an opening is low, that is, lower than the diffusion rate of the particles, contamination can move upstream. This is particularly evident at large openings where the outflow is likely to be relatively small.

Some rooms use negative pressure to keep substances in rather than out. These are used in the nuclear and genetic engineering industries so that any hazardous material is sucked in rather than blown out of the room when an opening occurs. Can we have cleanrooms which utilize negative pressure for research or the production of hazardous but contamination-sensitive materials? It is possible, but it would require an outer cleanroom which is at a higher pressure than both the inner room and the outside to act as a clean buffer.

EQUIPMENT AND MATERIAL DECONTAMINATION

Even though we have considered clean materials and how to clean the air fed to our room, we still have the problem of how to decontaminate equipment and materials going into the room and how to clean them once they have become contaminated by normal everyday use and by accidents. We will discuss some aspects of cleaning in this section.

Bringing Items Into the Cleanroom

Proper decontamination of equipment and materials going into the cleanroom is an important procedure which is, unfortunately, frequently not given the attention it deserves. We cannot afford complacency in this area, as we can seriously and permanently injure a controlled environment by bringing in items, including equipment and materials, contaminated with particles and/ or volatile fluids.

Most items destined for cleanroom use should be manufactured under clean conditions or cleaned by the manufacturer before they are packaged and shipped. We strongly advise that a microcontamination audit be made of any vendor you deal with to ensure that they are making a serious effort to make and/or keep their product clean. Do not expect to see equipment being put together in conditions as clean as your own facility; this is sometimes impractical for the manufacturer due to the types of assembly techniques required. However, you should demand that every reasonable action and precaution is taken during and after manufacture to reduce contamination. This means that all unnecessary dirty activities should be banned from the manufacturing area (smoking, eating, machining of components which will be installed later, and so forth) and decontamination procedures used after every step. We will discuss decontamination techniques later in this section.

As a general point regarding equipment, the buyer should be completely aware of what he or she is buying, not only in terms of how the equipment performs in a manufacturing sense but also with regard to how clean it is likely to be in service. Running witness wafers through to assess internal contamination is not enough, as in many cases the entire machine may be placed in your room and subsequently contaminate the whole process environment if it is dirty. Examine the outer casing. Is the steel coated with a paint which is likely to flake after some time or is the coating scratch resistant? Examine the bottom. That is where people generally forget to clean, and you may also find leaks, uncoated metal, or other hazards which could cause microcontamination problems. Does everything you see inside the equipment, especially dirty subsystems, really have to be built in or can it be located outside the cleanroom? It is up to you to ensure that no corners have been cut when it comes to making the equipment environmentally compatible.

Once the equipment or materials are ready for shipping, they must be packaged in a minimum of two separate layers of wrapping. This rule applies to small items such as chemical bottles right up to ion-implanters. The shipping process is certain to be dirty, so when the item gets dumped on your loading dock, the outer wrapping will be heavily contaminated. There is also a chance that it has been punctured, allowing some ingress of contamination onto the inner covering or even to the item itself. The item is therefore not ready to be brought into the cleanroom immediately, as we must first assess how clean it is. This step can cause problems because the people who will use the item may have been waiting some time for it and are desperate to get it on-line. Your incoming quality control or microcontamination specialists must stand firm.

Before the outer wrapping is removed, a particle count should be taken of the air between the wrapping layers. A large surgical syringe may be attached to a particle counter for this step and inserted into the first wrap. If there is too little air in the package to sample, supply some clean air of known quality into the package; this should flush out enough contamination to allow a determination of how clean the inner layers are. We can also repeat this procedure for the air surrounding the item within the inner wrapping. If you are concerned about contamination sticking to surfaces within the package, a wipe test can also be performed once the outer wrap is removed. A wipe test involves taking a clean filter pad lightly moistened with ultrapure water and wiping across the test surface. The pad is then examined under a microscope to see if it has picked up any gross particulate contamination. It is not a particularly quantitative method, but it does give you a good idea of the condition.

As far as the complete procedure for bringing items into the cleanroom is concerned, it is hard to give specific instructions for every piece of apparatus or type of material. However there are some general rules which may be followed. Take off the first layer of covering out on the loading dock or in an inside area off the dock if you have one. It is much too dirty to bring into the cleanroom. If you have found by sampling that the inner packing is also contaminated, wipe this down with ultrapure water-dampened wipes. If you try to remove the inner packing without cleaning it first, contamination will almost certainly be transferred to the equipment. Many facilities have a clean staging area adjoining the cleanroom for the unpacking of equipment and supplies. If this exists, the item may be moved here for the removal of the inner wrap and final decontamination. If the item itself has been contaminated during its journey, you still have to unpack it in your clean staging area where you have a better chance of keeping further contamination off it. If a clean staging area does not exist, the item in its (cleaned) inner wrapping can be moved into the cleanroom. However, in the case of equipment, it is advisable to erect a temporary vinyl curtain around the installation area to prevent the rest of the room from becoming contaminated while the final unpacking, decontamination, and installation take place.

Cleaning Techniques

We have mentioned several times in this chapter that we should decontaminate materials and equipment before we put them into our cleanroom and keep them clean while they are in use. Exactly how do we do this? Cleaning techniques tend to fall into two categories; those which may be applied to large surfaces, such as the casing of a piece of equipment or a benchtop, and those which are intended for small items such as equipment components and production materials. We will look at both in the next two sections.

We might begin by recognizing that surface contamination usually consists of various substances arranged in layers, as shown in Figure 4.9. For metals, the first layer is an oxide with perhaps some sulfides and other compounds such as salts of trace impurities. Depending on the metal, this layer can be very hard and well attached or can be mechanically unstable, becoming detached very easily. On top of this layer we typically have adsorbed moisture which helps to form a base for oily material (hydrocarbons) absorbed from the air, skin oils from employees, and residual oils from the manufacturing process. We will also find these on nonmetallic surfaces. These oils will help to hold gases on the surface, which is important for vacuum systems, but of more interest to us in the cleanroom is the fact that the oil layer will bond particulate material to the surface so that the total contamination is a mixture of oils and particles. The fact that only the particulates can be seen with a laser surface scanner is a problem.

Now that we know more about the surface, how do we clean it? One approach is the use of air- or nitrogen-blowing in the hope of sweeping particles off the surface. Many studies have shown that while blowing is effective for large (over 20 micron) particles, it is almost useless for smaller entities. Liquid spraying is somewhat more effective, but only if high pressures are used. However, high pressure liquids can erode delicate surfaces and induce electrostatic discharge problems associated with the breakup of large drops into smaller ones. In order to remove small particles, the cleaning media must get right to the surface and break the particles loose. If gases or liquids are used there will always be a boundary layer near the surface where the velocity is low or even zero. This reduces the cleaning effects unless very high velocities are used.

By far the best method of particle removal is perhaps the lowest tech one.

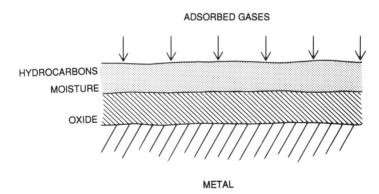

Figure 4-9 Typical surface contamination layers on metal.

Wiping with a cleanroom wiper, that is, one which is made of synthetic low-linting materials, will remove particles with the best efficiency of practically all known methods. Many cleaning operations require the use of a wiper material which is non-contaminating itself. One of the questions that frequently arises is, "Which clean room wiper should we buy?" This is a bit like asking "Which person should I marry?" In principle, we all want a person who has: wealth, beauty, good nature, talent, affection, and a lack of annoying relatives. Needless to say, most of us settle for a bit less. The same problem exists with wipers. We all want good absorption, low particulate generation, effective wiping, and the lowest possible cost.

The IES Recommended Practice (RP) calls for flex testing to look for particles and a solvent test to check for extractables. An extension to this test would be to quantify wiper particle removal performance using a fluorescent powder. The powder is spread on the test surface, and an ultraviolet light source is used to excite fluorescent light emission. The level of emission is measured, and then the wiper under test is drawn over the surface at a known speed, under a fixed load. The reduction in fluorescent emission, after wiping, is taken as a measure of the wiping process. In Figure 4.10 we show some

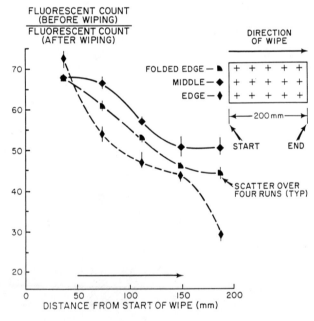

Figure 4-10 Experimental results of removal of fluorescent dust (Anthracene) from a polished stainless steel surface with a cleanroom wiper. Wiper folded once and used for a single stroke. Fluorescence measurement with an ES Corporation, Knoxville, TN, model 107-1000 Luminoscope.

data taken on a particular wiper. The figure indicates that wiper efficiency is quite high (removal factor of 70) at the beginning of the stroke but, after some 200 mm of travel, the wiping efficiency has dropped by some 29 percent. Examination of the wiper after testing indicated that most of the particles are collected by the leading edge of the wiper, as it is the edge which shears the particles from the surface. When the leading edge is loaded with dust, wiper efficiency drops very rapidly. We have observed significant differences between various wipers and in the use of dry versus wet wipers. Thus, when using a wiper to clean a surface, it should be dampened and refolded to expose a new edge after each stroke. Care should be taken not to put a hard crease in the material when refolding, as this could release contamination from the wiper. Unfortunately, a dry wipe will do little to remove the organics from the surface. Poorly manufactured wipers can release particles from their cut edges and contaminate whatever is being cleaned. Quality control is important to ensure that the wipers are of adequate quality, for example, by sampling the air in the wiper packages.

Brushes may also be used to scrub surfaces clean, but they should never be used dry as even nylon bristles can scratch sensitive surfaces if excessive pressure or sweep speed is used. A small amount of liquid will provide lubrication for the tips of the bristles and will help to prevent damage while sweeping particles off. Brushing is a popular technique for scrubbing substrates such as silicon wafers. The liquid here is usually ultrapure water but, once again, this does little to remove organic films.

The general theory of cleaning suggests that organic materials will be removed by organic solvents or fluorinated liquids (for example, Freon). These chemicals break the bonds that hold the contaminant greases or oils together and, in some cases, can interact with the substrate thereby displacing the organic contaminant entirely. In general, solvent cleaning by itself (for example, by immersion) is not totally satisfactory in removing particles unless high pressure spraying is used (with its inherent problems). The reason is that the particles seem to shield the underlying organics from the solvent cleaner, perhaps using surface tension to retain the organic material at their base. One way to increase the effectiveness of solvent cleaning processes is to use ultrasonic agitation in a solvent bath. This helps to reduce any surface tension effects and provides a mixing force. This technique is particularly good for intricate components with hard to reach internal surfaces, but would not be appropriate for a 5-ton ion-implanter. In the final analysis, for exposed surfaces, solvent dampened wipers are an efficient way to both remove particles and dissolve the organics, but the use of solvents is not without its problems.

The choice of the solvent is critical if cleaning is to be effective at all. There is no universal solvent. The choice of solvent must be based on a knowledge of the chemistry involved. Be aware that solvents are not very effective on mineral salts. Some of the other problems with solvent cleaning include:

1. Disposal of the contaminated solvent or solvent-dampened wipes in an environmentally approved way
2. Avoiding operator exposure to the more hazardous solvents
3. Cleaning or recycling the liquid for reuse in solvent baths
4. Monitoring the solvent to determine if the recycling process is effective
5. Removal of the solvent and solvent residue from parts after cleaning. This can be difficult if the solvent interacts strongly with the substrate
6. Inspecting the parts after cleaning to demonstrate that cleaning has been effective

As an alternative to solvents, we may perform wipe-down operations with a variety of detergents in ultrapure water. Detergents interact with oils, remove them from surfaces, and surround them with a layer of detergent molecules that allow them to float away with the water. The big problem here is that there is currently no good and simple method to evaluate the effectiveness of this process and how well the residue of the detergent itself is removed from the surface. If the residual detergents are not removed, they can dry on the surface and generate dust. If the residue is oily, the oil will hold any dust particles that fall on the surface. At the moment, many facilities simply wash down with ethanol and/or ultrapure water while hoping for the best. One major factor in favor of water is that, although it will not attack organics, it will dissolve mineral salts very effectively.

Perhaps the most vital part of any cleaning operation is verification that the job has been done right. We will finish this section with some ideas in this respect. One way of efficiency verification might well involve seeding the article to be cleaned with a fluorescent material that is similar in nature to the other contaminants. Fluorescent powders, containing various sizes of particles, are readily available and glow when illuminated with ultraviolet light. If the fluorescent material is removed, which will be obvious under uv illumination, you know the particle removal technique is efficient. Similarly, if we want to assess a grease removal technique, fluorescent silicone grease may be a good test material (if there is anything harder to remove, we haven't found it). Equipment is available to give a quantitative measure of the intensity of fluorescence, removing much of the subjectivity of the technique. Of course, we do not advocate the use of these test materials in the cleanroom itself; these materials should be kept in the lab.

New Cleaning Technologies

In view of the problems with solvent cleaning and in the knowledge that solvents are almost totally ineffective on salts, other techniques have been investigated. Among them is the application of ozone, water, and ultrasonic energy for cleaning. It appears to be quite practical to remove tightly bonded materials (for example, burnt-on photoresist) with the ozone/water/ultrasonic system. The ozone/water/ultrasonic system has a number of advantages, including: (1) The solvent (water) can be disposed of quite easily. No elaborate recycling system is needed, and (2) the ozone decays naturally into oxygen after a short period of time, particularly in the presence of liquid water or water vapor. Exposure of personnel to ozone during the cleaning process is avoided by doing the work in a ventilated hood.

The experimental ozone/water/ultrasonic system is shown in Figure 4.11. The water is saturated with ozone using a pressurized vessel cooled by ice. (The solubility of ozone in water goes up with increasing ambient pressure and a reduction of ambient temperature.) As the ozone charged water flows over the area to be cleaned, an ultrasonic generator is used to increase the rate of reaction. One important point here is that the ultrasonic beam is not in a wide format such as that normally used in cleaning equipment. It is a very concentrated beam that seems to be able to strip off contaminants quite easily, as is demonstrated by the fluorescent seeding technique. Unfortunately, at this stage in the development of the technique, the item to be cleaned still has to be immersed in water, and therefore only relatively small components can be treated.

An extremely promising technique for surface cleaning, which does not rely on liquid immersion, involves the use of dry-ice snow as a surface cleaning

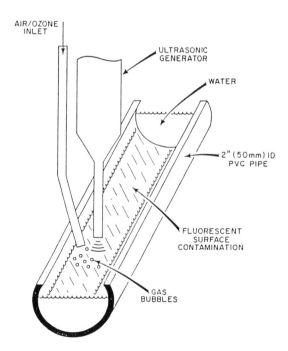

Figure 4-11 Schematic drawing test assembly for ozone/water/ultrasonic cleaning.

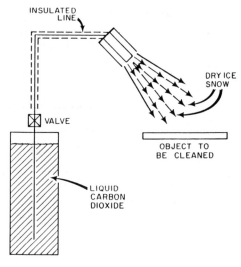

Figure 4-12   Scehmatic drawing dry-ice snow cleaning technology (scale none).

medium. The snow is generated from class 4 (ultraclean) liquid carbon dioxide using the system shown in Figure 4.12. The dry-ice snow particles slide across the surface on a layer of gaseous carbon dioxide and push dust particles off the surface. To date there has been no evidence of any surface damage due to the cushioning effect of the gas. The system has been used to clean Winchester hard disks and germanium mirrors. The disks had been lightly oiled at the factory, but there was no problem in removing metallics and nonmetallics from the surface. The same system has been used to clean finished semiconductor integrated circuits that had become contaminated by silicon dust. The dry-ice snow can also be used to remove oily contamination if the flakes are allowed to impact the surface at an angle close to 90°. We suggest that this process involves local melting of the dry-ice snowflakes as they hit the surface. Liquid carbon dioxide is well known to be an excellent solvent for organic materials, and this may explain why fingerprints and light organic contamination may be removed with dry-ice snow. If stronger cleaning methods are needed, the dry-ice system can be modified to generate very hard pellets that can be used for sandblasting. Cleaning of deposition residues in LPCVD systems and vacuum evaporators are possible applications in this area.

We met electrets earlier in this chapter in the context of removal of dust from the air. These remarkable materials may also be used in decontamination applications. For instance, the electret material may be cut into strips to form a brush which will attract particles. A version of this brush is currently in use in the magnetic disk and tape industry. Here, the problem of concern is magnetic particles as well as other particulate materials which can deposit on the read/write heads and interfere with the data storage and retrieval process. To solve this problem, the electret is combined with a magnet to help to retain magnetic particles once they have been removed from the surface being cleaned. After the electrostatic/magnet brush has been used, the magnets are cleaned by stripping off a thin layer of tape.

SUMMARY

In summary, we require certain things of our cleanrom construction materials. Most of all they must not contribute significantly to the particulate or vapor contamination levels within our environment. If we are using cleanroom materials that have a core capable of releasing contamination, it is extremely important to seal holes and cuts made during construction, after renovation or additions, and as a result of accidents. In addition, the cleanroom should be considered clean as soon as the internal surfaces are put in place, and materials and equipment brought in at any time after this should be decontaminated very carefully. High-efficiency air filtration is possible for particle reduction, but gases and vapors may not be removed so readily. Contaminants within the room are more readily swept out by unidirectional airflow than turbulent flow, therefore control of turbulence is important for open cleanroom areas.

We will return to issues regarding the decontamination and the prevention of contamination of production materials and the product in Chapter 10.

5

Cleanroom Layout

Since we now have some understanding of the problems of microcontamination and have acquired some elementary knowledge about how to create basic clean areas, it is time to tackle the problem of how to put together an actual cleanroom. We have a number of options in this respect, some of which we will describe in this chapter.

MIXED FLOW ROOMS

Before we start our description of the mixed flow cleanroom, we must make one (rather obvious) point abundantly clear. Assuming that we have constructed our room out of clean materials in a clean fashion, and that whatever is in the room is clean or acts clean, the way to make and keep the volume clean is to supply clean air. Therefore the air has to come in somewhere, at the inlet, supply, or delivery points. However, unless we want to blow our room up like a balloon, we must remove the same volume of air from the room as we put in. Therefore we must have removal or exit points for the air, too. These are typically called air returns, for the simple reason that this air is actually returned into the system for re-use (see Chapter 6). All rooms must have supply and return features.

Air Delivery

In mixed flow rooms, as the name suggests, not a great deal of attention is given to airflow patterns, at least compared to rooms with unidirectional flow regions. The supply air is fed from ducts which have HEPA filters at or near their ends, as shown in Figure 5.1. Care must be taken not to exceed the air velocity rating of the filters in trying to supply large volumes of air to the room, otherwise they will become stressed and particles may literally get blasted through.

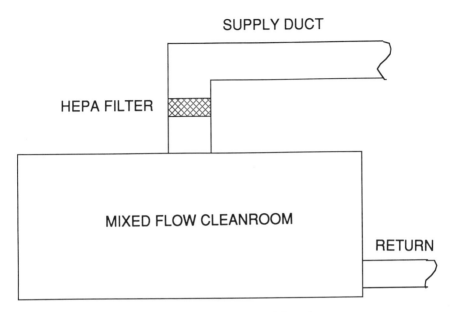

Figure 5-1 Schematic of a mixed flow cleanroom.

The air entering the room is therefore relatively clean. However, once in the room volume, it will spread out under the force of its own push. In occupying the larger volume (compared to the duct), the air will tend to slow down and move in all directions, eventually being deflected by solid objects (equipment, people, walls, and so forth). The air therefore mixes in the room, becoming highly non-unidirectional. Turbulent vortices will exist in a number of areas in the room, especially at the corners, and contamination reduction by sweepout will thus be poor. Because of these airflow patterns, a mixed flow room cannot be expected to maintain better than class 10,000 (best case). However, since the design is simple, with little in the way of expensive HEPA filters, the cost of this type of room is minimal.

A variation on this theme involves the use of a clean perforated ceiling, suspended on inverted T-bar supports, below the outlet level of the duct in the actual ceiling (Figure 5.2). This perforated layer acts as an air diffuser, and the entire structure effectively forms a *plenum* region. This plenum contains air at slightly higher pressure (0.2 inches of water maximum) than the room below so that the air enters the room evenly across the area of the diffuser. We will meet the concept of the plenum again later in this chapter, but the principle is fairly simple. A plenum should always be used to equalize the pressure across the area of a filter (or in this case across the area of a diffuser) when the air enters at single points. If a plenum is not used, there will be local pressure variations and uneven flow through the filter(/diffuser) region.

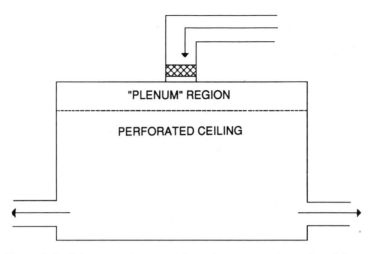

Figure 5-2   Schematic of a mixed flow cleanroom with a ceiling diffuser.

Since in the case described above, a small number of ducts and HEPA filters are used to feed this plenum (again to minimize cost), the exit velocity of the air through the perforations into the room will be too low to create true laminar flow. This is merely a consequence of having a limited airflow through the HEPA filters so that the corresponding velocity through the perforated ceiling, for the same air volume, has to be lower. Hence, this is still really a non-unidirectional flow room, although there should be less turbulence than before and class 10,000 will be easier to attain.

Air Return

In order for a room of the mixed flow type to be successful at all, the return vents have to be positioned with some care. These will take the air away from the room and therefore must be at reduced pressure with respect to the room. This means that the air will be attracted to these vents and will move preferentially in their direction. We have seen mixed flow rooms that have return vents placed in the ceiling, which is actually common practice for office air conditioning systems but is a complete disaster as far as cleanrooms are concerned. This is a cost-effective way of installing returns as all the air ducting can be hidden above a (solid) false ceiling. In this case, the clean air coming into the room from the inlet duct will tend to loop around to some degree and exit without getting anywhere near the work areas. Particulate contamination will exist happily at the lower levels of the room, being stirred up by the movement of personnel and doubtless landing on whatever is to be protected. If the velocity of the rising air is high enough, it will actually take particles upward, potentially allowing them to rain down on the work areas.

The proper position for air returns is actually as near to the floor as possible, if not in the floor itself. This is so that the air is ultimately persuaded to flow down through the room, taking some contamination with it. Rising air is therefore reduced somewhat. It is a more expensive design, as ducts have to be run in the walls or floor, but it is a necessary expense. The best rooms with sidewall returns are those which are narrow, for example, around 12 feet wide. They tend to perform better as the air is more likely to flow down to the region near the floor before it turns to move across to the return vents. Cross-flow air, which turns at higher levels in wider rooms, can be a problem as it may pass by relatively dirty employees, carrying their contamination onto the work bench. We will return to this idea later.

LOCAL UNIDIRECTIONAL FLOW

Mixed flow rooms are generally low cost but by no means provide a very high standard of cleanliness, at least as far as the high-technology industries are concerned. However, we can alter the nature of these areas by providing local areas which are much cleaner than the rest of the room. We will discuss how to do this in this section.

Fan Units

One of the simplest methods of providing local high cleanliness levels is by the use of self-contained fan units or laminar flow cabinets (LFC). These are essentially boxes which contain a small electric motor-fan set and a HEPA filter. Air is either taken directly from the room, through an integral prefilter, or preconditioned air may be ducted directly to the fan unit by way of a flexible duct. Some systems even go as far as to condition the exit air to some degree, that is, the temperature and humidity are controlled.

LFCs are extremely versatile and allow a great deal of production/research area flexibility. They may be ceiling mounted as single units or in banks if a structure exists to attach them to, or may be supported by C-frames as shown in Figure 5.3. A C-frame is an open-fronted frame which makes it relatively easy to place the LFC where it is needed. Many fan units have the HEPA filtered air discharging at the bottom of the cabinet to create a vertical laminar flow (VLF) region. Other configurations have the filtered air exiting horizontally as shown in Figure 5.4, to form horizontal laminar flow (HLF). The units come in a variety of sizes, the most common being 4 and 6 feet long. Thus the fan units may be positioned to create clean work spaces at assembly, inspection, extracted chemical benches, or equipment loading points.

The main thing to keep in mind with these systems is that they are not infallible. In other words, do not expect them to create a super clean, highly controlled environment under any ambient conditions. A rule of thumb which is often used is that the ambient has to be no worse than a factor of 100

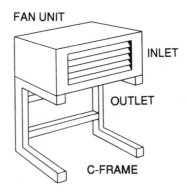

Figure 5-3   Fan unit (LFC) on a C-frame.

lower in class than what you are trying to achieve under the cabinet. Therefore, if you have a unit which is designated as a class 100 system, it should be operated in a room no worse than class 10,000. For higher classes, for most practical situations, it is best not to be more than one class away. The reason for this is that any activity in the clean work space will tend to mix the LFC air with the ambient air, and local contamination levels will rise as a consequence. Therefore, we frequently find class 100 LFCs in class 10,000 mixed flow rooms and class 10 LFCs in class 100 laminar flow rooms. Of course, any materials, tools, or clothing which enter the clean space must be compatible with what class you are trying to achieve, otherwise it would be nearly impossible to maintain this class.

LFCs are very economical in small manufacturing or research areas but become an expensive option if many units are required. Therefore, in large cleanrooms, the air is prefiltered, conditioned, and driven by centralized units. This aspect is discussed in Chapter 6. One other disadvantage of fan units is that since they contain their own motor-fan sets, they are generators of vi-

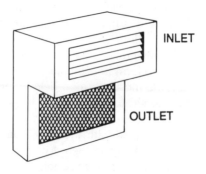

Figure 5-4   Horizontal LFC.

bration and therefore have to be isolated from sensitive systems. This is discussed in Chapter 7.

Workstations

The workstation is an extension of the LFC in that it is essentially self-contained, but we now also have an integral workbench. There are three main varieties of workstations:

1. *Horizontal.* The air enters through a vent above or below the worksurface, is prefiltered and driven through a HEPA filter to exit horizontally across the work surface. This effectively blows out any contaminants which attempt to enter the work space, but it will also blow any particles, fumes, gases, and so forth generated in the work area out toward the operator. This precludes the use of these units with hazardous materials.

2. *Vertical.* The air is taken as before through an inlet vent, is prefiltered, and then blown through the HEPA filter to exit vertically down onto the work surface. The work surface may be designed to deflect the air in a unidirectional fashion out the front of the bench, or the surface may be perforated to allow most (around 60 percent) of the air to pass through and be discharged by way of a low-level vent under the work surface.

3. *Exhausted vertical.* An exhausted workstation will take all or a significant portion of the air which is blown into it through vents in the work surface or back and deliver it to an appropriate chemical exhaust system. These systems are used with hazardous materials such as acids and solvents. The design of this type of workstation is not as simple as one may expect because careful air balancing is necessary, that is, the air passing by the areas of the bench which will have the hazardous material (usually near the back) must be extracted using negative pressure produced by an exhaust blower. However, if more air is exhausted than is blown in, air will be drawn from the ambient and may contaminate the work space. We therefore have to have a small amount of spill-out near the front of the bench with the rest of the air being exhausted. This type of system is quite wasteful of air as most of the precious filtered (and conditioned) air is discarded. This has prompted the design of alternative systems such as that shown in Figure 5.5. Here a local high velocity air knife or curtain is used to direct hazardous vapors to the exhaust with a reduction in air volume lost over conventional systems. Note that exhausted workstations differ greatly from traditional chemical fume-hoods in that the latter systems draw in air directly from the ambient without filtering and are therefore unsuitable for contamination-sensitive applications.

Spot VLF

Spot vertical laminar flow is similar in some respects to the distributed LFC approach, as the higher cleanliness laminar flow regions are only placed where

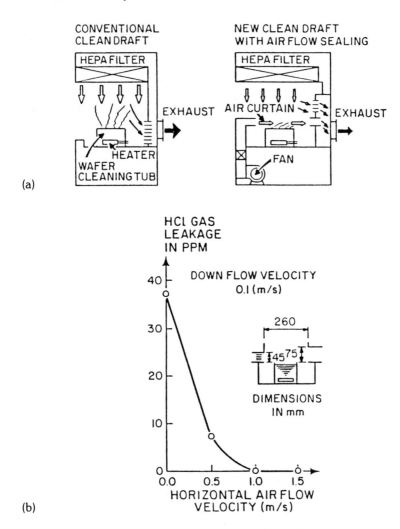

Figure 5-5 (a) Schematic drawing of air curtain system for acid fume control. (b) Figure shows the reduction in HCl gas leakage with the new system (data provided by Prof. T. Ohmi, Tohoku University, Sendai, Japan).

they are required within the room. In the case of spot VLF, however, the air may be supplied by LFCs but is more likely to be ducted by means of flexible ducting hose into passive HEPA filter units, as shown in Figure 5.6. The air in this latter situation is preconditioned and driven from one or more central air handlers (see Chapter 6), and hence there are no fans in the cleanroom as in the LFC case. One type of passive HEPA filter unit mounts below the ceiling level to bring the clean air outlet as close to the work area as possible.

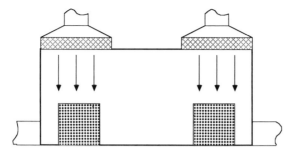

Figure 5-6 Spot VLF room with ducted supply air and side returns.

This is important, as when the airflow spreads out, the velocity drops as a function of distance below the discharge point and hence turbulent air will exist at large distances from the HEPA filter face. In this case the HEPA filters are mounted in a box unit which acts as a pressure equalization plenum, and the filters may be thicker than in the LFC case since there is no blower to take up space in the unit. (An 8-inch-thick HEPA filter is typically twice as efficient as a standard 6-inch filter). An alternative design of passive unit mounts flush with the ceiling level, having a relatively small plenum above and a sloped reducer neck for the connection to the supply duct.

Rooms which use spot VLF are really a cross between mixed flow and laminar flow. Unidirectional flow only exists under the HEPA filter outlets. The rest of the room is effectively fed with this same air, which will spread out laterally to mix in the room. A typical example is shown in Figures 5.7 and 5.8. In this example the HEPA filters are placed next to the walls of the room so that the air is partially confined and therefore spills out toward the middle of the room only. In many cases, the airflow in the regions not under the HEPA filters can be virtually zero, leading to stagnant air and a corresponding build-up of trapped contamination (a veritable Sargasso Sea). Additional HEPA filtered air inlets, which supply clean but non-unidirectional

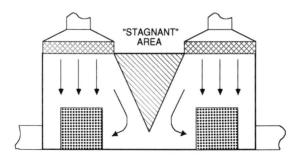

Figure 5-7 Air patterns in a typical spot VLF room.

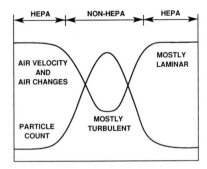

Figure 5-8   Particle counts and air velocity at workbench level for the room cross-section of Figure 5-7.

air, may also be included to reduce this effect by supplying a downflow or a transverse flow of air (depending where the return ducts are located). These rooms may therefore be considered as mixed flow rooms with local laminarity.

The air is returned by way of low-level ducts as in the mixed flow room case. The path may be through floor or wall vents, or a perforated raised floor (discussed more fully later in this chapter). In this latter case, not all the tiles need be perforated, only a sufficient quantity to handle the return air volume. A further alternative may be used if zones in the cleanroom are to be separated to create narrow "bay" structures. Hollow walls may be dropped from the ceiling, stopping short of the cleanroom floor by a few feet, and these can act as a return path to an overhead return plenum (Figure 5.9).

The main objection to this design of cleanroom, as we mentioned earlier, is that the level of cleanliness is highly nonuniform across the room (Figure 5.8). While it may be possible to maintain, for example, class 100 within the laminar flow regions, the rest of the room will have fairly non-unidirectional airflows and will therefore be less clean. Depending on the amount of spot VLF in the room, the rest of the room will typically vary from around just

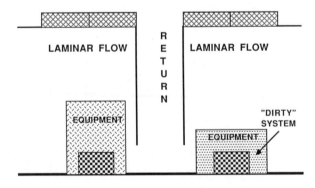

Figure 5-9   "Bay" layout with air return walls.

below class 100 (for 80 percent or more spot VLF) to near class 10,000 (for less than 10 percent spot VLF). Most rooms of this type have between 30 and 50 percent spot VLF, and the non-unidirectional regions tend to be around a factor of 10 worse in class, assuming, of course, that clean materials and practices are employed which are commensurate with the highest class areas.

MODULAR CLEANROOMS
Self-contained Systems

The first clean "rooms" were really just clean workstations. These offered an advantage to users in that they could be purchased off the shelf, delivered ready to go. They provided islands of acceptable cleanliness in relatively dirty environments but could not be expected to provide high cleanliness workspaces unless the ambient was controlled to some degree as discussed on page 83. We now can purchase entire rooms in the same off-the-shelf fashion, which enables us to set up total high cleanliness environments very rapidly.

A typical modular cleanroom unit has at its heart a portable cabin type of room which may readily be shipped by truck in much the same way as a mobile home unit. The internal walls are of a suitable cleanroom material, and some services such as power, ultrapure water, nitrogen, drains, and so forth may already be pre-plumbed. A roof- or side-mounted air conditioning/ air handling unit is attached to the room, providing pre-conditioning of the air (see Chapter 6) and a fan to drive the air through the ceiling- or wall-mounted HEPA filters. An integral control system is also included. The air returns may be in the floor or low on the side walls, and there may be a return plenum and ducts to take the return air back to the air handling unit. The alternative to the return plenum is to discharge the air into the room in which the modular cleanroom is placed. This may sound a bit odd, but many of these units are designed to be put inside an existing building to provide a low cost controlled environment capability. The supply air is taken from the room and the return air goes into the room. Air to replace air lost from exhausted hoods is supplied by the external room's "make-up" air system.

These rooms are generally designed in such a way as to allow them to be interconnected in some fashion in order to maximize flexibility, for example, a number of fairly standard units may be fitted together to create the desired size, and to some extent shape, of room desired.

Retrofit Systems

There are many times when a company wants to gain some cleanroom capability within an existing building or upgrade an existing room to allow them to produce a product which demands a higher degree of environmental control. If the room size is to be relatively small, a modular unit like those described above may be ideal. Unfortunately, in the case of the semiconductor industry, most processes require many large pieces of very heavy equipment (ion-implanters and the like), and this generally means that we need a spacious room with a good stable floor.

If this is the case, we would probably be better off by retrofitting the space with a cleanroom. This is not quite the same as buying an entire room off the shelf, but standard parts are available to some degree. For instance, there are manufacturers who supply wall kits which include a bolt-together structural frame, wall panels of an appropriately clean material, and joint seals. A structural or suspended ceiling may also be installed from a kit, and a solid or raised perforated floor is also quite readily available off the shelf. We then put in standard LFCs or ceiling-mounted HEPA units, provide an air handler and ducting for supply and return air, and we basically have our cleanroom.

If you think all of this sounds too simple, you are correct. In retrofitting a cleanroom we still have to do a great deal of engineering design to make sure we get it right, for example, how much air do we supply to the room and how many HEPA filtered outlets will be necessary to maintain the desired class, and so forth? In many cases, retrofitting a cleanroom can actually be considerably more difficult than designing and building one from scratch. When we build an entirely new building, we can design the cleanroom to suit the process and then design the building to fit the cleanroom. When we retrofit, we have to work with what we have, and this can lead to difficulties, for example, our room which will contain the cleanroom may be large enough but it may have support pillars every 15 feet, it may not have a basement when we need one for services, and so forth.

TUNNEL CLEANROOMS

We will now discuss some different cleanroom configurations. There are basically two main schools of thought as far as the macroscopic layout is concerned: (1) complete openness, with spot VLF where required or total laminar flow; (2) a tunnel configuration, which is a completely different use of cleanroom space and is discussed in this section.

Tunnel Layout

The basic tunnel layout is shown in Figure 5.10. A central aisle provides access to a number of work areas or bays which contain the process equipment, wet benches, and so forth. Each bay may contain equipment which performs similar tasks, for example, plasma etchers, or may contain a complete work cell, for example, a lithography and etch group, but the bays are

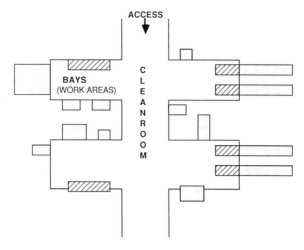

Figure 5-10   Typical tunnel cleanroom layout showing examples of equipment placement.

only as large as they need to be for these purposes. Equipment integration in these rooms is discussed later in this chapter.

The principle behind the tunnel configuration is simple. The area with the most traffic and movement is the access aisle, and this will also be the dirtiest region in the cleanroom. However, in this layout, no processing is performed and no work is exposed in the aisle; all contamination-sensitive materials and processes exist in the bays only (at least this is the plan). The relatively small cleanroom area in the bays helps to reduce cleanroom costs by lowering the cost of filtration equipment and reducing running (energy and maintenance) costs. It also makes it easier to control certain aspects of personnel work practices, in the respect that maximum numbers of people in work zones can be checked more readily; some facilities use an electric eye detector to count the number of people entering each bay. This is an extremely critical aspect which is often overlooked, but the contamination in an area of a cleanroom will rise dramatically with the numbers of people present. The tunnel layout also means that employees rarely have to pass through another person's work area, contaminating it in the process, in order to reach their own.

The bays may be left open (as shown in the figure) or closed by some sort of door or curtain. A physical barrier allows a higher degree of contamination control in each bay, as it helps to isolate them from the service aisle and each other. The disadvantage of solid doors is that many doors, whether swinging or sliding, generate contamination themselves. It also can be quite difficult to control the air balancing in a dynamic fashion if doors are constantly being opened and closed. Non-sealing soft curtains may be a better solution, although these will tend to get dirty as people brush past them going in or out of the bay, and therefore they should be cleaned regularly. A further alternative may be an air curtain or air knife across the entrance to the bay. Great care should be taken here as this may end up blowing contamination off personnel into the work area as they pass through. The alternatives should be studied before a scheme is decided upon.

Service Chase Configurations

The tunnel cleanroom configuration also has another advantage. Since the clean work areas are long, relatively narrow regions, we may space them in such a way as to allow access to the back of equipment placed in the bays (see Figure 5.10). Alternatively, relatively dirty subsystems may be taken out of the processing equipment in the cleanroom and placed in the space behind the cleanroom wall (Figure 5.11). Some equipment is designed so that only the loading areas protrude into the cleanroom and the rest is placed on the other side of the wall, as shown in Figure 5.12 (also see pp. 101–103). In all cases the major advantage of this layout is that maintenance may be carried out on the equipment without going into the cleanroom itself. Maintenance procedures generate a great deal of contamination, as actions such as driving screws and bolts, soldering, replacing vacuum pump oil, and so forth, will produce a shower of various contaminants. It is much better if maintenance can be carried out in the non-cleanroom zone, commonly called the *service chase*.

The service chases themselves may be arranged in a number of ways. In small non-tunnel cleanrooms, the chase may ring the cleanroom, providing service access to equipment at the walls. In tunnel rooms, the chase may be entered from the cleanroom access aisle, the area outside the cleanroom, or both. This latter arrangement could be a bad idea, as it allows the mixing of cleanroom and non-cleanroom items and clothing. Much also depends on

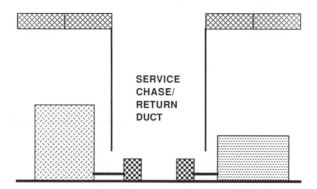

SERVICE CHASE/ RETURN DUCT

Figure 5-11   Tunnel cleanroom section showing service chase/return duct configuration.

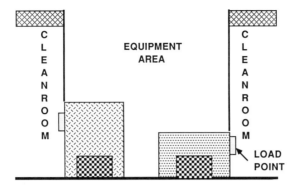

Figure 5-12   Tunnel cleanroom section showing through-the-wall equipment integration.

how the chase is used in other respects. For instance, many cleanroom designs call for the chase to be used in the air return path, as shown in Figure 5.11. This is an economical use of space in the facility as it incorporates the air return configuration of Figure 5.9 with the chase. If the chase is used in the air return, care should be taken not to contaminate the chase air. This is an important fact which deserves further discussion.

In principle, as we mentioned previously, the chase is a good location for things that need to be near but not in the cleanroom. Equipment repairs, electrical and plumbing services, vacuum pumps, and the like are frequently associated with or located in the chase. All of the above does not imply that the chase is a good location for rusty and dirty gas cylinders, old mops and brooms, leaking oil pumps, broken quartz, and general trash. The reason should be quite clear; any dust or oil vapors that get into the air will be carried back to the HEPA filters. The filters will remove only some fraction of the particulates and none of the vapors before dumping the air back into the cleanroom.

Some companies try to keep the chase as clean as the cleanroom or at least class 1000. If the chase is the same class as the cleanroom, it may be entered from the cleanroom side. If the chase is a lesser class, it is better if it is accessed from a separate maintenance clean access corridor (this takes up a considerable amount of space in the facility but is well worth it). If maintenance employees have to enter the chase, they naturally have to be gowned to reduce dust generation, and the use of clean tools, such as "exhausted" torque drivers, which remove generated particles by suction, is advocated. The walls and the floor of the chase are treated like cleanroom walls in order to reduce dust and vapors. The chase should certainly be cleaned as often and to the same degree as the cleanroom itself.

Does all this mean that it is a completely bad idea to locate dirty equipment subsystems in the chase/air return? Not if care is taken not to abuse the privilege. If an item absolutely must go in the chase as opposed to having it in the cleanroom, it should be modified to prevent excessive contamination. Examples of such modification include jacketing gas cylinders (with their own cleanroom suits) and exhausting cabinets around vacuum pumps, taking care to supply enough cooling air to the pump.

There is a further point to consider regarding the airflow in service chases. By virtue of the design of the chase/air return, the air flows upwards and is therefore exerting a force on particulate matter opposite to that of gravity. Therefore, some particles, particularly the larger varieties, may not be swept

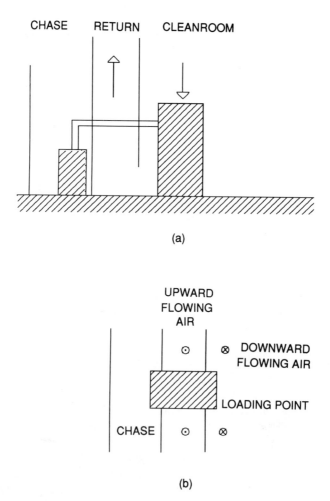

(a)

(b)

Figure 5-13   Cleanroom with separate air return and service chase. (a) Side view of equipment with remote pump in chase. (b) Plan view of through-the-wall equipment showing rear service access from chase.

out of the chase area and will be tossed around by any turbulent air present. This could have some far-reaching consequences for equipment maintenance if the sensitive innards of equipment are exposed to this air.

Of course, the return air does not always have to go through the service chase, as other return configurations are possible. For instance, a double-walled system may be used to create a chase behind a vertical return duct, as shown in Figure 5.13. In this case the chase has to be fed with clean air, fed from the top and ducted out at low level, to keep the inside of equipment reasonably clean during maintenance. The chase air is not mixed with the cleanroom air at any time. It is probably not worthwhile going to any great length to make the chase air unidirectional, as it will probably go turbulent over the somewhat non-aerodynamic surfaces of stripped-down equipment anyway.

Air Supply and Zoning

A further advantage of the tunnel layout is that it is relatively easy to create zones in the room which may be supplied with air which is "fine tuned" for each zone. An example of a zoned scheme is shown in Figure 5.14.

Bay 1 is a general equipment bay with two banks of HEPA filters on either side (a & b) being fed with 3500 cfm of air each. The filter banks in bay 2 (e & f) are also fed with 3500 cfm each, but the air going into bank f is totally extracted, as there are chemical benches underneath. The aisle units (c,d,g,h) are fed with 1500 cfm each. Bay 3 is a lithography bay with both banks (i & j) being fed with 3500 cfm each. The air returns for the bays are placed low in the sidewalls; in many cases the sidewalls stop short of the floor by some 2 or more feet to allow free passage of air. The aisle returns are evenly spaced along the lower parts of the aisle walls. The returns are balanced to prevent aisle air from entering the bays.

This cleanroom could be zoned as follows:

a + b + c + d—fed from one 10,000 cfm source

e + f + g + h—fed from another 10,000 cfm source, for example, a slave air handler linked to the controller in the unit that supplies bay 1

i + j—fed from an independent source which has a tighter control over temperature and humidity

In this case, most of the cleanroom does not need a high degree of control over air parameters, but the lithography area does. It makes good sense to zone the cleanroom in this way as tight tolerance levels are expensive to attain in terms of capital equipment costs and energy bills, so we should only have supercontrol where we need it.

Of course, this is not the only way we could zone this room. For instance,

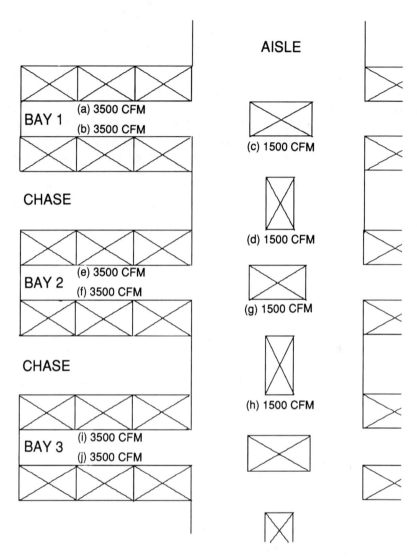

Figure 5-14   Tunnel cleanroom plan showing air supply zoning.

we could put the extracted benches on a completely different system to reduce the chances of dispersing harmful chemical vapors to the rest of the room, that is, if some of the vapors escape, they would be returned to the same area so that the problem is contained (vapors should never be allowed to escape, as it is bad health practice as well as poor microcontamination practice).

Besides zoning the room, we can create extremely clean bays by innovative local zoning. Figure 5.15 shows a tunnel system pioneered in Japan for class 10 work areas. As we can see, we have much the same equipment config-

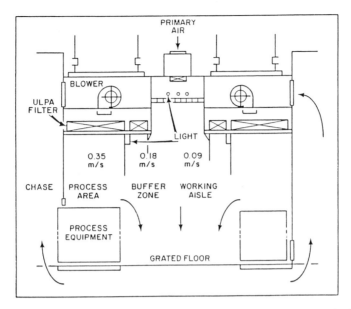

Figure 5-15   Japanese clean tunnel (from paper given by N. Jinno at the Institute of Environmental Sciences meeting Dallas/Fort Worth, TX, May 6–8, 1986).

uration as before, with the process areas running along the walls of the bay. However, the LFCs have a split air system, allowing a buffer zone to be created between the process area and the center of the bay where the personnel will be. A transparent sash maintains the laminarity of the 90 fpm air over the work area, and this air is allowed to spill out after reaching the work surface. The average airflow velocity in the working aisle is low (around 25 fpm) for economy, and therefore this area will be dirtier than the process area. The buffer zone helps to prevent mixing between these two areas, especially when personnel are working at the equipment. In this particular scheme, the air return is through a grated floor; we shall meet this again in the next section.

TOTAL VERTICAL LAMINAR FLOW

Until this point we have concentrated on having HEPA-filtered air outlets only in the most sensitive areas. This, of course, leads to local variations in levels of cleanliness. In many cases, particularly in cleanrooms which must be uniformly clean and open to allow maximum flexibility, spot VLF is unacceptable and thus we use total vertical laminar flow. This is sometimes called 100 percent VLF, but it is a bit of a misnomer. It is very difficult to make HEPA filters cover 100 percent of the ceiling, as we have to have supports for the filters themselves, lighting, sprinklers, and so forth. It would not be a good idea in general to hang these items below the HEPA filters as they could introduce contamination into the airflow.

Total VLF rooms are naturally cleaner than spot VLF rooms of the same size, as we can put much more clean air into them through the larger number of HEPA filters. Also, since there should be no zones of lower cleanliness as such, we do not have the problem of operator movement mixing the clean and dirty air. It is therefore more common to see total VLF in the highest standard cleanrooms. However, it must be kept in mind that this configuration is an expensive one, requiring higher capital investment (filters, and so forth) and higher running costs (energy, and so forth).

The simplest total VLF room configuration, used in small facilities, is shown in Figure 5.16. The air supply is ducted to the ceiling-mounted HEPA filters and air return is by low side wall vents. It is important to realize, however, that this configuration is not as flexible as one may first think. Since we are removing the air from the room at the walls, there will be a tendency for the air flowing down from the ceiling to change direction, that is, it will develop a horizontal component of movement. This can lead to a serious effect called *tenting*. The air entering the room near the middle begins to change direction at a high level. This can lead to a tent-shaped "dead zone" in the middle of the room in which the airflow is minimal. We have, therefore, created much the same kind of problem that exists in spot VLF rooms because we now have a dirty area next to the clean zones. If side returns are necessary due to budget or other factors, the best way to minimize this problem is to keep the width of the room to around 12 feet (15 feet maximum). This makes the room rather bay-like, and indeed this approach is used for total VLF tunnel cleanrooms.

In rooms which cannot be so narrow, we must rethink both the supply and return configurations. In large total VLF rooms, it is not particularly economical to duct air to all the HEPA filters. In this case it is better to seal the HEPA filters into the ceiling grid and use a supply plenum. As discussed

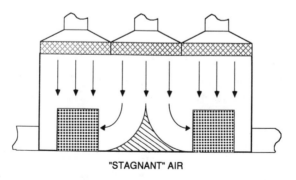

"STAGNANT" AIR

Figure 5-16   100 percent VLF (ducted supply, side return) cleanroom bay showing dead air region in the middle of the work area.

earlier in this chapter, the plenum is really like a single large-volume duct, supplying all the filters. The size of the plenum is chosen such that any internal pressure differences are reduced in order to create an even flow of air through the filters across the area of the ceiling. In zoned rooms, there are usually a number of independently supplied plena.

Air return is more of a problem in large rooms. Since we cannot use sidewall returns, we have to return the air through the floor. There are a number of ways to achieve this, but we will examine only two examples here. The first configuration, shown in Figure 5.17, shows the cleanroom as the middle story of a three-story structure. The work area is placed on top of a false or raised floor, which is perforated to allow the air to pass through and be carried out by way of ducts. This raised floor consists of rigid panels, held up by jacks at all four corners for stability (heavy equipment rests on separate supports). The panels can be drilled thick laminate material or cast metal coated with vinyl or anodized to reduce contamination. Depending on the design, the amount of free space for air transmission can vary from 30 to 60 percent. Unfortunately, the most open types, although being the best for getting air out of the room, are usually the worst in terms of comfort and wear and tear on cleanroom boots. In Figure 5.17, a service basement is also shown. This equipment service chases can be set up in the cleanroom to segregate maintenance activities from production. These chases can be defined using moveable wall sections to retain the flexibility of this configuration.

If a raised floor is undesirable for reasons of strength or stability, a purpose built structural return floor may be the solution, if perhaps an extremely expensive one. Figure 5.18 shows this "king" of cleanroom structures. We have, in effect, a four-story structure, crowned by the supply plenum. The

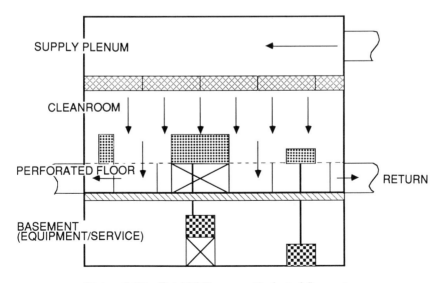

Figure 5-17 Total VLF room with ducted floor return.

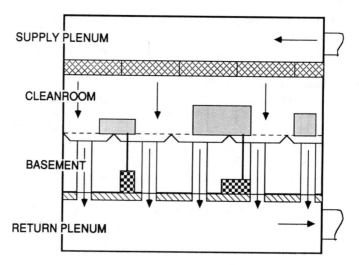

Figure 5-18   Total VLF room with return air ducted through basement.

cleanroom itself is next, resting on a structural floor rather than a raised false floor. However, this structural floor is built out of sections, each one allowing air to pass through its top surface to be collected by sub-floor ducts which carry it down through the basement to a return plenum region. This way the basement is as close to the cleanroom as possible, an important point as vacuum systems need to be close to the machines they serve. The air is ducted through the basement to prevent its becoming contaminated by basement systems. In this configuration, the air handlers may be located in the return plenum.

HORIZONTAL LAMINAR FLOW

In most of what we have said so far, we have taken the direction of the incoming air to be vertical, so that the workspace is bathed in clean air from is important, as pumps and other dirty systems can be placed here. In addition, the top and contaminants are driven out towards the floor. This seems a fairly sensible approach as we effectively create a contamination concentration gradient from the top of the room to the bottom, and as long as we do not do any sensitive work on the floor, this generally works. However, this approach does require a fair amount of "overhead," in that our ducts and filters have to go into or at least near the ceiling and this takes up space (in the form of another floor in the worst case) and produces maintenance problems. This latter factor should not be overlooked as it is awkward for we humans to work on things above head level, and anything we do to the cleanroom ceiling will cause a rain of trouble below. Therefore, if vertical laminar flow is not absolutely necessary for the application in mind, horizontal laminar flow may be a good option.

In horizontal laminar flow, one wall of the room is made up of HEPA filters which may be supplied by a plenum or individual ducts. The air sweeps across the room laterally and is removed by a perforated wall opposite to the inlet wall. Ducts carry the air back to the air handler and from there to the filters again. Of course, in an operating room with equipment and personnel, the contamination concentration gradient will run from the clean filter end to the relatively dirty outlet end. The room is therefore of constantly variable class along its width. This would be unacceptable for many applications but not all. Take electron-beam lithography for the production of photomasks, for example. The equipment has a load point which must be clean, but the rest of the machine consists of computers and vacuum systems. If we place the loading point at the clean end of the room and the pumping stations, control console, and so forth near the outlet end, we are achieving our goal; it really does not matter to the computers if they are the least clean part of the cleanroom.

One vital point should be remembered with these rooms, however. It is important not to place items upstream of the areas which require the most protection, as this may disturb the airflow and perhaps even introduce contamination into the airstream. This includes personnel; they should never be allowed to stand upstream of the loading point while loading. It is best to lay out the equipment in such a way as to prevent people from moving into critical airflow areas while still allowing them to perform their functions. This can be a difficult task at times, but it is one of the drawbacks of the horizontal laminar flow room.

PROCESS EQUIPMENT

Equipment Integration

Throughout this chapter we have seen examples of cleanroom configurations. In showing the cleanrooms, we have frequently shown equipment in place to help the reader appreciate some of the reasons for the particular layout. We will discuss further how equipment fits into cleanrooms in this section.

The first thing to appreciate about equipment is that it can have a dual nature, being both super-clean and reasonably dirty at the same time. For instance, a plasma etching system may have a clean chamber, compatible with the processes used in the manufacture of leading edge microelectronics components, but can also have a relatively dirty rotary vacuum pump/blower set. This leads to a dichotomy, as we want the clean part in the cleanroom so that we can always keep the wafers in the cleanest environment possible, but we would ideally like to get the pumps and so forth as far away as possible.

Returning to some of our previously discussed layouts, we can see how we may resolve this dichotomy. For instance, in looking at Figure 5.10 we can see that some equipment is fully in the room, with the back against the bay wall, some equipment is fully out of the room, with only the loading point being seen from the room, and some equipment is somewhat half in and half out. How we arrange the equipment in this sense depends greatly on the nature of the equipment itself. For instance, inspection stations, assembly benches, and the like are inherently clean and need to be fully in the room. However, modern ion-implanters, coating systems, and so forth need only have a loading point protruding into the room, the rest being in the service area. Furnaces, on the other hand, have to have a relatively long loading bench in the cleanroom while the element sections and gas jungles can be outside the room. This sounds simpler than it actually is, as it is only fairly recently that equipment has been built so that this separation can be made. For instance, all the controls have to be moved into a position where they may be accessed by cleanroom personnel. This means a high degree of remote control of elements such as valves, and remote reading of measurement instruments such as mass flow controllers. This has added a great deal of expense to process equipment.

Even equipment which has been modified to allow remote or "through the wall" operation may still require some regular or semi-regular operations to be performed at the back end. Ion-implanters, for instance, will require source change-outs (in addition to the huge amounts of maintenance behind the scenes). There may be a temptation to put a door between the cleanroom and the service area, but this could lead to some degree of cross contamination and should be avoided if possible. An intercom at an appropriately placed window will be better in most cases. If parts have to be transferred for any reason (unlikely) then a double door.(airlock) pass-through may be installed.

All the above assumes that you are in the position to design the cleanroom around your process and process equipment. What if you already have a running line and would like to clean it up by segregating clean from service operations? In this case you probably cannot go too far wrong by placing a cleanroom wall in such a way as to block off the backs of furnaces, maintenance areas, and so forth. If access is required by operators to the back of equipment, a door may be used from the cleanroom to the back, but this means that every effort must be made to keep this access zone as clean as possible. One option which may be considered is to create clean corridors in these areas with doors at either end so that access is possible from either end (cleanroom or service chase) but the double door system does not allow the intermediate area to become too dirty. Another very important point which should be kept in mind is that retrofitting an equipment wall in this fashion may severely interfere with airflows in the room. If the air return vents are at the original cleanroom boundary walls, vents should be placed at low levels in the new walls and doors with ample area so as not to significantly reduce the airflow. Care should also be taken to ensure that the pressure difference is not too great between cleanroom and service area, otherwise air may be blown through equipment. This is fine for some types of equipment. In fact, some machines are designed this way so that air can flow through the body of the unit to exit at a wall vent (the equipment is transparent to the airflow and it therefore does not obstruct the return ducts). However, other systems, for example, furnaces, cannot have air flowing through them as it will interfere with functions such as temperature measurement.

Services

In addition to equipment, central plant services must also be integrated into the cleanroom environment. These services include electricity, water, compressed (clean dry) air, technical vacuum, nitrogen and other house gases, piped chemicals, fume exhausts, and chemical drains. These are all run in pipework and ductwork, and many pieces of processing equipment require more than one service. It is something of a headache to supply cleanroom equipment with services because we cannot run a mass of pipes and ducts within the room as they can be awkward to clean and keep clean and will obstruct normal operations.

The ideal place to run services is, of course, the service chase. Since the equipment backs on to these regions for ease of maintenance, it is relatively straightforward to connect into services which run the length of the chase and feed the connections through the wall to the equipment. If holes have to be cut in the wall material for the feed-throughs, a hole cutter exhausted with house (cleaning) vacuum is used to minimize contamination generation. If the wall material is a sealed substance, such as vinyl sealed sheetrock, this must be resealed, after cutting, with an appropriate cleanroom sealant.

The alternative to the service chase approach is to run the services in the basement and feed the equipment connections through the floor. This is more flexible in terms of positioning the equipment as it does not have to be near the walls. However, it is less accommodating for the later addition of equipment as it is considerably more difficult to core drill a concrete floor and reseal it than to cut a hole in a wall. Therefore, cleanrooms of this type tend to have the compromise of pre-cored sections which are plugged when not in use. A further alternative for rooms with raised floors is to run the services under the false floor. This only works for situations where there is sufficient headroom under the floor to enable installation and maintenance without tearing up the floor sections. Finally, some facilities have the services run overhead and dropped through the non-filter ceiling sections. This is not very popular in modern facilities with many HEPA filter outlets as there is already a large amount of ductwork in the roof space, and installation and maintenance (leak fixing and additions) become a problem. If the air is supplied by way of a plenum instead of individual filter ducts, we still have a problem with overhead services as, although we now have plenty of working space, it is not a good idea to have people working directly above the filters. Of course, in the case of gravity drains, these must be below the level of the outlet on the equipment.

In all cases, accessibility is not the only consideration as we must be careful not to set safety traps, for example, running piped chemicals such as acids over an electrical conduit in a service chase sounds like an accident waiting to happen. Also, if a hazardous material is piped in an area which is included in the supply or return air paths, a leak could be disastrous to employee health. In many cases, regulations prevent these situations from occurring in the first place. These aspects are discussed in Chapter 8.

There are also other services in the cleanroom which are not directly related to equipment. Lighting and sprinkler systems must be integrated into the environment. These generally have to be in the ceiling to provide coverage with as little shadowing as possible, for both light and water. This poses a problem in total laminar flow rooms as most of the ceiling is made up of HEPA filters, and this leaves little room for anything else. In the case of lighting, some room designs have sacrificed HEPA filter space for flush fitting fluorescent lighting units. This leads to a dead space beneath the lights which extends a distance of three times the width of the unit beneath the ceiling level. An alternative to this is to mount single fluorescent tubes along the lower surface of the T-bar ceiling supports. To reduce the turbulence below these tubes, a teardrop-shaped cover is usually placed over them. The T-bar supports are also wide enough to include sprinkler heads fed from pipes above the ceiling. This brings us back to the problem of having services where there is air ducting or a supply plenum, but in the case of sprinkler pipework, it is generally installed and not added to, so maintenance should be minimal. There are some cleanrooms which have had the sprinkler pipework installed under the HEPA filters after the cleanroom was built. The air will return to laminarity within three pipe diameters below the pipes. However, since cast iron is used for these pipes, they must be coated with a reduced contamination material such as vinyl, otherwise particle shedding will occur.

DESIGNING FOR CLEANROOM CLASS
Determining Class Requirements

There are two related questions which we must now ask ourselves: how do we determine which class of cleanroom we need and how do we attain this class through the design of the cleanroom? The answer to the first question depends very much on what we are trying to do in the room and is generally driven by economic considerations. We will tackle this question first.

Say we have a specific product type in mind. We obviously wish to have the highest yield possible, that is, 100 percent, so that there is no material, labor, and overhead wastage. This is a sound way to maximize our profits.

However, we should ask ourselves the question, "How much will it cost in terms of cleanroom facilities (and practices) in order to attain a yield goal, as capital costs are also a factor in profits?" This is a tricky question, as we must understand the relationships between product yield and the mechanisms of yield reduction. If we assume for this discussion that the only yield-reducing factor is microcontamination, then we must know how point defects, for example, will affect our yield. This may be determined from work done in existing facilities or from existing mathematical models (or both). If the product is at all sensitive to microcontamination, then it will be impossible to attain 100 percent true yield 100 percent of the time, as it is virtually impossible to eliminate all particles from the environment. We may get very close to this, but the cost would be vast. In fact, the cost of our ultra-clean facility could be so large that we could never hope to make it up with product sales during the sales life of the product. We must therefore settle for a target yield lower than 100 percent to reduce our facility cost and leave us with some profit. In the case of research operations where profit is not a motive, the class of the room is frequently determined by how much funding is available or by how much contamination we think we can tolerate.

With regard to the question of how do we actually create a cleanroom of a particular class by design, the answer is simple in principle but rather more complex in practice. The basic premise for any cleanroom is that particle contributions from materials, personnel, processes, and filter penetration (what we shall call "particle generation" in this section) should be negated by the room airflow. At any instant in our controlled area, we should never have more particles than the class allows for. Therefore, all we have to do is ensure that there is enough clean air entering our volume to sweep out any particles which enter. For instance, if we have a generation rate within a class 10 controlled volume of 20 0.5-micron particles per cubic foot per minute, we must adjust the airflow so that we change all the air in the volume within 30 seconds to maintain the class specification. In other words, we must have a minimum of 2 airchanges per minute so that the particle count can never exceed 10 per cubic foot within the volume. In arithmetic terms,

$$\text{minimum airchanges} = \text{generation rate (at 0.5 micron)/class}$$

This may sound simple but closer examination reveals that this problem is actually rather complex. For one thing, our air does not travel at infinite velocity. In fact, the largest air velocities we are likely to find will be around 90 feet per minute. This means that in a total VLF room (100 percent VLF for this calculation; it will be less in practice) with a 9-foot-high ceiling, the maximum number of airchanges is set at 10 per minute, regardless of floor area, that is,

$$\text{room airchanges} = \text{air velocity/room height}$$

In our hypothetical class 10 volume, this sets the maximum allowable particle generation rate at 100 0.5-micron particles per cubic foot per minute:

$$\text{maximum generation rate (at 0.5 micron)} = \text{airchanges} \times \text{class}$$

We cannot tolerate a higher rate under any circumstances, which means that we have to strictly control the numbers of particles entering the volume. We do this by using filters of appropriate efficiency to reduce this contribution, by using low particle transmission cleanroom apparel to reduce this contribution, and so forth. The numbers become more frightening when we do not have total laminar flow, as average air velocities will be reduced according to the percentage of HEPA filter area present, for example, 50 percent HEPA filter area will give an average room air velocity of around 45 feet per minute:

$$\text{room air velocity (average)} = \text{inlet velocity} \times \text{HEPA \%}/100$$

The maximum number of airchanges possible here is 5 per minute, which would mean a maximum allowable generation rate of 50 0.5-micron particles per cubic foot per minute if we had to maintain class 10 conditions. This is, of course, why we are forced to go to a larger percentage of HEPA filter inlet for better class rooms.

The problem of fixing class is more terrifying yet. The above discussion considered the steady-state, assuming a constant generation rate from all sources. This is rarely the case in practice, as activities in the cleanroom will tend to create bursts of contamination. Our cleanroom should be able to recover from these bursts within a reasonable time or it will be difficult to maintain class specifications. For instance, in our total VLF class 10 volume above, if a particle enters the room at ceiling level, it will take a minimum of around 6 seconds to leave the room. Therefore, if a burst of 20 particles per cubic foot suddenly enters a class 10 room near the ceiling, for 6 seconds the room will be out of specification. We may say that our recovery time is 6 seconds. It is important that these bursts do not add significantly to the recirculated air particle concentration after refiltering, otherwise we will get a "memory effect" with the burst, in reduced form after filtration, coming back to haunt us. Once again, for less HEPA filter coverage, the recovery time is slower, for example, 12 seconds for a 50 percent coverage room.

Lower-, Middle- and Upper-class Rooms

What can we expect to see in cleanroom designs for lower-class (10,000 and worse), middle-class (better than 10,000 but worse than 10), and upper-class (10 and better) rooms? This question is difficult to answer specifically, so we will attempt to generalize and supply examples.

Lower-class rooms tend to be of the mixed flow type with some spot VLF where required, the air being returned through evenly spaced low level wall vents. HEPA filter efficiencies will be in the 99.95 percent or better region The airchange rates are typically in the order of 10 per hour. The rooms are generally open plan, making it easy to place equipment and workbenches. Typical applications are non-critical device packaging and assembly operations.

Middle-class rooms will also have spot VLF but using more HEPA filtered air outlets so that the HEPA filter coverage is greater than 25 percent or so. HEPA filter efficiencies will typically be around 99.995 percent at 0.3 microns or better. The larger volume of air will be returned through larger capacity low wall vents or in some cases through a partially perforated floor. The airchange rates are typically in the order of 1 or 2 per minute. The rooms can be open plan or arranged as bays by air return walls or full service chases. Typical applications are MSI to LSI semiconductor components or critical subsystem assembly.

Upper-class rooms are almost always total VLF, except in the cases where the work areas are physically separated from the rest of the room (there is total VLF in the work areas in any case). The highest efficiency HEPA (or ULPA) filters will be used here, with a minimum rating of 99.9995 percent. Most use full perforated floor returns, unless very narrow bays (12 feet or less) are employed, in which case large side return vents are used. The tunnel cleanroom configuration is favored here although a small number of rooms of this type are open plan. Typical applications are VLSI or ULSI integrated components and systems.

Cleanroom Models

One of the best design aids you can have when putting a cleanroom together is a good mathematical model. Even if you already have a cleanroom, a model is extremely useful as you can vary parameters in the model to see what happens without disrupting the real system. In mathematical models, we represent each element of the physical situation by using rules or equations. We then supply the required constants and input variables to attain the response variables. Unfortunately, as far as cleanrooms are concerned, there is no such thing as a totally accurate model as there are simply too many variables involved. However, a first order (simple) model can still be very helpful. There are a number of things you can model at a first order level. Our example given in this section models a cleanroom with recirculated and make-up air supplies as shown in Figure 5.19. Note that the principles of recirculation and air handling are covered in the next chapter, but you do not have to worry about the details for the model.

Our first order model treats the cleanroom and air handler from a system point of view. Instead of describing the detailed spatial and time dependence of the particles, the components of the cleanroom system are modeled as

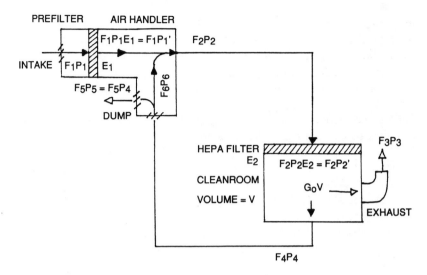

Figure 5-19   Schematic of cleanroom system for sample model.

black boxes affecting air flow and particulate densities. This is analogous to modeling a transistor as having certain terminal characteristics without concerning oneself with the details of electron and hole distributions within the device. The components of the system are (1) an air intake for make-up air, (2) a prefilter, (3) ducting, (4) HEPA filters, and (5) the cleanroom itself. Air flows (ft³/min) are labeled F_1 through F_6, and particulate densities (particles/ft³) are labeled P_1 through P_5. A filter is modeled as a box which does not affect air flow, but removes particles from the system. This removal is modeled by the dimensionless particulate transmission efficiency (labeled E_1 and E_2) of the filter. The number of particles entering the filter per unit time is F_nP_n, but the number leaving per unit time is F_nP_n', where $P_n' = E_nP_n$. Thus, the filter removes particles from the system at a rate of $F_n(P_n - P_n') = F_nP_n(1 - E_n)$ particles per unit time.

The cleanroom is modeled as a box with floor area A, height H, and volume $V = AH$ within which particles are generated at a rate G_0V particles/minute. The parameter G_0 is representative of a generalized particulate source in the room generating G_0 particles/ft³-minute. There is also a loss of particulates and air flow from the room through unintentional leakage or intentional dumping of air which is modeled by the flow F_3 out of the system, carrying with it particulate density P_3 for a loss of F_3P_3 particles/minute. The air intake is modeled as a box which adds an air flow F_1 and particles at a rate of F_1P_1 to the system, while the ducting circulates this and the cleanroom air, but also removes air flow and particulates from the system through controlled dumping via the flow F_5 and F_5P_4.

With the system described, it is now possible to apply two basic physical principles which will allow equations to be developed for calculation of the particulate densities at the entrance point (ceiling), P_2', and exit point (floor), P_4, of the cleanroom. Time dependence of air flow and particle densities is not addressed by the model, so steady-state conditions are assumed. Under this condition there must be (1) a balance of air flows and (2) conservation of particle flow in the system. Looking at Figure 5.19, it is clear that the balance of air flow requires that:

$$F_2 = F_1 + F_6$$
$$F_4 = F_2 - F_3$$
$$F_6 = F_4 - F_5$$

Combining these, it is not surprising to find that:

$$F_1 = F_3 + F_5$$

That is, make-up air must equal losses from the cleanroom and ducting in steady-state or there would be a pressure decrease or increase in the system. Conservation of particle flow requires that particles entering and leaving both the air handling system (intake, prefilter, and ducting) and cleanroom (HEPA filters and the room itself) balance, giving:

$$F_1P_1E_2 + F_4P_4 = F_2P_2 + F_2P_2$$
$$F_2P_2E_2 + G_0V = F_3P_3 + F_4P_4$$

To simplify these equations, the loss of particles from the cleanroom via F_3P_3 must be estimated. Since the distribution of particles in the room is unknown, perhaps the best estimate of particles leaving the room by flow F_3 is to assume that it is equal to the number of particles generated per minute in the room times the percentage of total flow out of the room represented by F_3 or:

$$F_3P_3 = F_3G_0V/(F_3 + F_4)$$

Now, using the fact that $P_2 = P_2'/E_2$ and combining the above equations, we can create two equations in two unknowns (P_2' and P_4). It is quite simple to solve for these two unknowns using the techniques of linear algebra. In doing so we find that:

$$P_4 = [G_0V(1 - F_3/F_2) + F_1P_1E_1E_2]/[(F_2 - F_3) - E_2(F_2 - F_1)]$$

and

$$P_2' = [F_1P_1E_1E_2 + P_4E_2(F_2 - F_1)]/F_2$$

As we can see, even in our simple model the expressions for our parameters of interest are not mathematically complex but are rather unwieldy. The best way to handle such expressions is to write a simple computer program to perform the calculations. Such a program, written in Basic, is included in Appendix III. The inputs to the program are the air dumped (F_5), which is usually zero, cleanroom exhaust (F_3), make-up particle density (P_1), prefilter efficiency ($EF_1 = 1 - E_1$), cleanroom airflow (F_2), HEPA filter efficiency ($EF_2 = 1 - E_2$), cleanroom particle generation (G_0), and the room floor area and height. This type of model is extremely easy to alter to fit most system configurations.

Note that the main shortcoming of this model is that since it treats the cleanroom as a black box, it cannot predict the particle distributions within the room. In order to do this, we have to involve differential calculus and/or numerical techniques. However, these are beyond the scope of this text.

SUMMARY

To summarize, there are a large number of cleanroom configurations, ranging from mixed flow rooms with side returns to toal vertical laminar flow in tunnel or open rooms with perforated floor returns. In many cases, spot VLF may be used to provide cleaner zones in a mixed flow room at low cost. The best cleanrooms tend to have only the equipment-loading points in the clean area containing the product and the dirtier systems in a separate service chase. A number of simple calculations may be used to help us determine which class our cleanroom will maintain. These calculations may also be used to tell us what our maximum allowed contamination generation rate will be for a particular class of room.

6

Preconditioning, Control, and Static

We have, until this point, concentrated on the clean part of our facility, namely, the cleanroom itself. However, our cleanroom system extends beyond the walls of the clean production or research area. Since we are attempting to create a controlled environment, we have to apply control to other parameters, such as temperature and humidity, and provide the necessary drive for the air to get it through our HEPA filters and into the room.

In this chapter, we will examine how the air may be preconditioned before it reaches the cleanroom (or during the recirculation phase) and briefly discuss the types of control systems which could be used to meet the exacting tolerances required. In addition, we will include a discussion of how we may attempt to control static electricity in the environment, much of which can be carried by ionized air.

AIR HANDLING SYSTEMS

The preconditioning of air is really the realm of HVAC (heating, ventilating, and air conditioning) engineering. The aim here is to provide control of temperature, humidity, and air velocity/pressure using fans, heaters, chillers, humidifiers, dehumidifiers, sensing elements, and control systems. We have already discussed sensing elements in the cleanroom in Chapter 3, and these same elements may be used elsewhere in the system, for example, in critical ducts and air plenum regions. The other components mentioned above can be combined in a single air handling system. We will discuss such systems in this section.

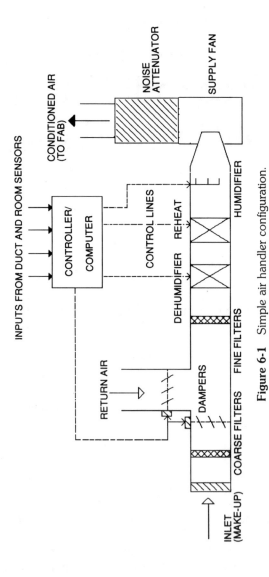

Figure 6-1 Simple air handler configuration.

System Configurations

A variety of air handling system configurations are available, the size and configuration being determined by the volume of air, degree of control required, external conditions (typical temperatures and humidities in a region), and economics. By way of providing representative examples, we will discuss two typical configurations, as shown in Figures 6.1 and 6.2.

The first thing we should note about these configurations is that they both have make-up and recirculated air paths. Recirculation of air from the cleanroom air return vents is important for economic reasons. As we will see in this chapter, a great deal of time and money is spent in engineering cleanroom

PLAN VIEW

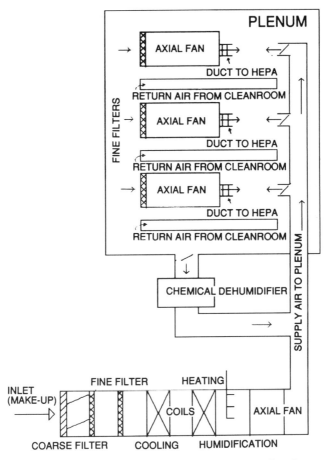

Figure 6-2   Alternative air handler configuration (as found in dry regions).

air to our specifications. It would be a great waste to throw away all this air once it had passed through the clean environment just one time. We therefore recirculate most of it back through the system. The make-up air is used to replace any air lost in the system through exhausts or leakage, and also to provide "fresh" air. Typically, 25 percent of the total air volume in the system is taken in through the make-up path, leaving 75 percent to be recirculated. Some facilities are more efficient at keeping their recirculated air in the system, but it is probably unwise to go below 10 percent make-up for health reasons, that is, airborne process vapors will not be as well diluted. Many facilities have such vast exhaust systems at wet benches dealing with volatile and hazardous chemicals that much higher percentages have to be made up (we have heard of 70 percent make-up in one facility).

In the system shown in Figure 6.1, the inlet air is mixed with the return air, the relative amounts being controlled by dampers. Some prefiltering is employed at this point. The mixed air then passes through cooling/dehumidification coils, heating coils, a humidification unit, and then through the fan to the cleanroom. This is an extremely simple system and is widely used in "off-the-shelf" stand-alone units for small and modular cleanrooms. Multiple units are typically used to supply cleanroom zones or as standby systems in the event of primary system failure or maintenance.

The configuration of Figure 6.2 is considerably more complex. The make-up air is prefiltered as before, but cooling, heating, humidification (not dehumidification—see later), and initial "injection" into the system by fan is only performed on this air and not the recirculated air. This assumes that the temperature of the cleanroom air remains fairly constant, which is a reasonable assumption as in many cases room heat loads can remain at nearly constant levels on average. If the room temperature does change, the make-up air is heated or cooled accordingly and fed into the system to compensate. (Additional cooling coils may be placed in the return path if the heat load in the room is particularly high.) The air is fed into a plenum above the cleanroom, where it is taken up by fan units in the plenum itself. These fans filter the air and drive it through ducts to the HEPA filter units in the cleanroom ceiling below. The air return from the cleanroom is by vertical ducts (which could form the service chases also), which take the air back into the same plenum where it mixes with the make-up air and may then return through the plenum fan units again. This system is typical of those found in dry regions and therefore the dehumidification system is perhaps a little unusual. Since little dehumidification is typically required in this case, only a portion of the mixed air is taken off to a small energy-efficient dehumidification unit and then returned to the plenum.

There are many other variations on these themes, but the system elements are generally the same. In the next sections, we will look at these system elements more closely.

Prefiltration

Prefiltration for the removal of gross particulate contamination is generally a multistep process. A wire mesh and a pad type filter generally come first to remove any debris such as leaves, feathers, litter, and insects from the make-up air. Some larger dust particles will also be removed at this stage. This is followed by a finer filter, such as an 80-micron prefilter and/or a *bag filter*. As the name suggests, this is really a large surface area fiber filter, formed in the shape of a bag. This helps to remove dust, grit, soot, and so forth by trapping the particles in the fibers. The best prefilters for cleanroom use can remove 0.3 micron particles with removal efficiencies in the mid-80s to mid-90s percent. The air is now of reasonable quality for areas intended for human occupation, but is still a bit too dirty to be HEPA-filtered as most of the finest particles will still be present.

The final stage in prefiltration should be designed to take out as much of the remaining particulate material as possible. The reason for this is simple; it is better to have some high-efficiency filters upstream of the HEPA filters which have to be changed out regularly than to have low-efficiency pre-HEPA filtration and have the HEPA filters changed more frequently. The latter option leads to maximum disruption in the cleanroom itself, which in many cases will prove to be more costly than installing higher efficiency prefilters in the first place. Many facilities now use HEPA filters as prefilters; for example, 99.97 percent at 0.3 micron to prefilter for 99.9997 percent efficient cleanroom HEPA filters.

Heating and Cooling

When considering heating and cooling systems for cleanroom facilities, it may be tempting to think of the systems found in our own homes. Domestic air conditioning systems generally use electric resistance elements for heating and an electric motor driven compressor type refrigerant expansion system for cooling. Such systems may not be the answer for clean facilities, especially where large volumes of air are concerned (millions of cubic feet per minute). A more energy-efficient approach is to use a hot or cold liquid (or gas) passing through an appropriate heat exchanger.

These heat exchangers are merely coils of tubing, which have an extended surface area created by the addition of fins to the tubes. Heating or cooling is thus achieved by the transfer of heat (due to the thermal gradient) from the coil surfaces to the air, or from the air to the coil surfaces respectively. To ensure efficient heat transfer, the tubes, typically 3/8"–1" in diameter, are spaced 1"–3" apart to form rows. The complete exchanger will consist of 8 to 12 of these rows in series. The coils will typically carry hot water or steam for heating, or chilled water or refrigerant for cooling.

For the coils described above, the air face velocity for heating will generally be between 500 and 750 feet/min. If water is used, the temperature will typically be between 170 and 190° F with a flowrate from 2 to 7 feet/sec. If steam is used, pressures will be between 5 and 25 psi (gage). The hot water or steam will usually be generated in another location, for example, a central plant building, by oil- or gas-fired boilers. For the above conditions, the air discharge temperature will generally be 70 to 120° F, depending on valve opening (flowrate). This may sound rather high, but a lower room temperature will result when the hot air is mixed with cooler air from outside. This is a much more energy-efficient way of getting the heat energy from the source to the air than are electrical heating methods, which would require high current capacity elements.

For the case of cooling, the air face velocity is lower, in the range 300 to 700 feet/min. If chilled water is used, produced by a central plant chiller (which generally uses the expansion of a refrigerant to produce the cooling effect), the water velocity is again 2 to 7 feet/sec for a relatively wide cold water temperature range of 40 to 55° F. Direct expansion of refrigerant, for example, Freon 14, can also be used to produce refrigerant temperatures of 25 to 55° F. The air discharge temperature from cooling units can be as low as 48° F, controlled by the coolant flowrate. Evaporative coolers (using water in the airstream), water sprays, and washers can also be used for air cooling, but these can increase the humidity of the incoming air to unacceptable levels, especially if the air is warm.

Humidity Control

The control of humidity is somewhat more difficult in practice than the control of temperature. In this case we have to physically remove or add a material, that is, water, in compatible form, that is, as a gas or vapor, to the air.

Dehumidification may be performed utilizing the same coils used to cool the air. If moist air is cooled, it can lose some of its water content due to condensation. In semiquantitative terms, air can only hold so much water in the form of water vapor. The limit to how much it holds depends on temperature; the higher the temperature, the higher the humidity of the air. If air at a particular temperature is saturated with water vapor, a reduction in temperature will reduce its water-carrying capacity and water will "fall out" by condensing into droplets. Alternatively, if air at 80° F has the percentage of water vapor which would saturate air at 50° F, it will retain this water unless it is cooled to below 50° F. In this case we say that the air has a *dew point* of 50° F, the dew point naturally being the temperature at which condensation (or dew) will begin to form. Therefore it is *easy to see* how we may use cooling coils to reduce the humidity of the air, the amount of water being removed increasing with decreasing temperature. Unfortunately, it also means that we usually have to reheat the air after we have cooled it; therefore, this is not a particularly energy-efficient process.

An alternative, more energy-efficient option is to use a chemical dehumidification system. In this case the air is not cooled, but a solid (or sometimes liquid) desiccant is used instead. The desiccant is extremely hydrophillic, absorbing water from the air on contact, and hence dehumidification is relatively simple. The control of humidity is achieved by mixing dried air from the dehumidifier with moist air from the inlet or return to achieve the desired humidity level. The main problem with this approach is that the desiccant will become saturated with water after some time and will no longer take in water to the same degree. These materials may be regenerated by heating to drive off the absorbed water and then placing them back in service. In an industrial process, this usually means the use of two identical units with one being in service while the other is undergoing regeneration. Newer systems, using liquid desiccants, can operate continuously as the liquid can be returned to a regeneration vessel and then sent back to the dryer without shutting the unit down.

Humidification is used when the inlet air is particularly dry, which is generally the case in desert regions or in a hard freeze situation. The traditional humidification method is to use a pan unit which is fed with water. A heater is used to evaporate water into the air. Precise humidity control is difficult with these systems and scale tends to build up in the pan, therefore this method is reserved for occupant comfort systems. A more controllable method is to use spray nozzle or spinning disc atomizers which break the water into fine droplets. This is not the most ideal way of getting water into the air as droplets will not be held in the air as readily as water vapor. Collisions between droplets, and between droplets and surfaces, cause coalescence and a loss of airborne water. In addition, the incorporation of the water from this form will depend strongly upon the air temperature.

One of the most popular methods for humidification is by steam spray. The water is already in the gas phase and will thus be readily incorporated. This method is particularly ideal for large airflows and volumes of air as steam sprays can supply 1700 liters/hour of moisture into the air (with only 2 psi [gage] of steam pressure). As is often the case with so-called ideal methods, steam injection can suffer from a rather nasty problem. If the water used to create the steam is not pure, there is potential for contaminants to be introduced into the airstream. Since these can be extremely well dispersed, it is difficult to remove them, and they could thus accumulate elsewhere. The obvious solution would appear appear to be to use ultrapure water. However, ultrapure water is actually an extremely good solvent and the act of heating it in a boiler and putting it through piping systems as hot water or steam tends to result in damage to all but the most resilient system materials. The answer could be the addition of corrosion-inhibitor chemicals, but we come back to the same problem as before. The best way around the problem is to meet it halfway and use RO water (see Chapter 9), which is much purer than city water but not as pure (and as much of a solvent) as ultrapure water. High quality stainless steel pipes may then be used for the system with reduced contamination and system damage. Unfortunately, a glass lining is still necessary for the boiler.

FANS AND AIR CONTROL

As we have seen in the last section, fans are used in air-handling systems to produce a continuous flow of air by aerodynamic action. They are generally found in stand-alone air handler units to provide pressurization for the make-up air and may also be present downstream to provide pre-HEPA filter pressurization. There are essentially two basic types of fan used for these purposes; the centrifugal fan and the axial fan. There are other types, but these are really just variations on these configurations.

In a centrifugal fan, the air enters the inlet axially, turns at right angles through the blades, and is discharged radially. The air is taken up at the center of the blades and is literally thrown off the ends by centrifugal action at great speed. This type of fan can thus produce high pressures at relatively low speed. It is therefore a quieter and somewhat lower vibration configuration than the axial type. The other advantage of the centrifugal fan is that only the blades are immersed in the airflow and the motor is outside the duct, the mechanical connection being by shaft or belt drive. This is good, as even induction motors can be a source of contamination from the bearings and coils. The disadvantage of centrifugal fans is that they take up a relatively large amount of space. However, due to their large pressurization to size ratio, they are the configuration of choice for laminar flow cabinets with integral fans.

In axial flow fans, the air enters axially and leaves axially. The motor and blades are placed in the airstream in line with the inlet and outlet ducts. These fans will develop less pressure than centrifugal fans at the same speed. Interestingly enough, axial fans are actually more popular for semiconductor facilities than centrifugal fans due to their compact nature. This aspect becomes very important if, for example, five fan units have to be placed in a plenum area. For those who are concerned about the motor being in the airstream, a variation on this theme has axial flow blades driven by an external motor.

Speed control tends to be standard on virtually all cleanroom air fans in modern facilities. Control of motor rpm is important as it allows dynamic air balancing to some degree, that is, if a door is open for a relatively long period while equipment is brought in, the air volume may be increased to compensate and maintain positive pressurization. There are also economic reasons why

speed control is used. For instance, during periods of inactivity in the production line, contamination generation will be reduced and the volume of air
being pushed into the room may be reduced without penalty. This can result
in major cost savings, as a lower fan (to drive less air) speed means less
power consumed, and electrical power is a major cost item in cleanrooms
(Figure 6.3). Since high current motor speed controllers tend to be expensive,
some fan units use variable attack angle blades which may be adjusted to
drive more or less air, increasing or decreasing the motor load and hence
power consumption, respectively. These are less popular in cleanrooms as
mechanical systems generally require more maintenance and higher levels
of generated contamination have been associated with them.

The air is directed to where it is supposed to go by ducting systems. The
main trunks are generally constructed of galvanized, or otherwise coated,
sheet steel. This can lead to vibrational problems due to its flexibility, as
turbulent air in the ducts will tend to hammer the sidewalls, creating booming.
This effect may be reduced by giving some thought to the aerodynamics of
the situation, for example, by putting in curved rather than right-angled bends
so that the air is not made to change direction suddenly and hence go

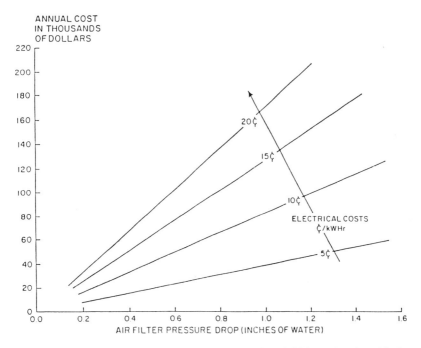

Figure 6-3 Annual electrical costs for operation of a 10,000 sq. ft., class 10 cleanroom as a function of air filter pressure drop and electric power rates (data provided
by Mr. F. Fichter of Comp Air, Inc., Grand Rapids, MI).

turbulent. In addition, the ducts should be suspended by vibration damping supports (see Chapter 7).

The main trunks will feed a plenum, if used, or will lead to a manifold which allows connection to a number of flexible ducts. These flexible ducts are generally wire-reinforced PVC (or similar material) tubes formed as concertina or spiral structures so that they may easily be bent and routed down to individual HEPA filter units. It is important to lag these flexible ducts to reduce heat loss or gain. The lagged flexible ducts can be a relatively expensive option if used in large numbers; therefore, this approach is more common in smaller cleanroom systems.

Dampers may be used within the ducts to throttle the airflow through branches and to control mixing. These dampers are generally butterfly or louvre elements which physically block the passage of the air through the duct, the amount depending on the angle of the vanes.

CONTROL SYSTEMS

Controlled Parameters

What we are aiming for with regard to control of cleanroom air parameters will depend on the requirements of the processes carried out in the room. A typical semiconductor facility may require, for example, temperature control at 68° F with a tolerance of ±1° F in the diffusion or etch areas. This temperature is chosen mainly for operator comfort; it may sound a bit chilly, but the operators are generally dressed in relatively hot cleanroom apparel. However, a tolerance of ±0.5° F is required for lithography rooms to reduce temperature-induced dimensional changes in sensitive equipment. For example, if the structural elements in a projection aligner were to expand and contract due to temperature changes, the focus could also change, and this would affect the system performance, that is, alter the repeatability of results. Many modern lithography systems are somewhat tolerant to temperature variations by the use of alloys which have a low temperature coefficient of expansion (for example, Invar), but due to the increased demands for resolution, overlay, and consistency, these systems still require strict control of environmental temperature.

Relative humidity may have to be controlled at 40 percent ± 5 percent for a number of reasons. Higher humidity can be extremely uncomfortable to operators in body-covering apparel. The loss of body heat by the evaporation of perspiration is reduced when the air is damp, as less evaporation from the surface of the skin occurs. Too little humidity can result in too much static in the environment; dry air prevents the natural leakage of static to ground. Dry air is also unpleasant for people, leading to dry throats and mouths. Finally, many chemical processes require a strict control over the amount of water in the air as too much or too little can alter the outcome of

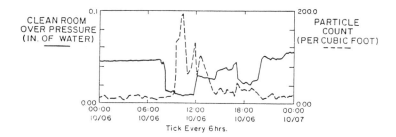

Figure 6-4 Effect of cleanroom pressure loss on particle counts in diffusion furnace area (from *Jnl. Env. Sci.*, Nov./Dec. 1986).

a reaction. Photoresists can be particularly sensitive to the relative humidity, using the water absorbed on surfaces in adhesion reactions.

Control has to be applied to air pressure and velocity, too. If our airflow is to be laminar, we require a velocity between 50 and 150 linear fpm in the flow area. We also have to overcome the pressure drops across filters, which can be as high as 1″ H$_2$O near the end of the filter life. We have to be able to do this while maintaining the room at positive pressure, perhaps around 0.05–0.1″ H$_2$O overpressure with respect to the outside, to prevent the back-flow of contaminants into the room through doors, passthroughs, and gaps in the ceilings, walls, and floors. The requirements above are not very easily met. The opening of doors turns the control of pressure into a dynamic problem; if a door is opened for any length of time, the overpressure drops and contamination can enter the room (see Figure 6.4). However, if the system responds by increasing the airflow into the room, when the door closes there is a danger of the sudden increased overpressure blowing out ceiling tiles or even the walls. Therefore, it is clear that we require a control system which is capable of providing accurate and rapid control of a number of parameters.

Types of Control

Control systems have three basic functions:

1. Control—maintain temperature, humidity, pressure
2. Safety—prevent parameters exceeding danger levels
3. Operational—provide economic operating cycles, match load conditions, interface with other factory systems

We have already discussed control. The safety function is extremely important as the system has to be able to automatically signal a dangerous or out of specification condition and take some predetermined action. For instance, a component failure could cause a constant temperature rise which takes the room temperature above the tolerance limit. The system should be able to sound an alarm and perhaps even switch over to a back-up system or shut down before damage is done to equipment or product. Sudden increases in duct temperature could suggest a motor overheat or fire, or sudden pressure drops could indicate a significant filter or duct leak. Ideally, the control system should be able to "see" these conditions and react accordingly. This brings us to the third aspect of control systems. Since our cleanroom is indeed an integrated system, the control system for the air-handling apparatus should be an integral part of this system. For instance, the safety functions on the control system could be connected to the factory-wide safety systems, for example, the fire alarm system. In this era of computer-integrated manufacturing, the information from sensors could be fed into a central database so that environmental data could be logged and later related to product factors such as yield. If the control system is flexible enough, it could also be programmed to save energy by reducing fan speeds and airflows during periods when operations are suspended.

How does a control system actually work? The answer to this depends on which type of control is employed, but all systems use some type of measuring device, a controlled device, a communications path, and a source of energy. The signal from the measuring device is ultimately fed to the controlled device, which acts in some predetermined manner to alter the parameter it is designed to control. The change which the controlled device makes can be detected (downstream) by the measurement device so that information feedback is employed.

Feedback is critical in control systems. Without feedback, the control system would have no way of assessing if its action was successful. For instance, if the temperature rises due to an increase in the heat load in a room, the control system may act to reduce the temperature (by closing a steam valve to heating coils, for example). The measurement device in the room will continue to sense the temperature during this action, and if the ensuing temperature drop is too great, different action may be taken to correct this (the valve is opened again). How feedback is used depends on the type of control employed, which is discussed later.

There are six basic types of control systems:

1. Self-acting. The measurement device and controlled device are in one unit. An example of this is a valve (controlled device) on a hot water radiator which is set by a bimetallic strip (measurement device). If the ambient temperature gets too high, the strip deflects and the valve is closed, reducing the flow of hot water into the unit and hence the heat input. The communications path is the mechanical junction between the elements, and the source of energy is the potential energy of the bimetallic strip/heat input from the environment.

2. Hydraulic. Pressurized liquid is used as the source of energy and communications path. The measurement elements change the pressure of this liquid. It may thus be made to act on rams and pistons to control dampers and valves.
3. Pneumatic. Compressed air is used in much the same way as the pressurized liquid above.
4. Electric. Electrical energy, communicated by wires, is used to act upon motors and solenoids. For example, the power to a solenoid valve may be switched by a bimetallic strip-controlled mercury switch.
5. Electronic. Small electrical signals from the measurement devices are processed and the outputs are amplified to control motors and solenoids.
6. Hybrids. Combinations of the above types of systems, for example, electronic/pneumatic, are also used.

A number of different types of control action may be used:

1. Discontinuous. An example is "on-off" control. The controlled device is switched from a nominal on position to a nominal off position, or vice-versa, when the measurement device detects a parameter outside preset high and low limits. An example of this is domestic air-conditioning systems, which will switch the fan/compressor off when the measured temperature gets below the "lo" limit and switch it on when it exceeds the "hi" limit. These limits cannot be set too close together, as the system would be switching on and off every few seconds, which would be extremely hard on the mechanical and high current elements.
2. Continuous. Here we have a control action which is not of the on-off type but has a range of controlled device settings.
3. Timed two-position. This is really discontinuous (on-off) control and provides the least degree of control as the controlled device is on for a preset time independent of what is actually happening to the controlled parameter. This type of control would only be used if other parameters were fixed and known in advance. (Feedback is not necessarily involved here.)
4. Floating control. The measurement device moves the controlled device at a constant rate to the on or off position. This is also not very accurate as, although it may be considered to be continuous to a degree, the rate of change is preset and not controlled in real time.
5. Proportional control. The controlled device is altered in direct proportion to the condition measured. This is a more precise and faster acting type of control; if the controlled parameter is off a little, the control action is small to avoid gross overshoot.
6. Proportional control with integral action. This is similar to proportional control but will react to changes in the controlled parameter regardless of the absolute value. The measurement system is continually reset to

remove any offsets. This allows the system to work around the optimum operating point.

7.   Derivative control. Corrective action is applied proportional to the rate at which the deviation in the controlled condition is taking place. This makes for a very fast acting and accurate control system.

8.   Proportional-Integral-Derivative (PID) control. This combines the best of all worlds by using integral offset nulling and derivative accuracy and speed.

As far as control of cleanroom air parameters in large facilities is concerned, PID control is the modern method of choice, although other systems are still in use. This was a very expensive option in the days when control systems used pneumatic logic. Now, computer programs may be bought off-the-shelf which offer PID control on a computer-based electronic control system. Signals from the measurement elements are led to analog-to-digital conversion units and fed, in numerical form, into the computer. Control outputs from the computer are reconverted to analog form and amplified, through semiconductor devices or relays, to alter the controlled devices.

STATIC

The final area we will consider in this chapter is related to the cleanroom air and cleanroom materials. Our need for clean polymeric materials and large flows of relatively dry air leads to problems of static electricity build-up, which may be considered to be yet another form of contamination.

Effects of Static Charge

When certain materials are rubbed by another material, which may be a solid or even a fluid, they may lose or gain electrons from the atoms near their surface, thus creating a charge imbalance. If the materials in question are nonconducting or isolated from ground, this charge cannot move away, that is, it is a static charge. Some very large potential differences, in the order of tens of thousands of volts with respect to ground potential, can be built up in this fashion. However, if a relatively conducting path becomes available, this charge can very rapidly move to ground, creating a sudden spike of current. This path to ground may even contain an air gap, the charge jumping across this gap by ionizing the air and creating a spark in the process.

Unfortunately, many electrical and even some micromechanical components are sensitive to static discharges. The best example of this is the very thin gate oxides used in MOS transistors. These are insulators, but will break down electrically or even melt if they are placed in the path to ground and are consequently subjected to a current spike as described above. An entire microprocessor or high density memory circuit may be rendered inoperative by just one gate being blown like this. Thin metal tracks may also be blown like fuses by current spikes. Integrated circuits now contain electrostatic discharge (ESD) protection elements on their input connections to help to protect them while in service, but these structures cannot help during manufacture as they are not complete.

The problems of static electricity are not confined to the microelectronics industry. The handling of small components becomes a nightmare if they acquire a static charge. This charge can induce an opposite charge in other objects, making the component stick to tools and surfaces. There is also a safety aspect, as a discharge across an air gap creates a spark which is locally hot. This could be a source of ignition for flammable vapors. Many large-scale disasters have been created by these conditions. In addition, static charges on surfaces will attract and hold contamination. In our discussion of charged materials in Chapter 4, we saw in Figures 4.6 and 4.7 how a charged substrate could reach out and grab a floating particle, whether the particle was itself charged or not. Therefore, charged materials are difficult to keep clean (think of how quickly a television screen becomes coated with dust), and the particles will also be tightly held to the surface, resisting removal by blowing.

We can see that static charge is not welcome in our facilities. We will now consider how to reduce the problem.

Sources and Detection of Charge

As with many problems of this type, the best place to start when tackling them is to identify the sources. As mentioned previously, many polymeric materials will generate a static charge when rubbed. Therefore, to help reduce the problem we could test our cleanroom materials to assess their generation capacity. In addition, we would also have to think of how these materials are likely to become rubbed in the first place.

The three most obvious ways a charge can be built up on a material mechanically are:

1. Operators. Personnel may brush past or lean over surfaces while working. There is plenty of opportunity for rubbing contact between the synthetic materials of the cleanroom suit and the materials in question. Very high charges may be built up in this fashion.
2. Equipment. Many pieces of cleanroom equipment have moving parts, gears and slides, made up of synthetic materials.
3. Fluids. Dry air passing over synthetics can produce quite large static charges. Charges may also be induced in various liquids, including ultrapure water.

There is also a fourth way that charges may be built up on materials. Ionizing radiation and ultraviolet light will charge materials by removing electrons.

Unfortunately, build-up of charge is not limited to solid surfaces. Cleanroom air and some process gases can also become charged when they come in contact with highly charged surfaces or ionizing radiation. In this case, individual gas molecules lose or gain electrons to create molecular ions. These ions are free to move with the other gas molecules and hence can carry charge around our clean facility. This charge can be donated to surfaces or even small particles.

The measurement of static charge is not a particularly easy task as it tends to be fixed and dispersed over a surface. It is easy enough to detect, using a gold leaf electrometer which takes charge from a small area of the surface and leads it via a conductor to thin gold plates which will repel each other when they receive this charge. However, to be more quantitative, we need some sensitive instrumentation with extremely high impedance input amplification and current spike protection (in case of sudden discharge). The static charge on the surface under test may be used to induce a charge on a metal plate of known area electrically connected to the input of our high quality electrometer. This induced charge can thus be measured by the instrument and the charge per unit area determined.

Reduction by Ionizers

Since we began this chapter by discussing how to precondition air, let us now describe how we may control static charge in the cleanroom air. The standard method is to use *corona grids* suspended immediately below the discharge points. These are merely arrays of sharp metal needles which are connected to a high voltage DC supply. The idea here is that such points will produce a very high local electric field, and this will effectively neutralize the ionized air. This actually works very well and can even reduce particulate contamination in very dirty environments by charging the particles and getting them to stick to surfaces. The same basic technique is also used to reduce the static content of house nitrogen gas. In this case a small point ionizer is placed at the nozzle of the blow gun to neutralize any charge present.

There is, unfortunately, a dark side to this technology. When the corona grids were examined in controlled tests in class 100 conditions, they did reduce static but actually seemed to add contamination. Some test results are shown in Figure 6.5, which shows the distribution of ions and particles around a high voltage ionizer. The contamination generation is thought to occur in two ways: corrosion of the ionizer tips due to contact with low levels of acid vapors in the atmosphere and condensation of organic vapors into polymeric solids, which may appear in particulate form. The latter mechanism has been investigated extensively, with various test vapors used, and the results seem to support the condensation theory. This makes a lot of sense, as organic mol-

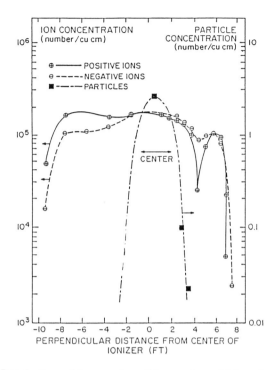

Figure 6-5. Distribution of ions and particles under a high voltage ionizer (data provided by Prof. Ben Liu, Univ. of Minn.).

ecules in vapor form will become charged near these ionizers and will thus tend to coalesce. What is worse still is that the particles so created are soft and sticky and will be difficult to remove from surfaces once they have landed.

How do we tackle this contamination generation problem? Some say don't use corona grids, leaving the control of static to other methods. If grids must be used, frequent cleaning of the tips to prevent build-up of material which will eventually fall off is vital. In addition, strict control of airborne organic vapors (which should be done anyway to protect the product) is necessary. Fortunately, the latest systems reduce the problems of particle coalescence by the use of voltage-switching techniques.

Reduction by Other Means

Reduction of static charge in synthetic cleanroom garments is discussed in Chapter 11. We will concentrate on static control in work areas in this section.

One of the best ways of controlling static charge in workbenches is to use materials specifically designed for this task. These materials are either rigid or soft in nature. In the case of the rigid materials, some of the actual bench components, for example, the worktop itself, may be made from a static control material. The soft materials are generally used in coverings and mats which are laid on the worktop or floor.

The rigid materials fall into three categories:

1.   Antistatic. These have an inherently high resistance, and it is extremely difficult to strip electrons to create a charge imbalance. Certain special polymers are very good in this respect.
2.   Static dissipative. These have a reasonably high resistance associated with them, but they control static by slowly bleeding the charge away. For instance, a permanently positively charged plastic film will neutralize negative charges in the environment.
3.   Conductive. These have a relatively low resistance and will conduct charge away to ground. It is actually better to have some resistance in the ground path to reduce the discharge currents, and so conductive polymers are actually better than low-resistance metals.

Where it is not economically feasible to have these relatively expensive materials used throughout, for example, in floors and walls, other building materials may be altered to give them static control properties. A good example of this is vinyl flooring materials which have carbon fibers, conductive polymer filaments, or fine wire grids embedded in them. This creates a composite material which conducts to some degree to create a continuous discharge condition. There is some concern regarding the contamination arising from old floors of this type, so careful monitoring should be performed to check their integrity. Of course, conductive vinyl- or enamel-coated metal raised floor sections will also control static well.

The soft static control materials are available in two basic types: (1) static dissipative vinyl, which will bleed charge, and (2) conductive polyethylene, which will remove charge by conduction. These may readily be formed into mats for personnel to stand on while working, or benchtop or equipment covers. Remember, however, that the conductive materials must be grounded so that the charge can flow to ground, and this can present a hazard for technicians working near exposed electrical apparatus.

Finally, apart from wearing suits with static control built in, the work force can also be grounded using conductive wrist and ankle straps. Some training and discipline is required here to ensure that these straps are worn properly and not allowed to fray or become dirty.

SUMMARY

We have seen in this chapter what happens to the air before it reaches our HEPA filters. Preconditioning of air involves prefiltering and the control of temperature and humidity. Recirculation of a large portion of the air keeps preconditioning costs down, but some make-up air will have to be introduced into the system to replace that which is lost to fume exhausts or merely to maintain a healthy atmosphere. The static charge carried by the air or by personnel may be controlled by appropriate flooring and benchtop materials or by ionizers, although care has to be taken with the latter method to ensure that the systems do not add to the particulate contamination levels in the room.

7

Site and Structural Considerations

We are at the stage where we could perform a reasonable job of designing and constructing some kind of cleanroom or clean space, as we now know something about microcontamination and controlled environments. However, we may still ask ourselves, "Where would be a good site for the facility?" This is a tricky question as the answer could depend on a number of diverse factors, including emotional as well as technological elements. However, assuming that economics and the whims of the company president do not present a problem, where is an ideal site? The air should be as free of particulate and gaseous contamination as possible. Non-ideal sites in this respect are adjacent to a heavy industrial area or right next to the ocean. However, the site should be close enough to population centers and other industries so that personnel and material supplies can reach it with ease.

We will study yet another factor in this chapter. It is, in effect, another type of contamination which we have not yet discussed. In many high-technology industries, *vibration* can be as deadly to the product as microcontamination and deserves consideration at every stage of the design and construction of the facility.

VIBRATION

Effects of Vibration

Vibration is the physical movement, both lateral and/or vertical, of objects and surfaces. The displacement generally has a periodic or near-periodic component. This movement can be devastating to precision mechanical and optical systems which rely on a tight control of the relative position of components for their operation.

A prime example of the importance of vibration control is high resolution lithography for semiconductor manufacture. In these processes, we typically are required to align a pattern on a photomask with a preformed pattern on a wafer with a tolerance of a few tenths of a micron. If a projection lithography system is used, the separation between these two elements could be several feet. Any relative movement of mask and wafer will result in a blurring of the pattern which is being formed in the photoresist on the wafer, and hence resolution is lost. This lack of image quality can have a profound effect on yield, especially on small geometry circuits.

Vibration is also unwanted in other technology areas. The micromanipulation of mechanical elements is made very difficult if vibration is present. For instance, a high-density magnetic read/write head could not be aligned properly, or a test probe could actually do damage to a circuit due to vibration effects. Modern analytical techniques would also be impossible without strict vibration control; scanning tunneling microscopy requires that a super-sharp probe be placed in close proximity to individual atoms, an impossibility when there is any relative movement.

It is clear in many cases that we do not want vibration any more than we want contamination. However, as in the case of contamination, we must understand what the sources and transmission routes of vibration are before

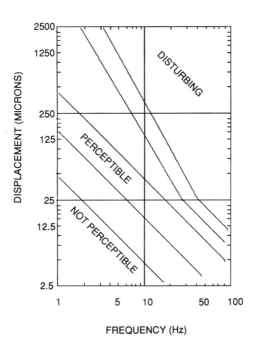

Figure 7-1   Human response to steady state vertical vibrations.

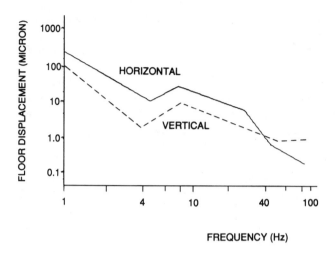

Figure 7-2   Maximum allowable displacements for a typical photolithography system.

we may tackle the problem effectively. One point which we will add at this stage is that, in many cases, vibration levels which may be considered as significant to an operation may be completely imperceptible to humans. Figure 7.1 shows the human response to steady-state vertical vibrations. When we compare this to the maximum allowable displacement requirements for a typical photolithography system (Figure 7.2), it becomes apparent that sensitive instruments, such as a high quality accelerometer (which measures movement), are required as humans cannot hope to detect these levels themselves. Another point to note from Figure 7.2 is that this equipment has already been made vibration tolerant, but it still requires a reduced vibration environment. As a general rule for sensitive lithography equipment, we should strive to keep ambient vibrational displacements to less than 5 microns below 10 Hz and less than 1 micron from 10 to 50 Hz both vertically and horizontally. We will return to this point later in this chapter.

External Sources

When considering sources of vibration, it is useful to split them into the categories of *external* (outside the facility) and *internal* (inside the facility). We will tackle external sources first. There are essentially four groups of external sources: tidal, seismic, meteorological, and man-made.

It is an unfortunate fact that there is no place on earth which does not vibrate to some degree. This ambient vibration is generated by the Earth itself. To understand why this happens, one must step back and look at the macroscopic situation. The planet Earth is not a solid body; the solid crust on which we live rides on a sea of molten rock or *magma*. When other celestial bodies such as our nearest neighbor, the Moon, exert gravitational forces on the Earth, the planet distorts, with mass being pulled toward whatever is doing the pulling. This is precisely why large bodies of water, such as the oceans, are tidal. When the moon pulls on the water at the middle of the ocean, it rises up, reducing the level at the shore, that is, the tide goes out. Unfortunately, the crust also feels these forces and literally creaks and groans as it deflects, leading to continual vibration, albeit at very low frequencies. The intensity of these tidal vibrations at any particular site tends to depend on factors such as the local geology and geography, the rock types and structure, and the type and depth of soil, the latter having an effect on the transmission of the vibration.

The fact that the Earth has a solid crust which resides on a liquid support also leads to other forms of vibration, this time in the seismic category. The continents which we live on are part of individual plates in the crust. These plates are actually still being slowly formed by molten rock coming up to the surface (under the sea) and cooling to form solid material. However, something drastic has to happen at the other end of the plate, otherwise it would continue to grow in size and the Earth would get bigger day by day (which it doesn't). The plates tend to slide over one another and, as one might imagine, this process is not a particularly smooth one. In fact, the friction between the plates holds them steady until sufficient pressure is built up at faults in the crust, whereupon movement occurs, resulting in what we call an earthquake and its attendant tremors. In this case the incidence of vibration is somewhat sporadic, arising in unpredictable bursts. This is the most drastic kind of vibration, as lateral and vertical movements can be as much as several feet in local areas. This movement has extreme safety implications as well.

The weather can also add to the vibration in an area. Turbulent air caused by a strong wind passing by a somewhat less than aerodynamic building will buffet the structure and cause it to vibrate. This action tends to be especially evident on large panes of window glass in high-rise buildings. The wind blowing through deep-rooted mature trees applies a force to them, causing movement which is transmitted into the ground by the roots. Even loud thunderclaps create vibration, as these are actually shock waves which travel through the air and are able to shake structures.

Significant sources of vibration in any populated area are transport and industry. The construction industry tends to be a great generator of vibration, as it frequently employs drilling systems, pile drivers, heavy machinery, and even explosives from time to time. Other heavy engineering industries such as steel mills and shipyards create their share of vibration. Heavy trucks on highways and moving trains create vibration which is readily perceptible to humans, and even the roar from a jet engine can add to the ambient levels. As a general guide, large earth-moving equipment will produce local vibration in the frequency range of 1 to 5 Hz with vertical ground accelerations approaching 1 g (this is equal to the acceleration produced by gravity and so it is possible for objects to be lifted off the ground by this movement). Highway trucks have a vibrational frequency range between 2 and 8 Hz, with much lower local vertical ground acceleration (around 0.2 to 0.4 g), and tracked vehicles tend to be in the 4 to 10 Hz range, with vertical acceleration similar to trucks.

Internal Sources

The most significant vibration source internal to the facility tends to be machinery. The buildings which house high-technology operations tend to be filled with vibration-producing machinery in the form of fans, pumps, compressors, electric motors, and so forth. Anything which rotates or reciprocates, slides, rolls, or moves in a variety of ways is capable of producing vibration. Even electrical transformers, which are essentially magnetic in nature, will vibrate slightly when an alternating current is applied, and thus will add to the volume of internal vibration.

Consider the case of an electric motor which is used to drive a cooling fan on a large power supply. It is virtually impossible to balance the rotor so precisely that all the mass is evenly distributed around the axis of rotation. Therefore, when the rotor spins around, a net centrifugal force is created due to the mass imbalance, and the motor assembly will move or vibrate. If a similarly designed (if somewhat larger) motor is used to drive air through ducts, the problem is worse, as any turbulent air created by the fan or poorly designed (nonaerodynamic) structures will vibrate the ducts, and this may be transmitted to the building structure by way of the duct hangers. Of course, the vibrational effects of reciprocating compressors and the like are easy to imagine, as there is a large mass being thrown back and forth in this case, leading to movement.

What is perhaps most distressing to those individuals engaged in the business of trying to control the effects of vibration is that a single vibration source may actually produce a wide range of frequencies. For instance, if an out-of-balance motor is running at 600 rpm, one might expect to see a vibrational frequency of 10 Hz. This fundamental frequency is seen in practice, but it is also accompanied by lower intensity multiples of this frequency. These are called *harmonics* and tend to become smaller in intensity as the frequency increases. They arise due to the nature of the vibration generation; if the vibration generator produced a perfect sinusoidal wave, then all we would see would be the fundamental. However, vibration sources rarely do this, and therefore we get a complex waveform which is actually composed of a number of different frequencies superimposed (this is obvious if you are familiar with Fourier analysis). The situation can actually be worse. If the rotor has multiple out-of-balance spots, some very odd frequencies can arise, with multiple strong fundamentals as well as a plethora of harmonics. We should not forget that many large machines also produce sound which is a type of airborne vibration capable of coupling to surfaces, especially if the sound is at the resonant frequency of the system.

The concept of resonant frequency, (sometimes called the *natural frequency*), is extremely important in vibration control. Systems which are able to oscillate have a natural frequency associated with them. This is the frequency at which the maximum energy from a vibration source can be coupled into the system. Therefore, if one of the frequencies created by a source happens to be the resonant frequency of, for example, a support that holds a piece of equipment, a relatively large vibrational effect will be experienced at the equipment.

Personnel are also a source of internal vibration. They walk on floors which may be suspended off the ground, allowing movement to occur in these structures as the body weight crashes down through the relatively small contact area of the heel. A typical footfall produces vibration which drops off exponentially with time and distance, but the immediate effect can be quite significant, the effect becoming worse with the weight of the individual. Personnel may also be pushing carts, or driving indoor vehicles such as fork-lift trucks, and these add to the vibration. Indeed, there are actual reported cases where people talking have had an adverse effect on high resolution electronbeam lithography systems, causing column vibrations at the resonant frequency.

SITE SELECTION

We now return to the question of site selection, but this time we will look at the problem from the viewpoint of vibration assessment and control. Because of the nature of the problem, vibration engineering has to start well before the facility is constructed. A vibration survey, performed by professionals in this area, should always be performed on the potential sites as part of the site selection process. It is not uncommon for some sites to be ruled out on the basis of the findings of these surveys.

The first and most obvious things to look for are the local human geography features. For instance, is the site near a major highway, a railroad, an airport or approach flightpath? Is there heavy industry or long-term construction nearby? Will people be blasting rocks in a nearby quarry or mining directly underground? These situations are all fairly easy to discover and should all be avoided if possible. What is slightly less easy to find out but nevertheless just as important is what is planned for the area in the future; will a major transport route be placed right by your facility three years after you start to build? Check with the local planning offices for information.

Next in the site selection process is consideration of natural geography. Is the site prone to strong prevailing winds or is it relatively sheltered from the weather? Are there likely to be resonance effects from even light winds? (This can become a very serious point as bridges have been destroyed by relatively light winds due to the natural landscape creating a resonance coupling situation.) What are the amplitudes and frequencies of tidal vibrations in the area and is seismic activity frequent or likely? (It is interesting that California and much of Japan are very poor in the latter respect, and yet both areas have large concentrations of high-technology industries. Because this is so does not recommend the practice.)

The geology of the area should also be examined carefully. For instance, what is the rock and soil structure like and does it readily transmit natural and man-made vibrations? Can we build a stable building on the site without coupling into the ambient vibrations? Generally speaking, grounding the building structure to the bedrock is usually worse than allowing it to float on an overlying gravel, as the looser material will naturally absorb rather than transmit vibrations. However, we do not want the building to subside several years after it is built due to a lack of stable foundations.

Unfortunately, since we have resigned ourselves to the fact that everywhere on Earth vibrates to some degree, and also since we will always have to make compromises as far as site selection is concerned, vibration engineering does not stop at this point; in fact it never stops until all the equipment is installed.

STRUCTURE

Foundations

The foundations of a building are extremely critical as the structure rests on these and will pick up any earthborne vibrations which they absorb. As mentioned in the previous section, we have something of a dichotomy as far as foundation design is concerned as we desire maximum structural stability with minimal vibration coupling. Since, in many cases, we want to avoid hard bedrock contact, we could design the foundations such that they have a large stable mass, or multiple large masses, at their base. The mass or masses in turn support the caissons, which are effectively the "legs" of the building. This design is generally a good idea, but we must be extremely careful not to fall into a trap here. For instance, by performing a vibrational analysis of the area, we will know which frequencies dominate. We can then design the foundations such that the resonant frequency, which depends on the mass and flexibility of the structural elements, is completely different from the frequencies present in the ground. A common mistake is to assume that the biggest foundations are the best. This approach can result in vibration amplification rather than vibration reduction.

Reducing Vibration

The building foundations are not the only way that external vibration can get into the building. The largest amplitude vibrations from man-made sources frequently travel very freely along the surface of the ground, especially if it is continuously paved with a poor absorbing material like concrete or hard tarmac (blacktop). For instance, vibrations from trucks passing along a nearby highway will create vibrations which can travel across a parking lot and into the building structure. An effective way of reducing this effect is to put a narrow deep trench around the building and back-fill with a loose or semi-liquid absorbing material. This technology has been used for many tens of years on buildings in large cities which have subway railroad systems running nearby. The isolating trench, a few feet wide, is typically filled with a slurry and capped with a nontransmissive solid material.

Vibrations caused by the wind can be reduced by appropriate building design. Many architects have produced wedge-shaped building designs to make the building more aerodynamic in the face of prevailing winds, thereby reducing the vibration-producing turbulence. Where this is not feasible, for example, when an existing building is being retrofitted for a high-technology application, wind breaks may also be used. A good and inexpensive way of doing this is to use a series of shallow rooting trees and shrubs to form the "wedges" in front of and behind the building. Care should be taken not to use deep rooting plants or to space the trees so that that turbulence is created around them. Acoustic vibrations from external sources may be dampened by the use of a thick layer of absorbent material within the outer walls of the building and double glazing on all outer windows. These devices are usually put in place for energy conservation anyway and hence do not represent additional cost.

Floor Isolation

Since it is almost always impossible to completely isolate a building from external vibrations, and since we will also have internal vibration sources to contend with, we have to employ other techniques within the structure to reduce vibrational effects. This is especially true of older, non-ideal buildings which are being retrofitted, as these may not have been designed with high vibrational control in mind.

One question we may ask is, "How does the vibration travel inside the building?" Any vibration which enters via the foundations or walls, or is generated by internal plant, will travel up or down the vertical structural elements, and from there will be transmitted across the floors supported by these elements. There is not much we can do about the columns which support the building, as these must have a high degree of structural integrity and therefore cannot easily have vibration-reducing elements incorporated. However, we can do something about isolating the floors.

Take the simple case of a slab on grade structure, where the floor is laid directly on compacted soil. If the soil is sufficiently absorbing, and it usually is, the main vibrational path will be from the structure/walls into the floor. To break this path, we can cut around the section of floor which is to be isolated so that there is no longer a structural tie between the floor and building, that is, a 1- to 2-inch gap is created (Figure 7.3). The gap can be filled with polystyrene foam and covered with a floating cap to prevent tripping. Any connections to equipment resting on the isolated slab should be by way of flexible couplings, for example, plastic hoses or stainless steel bellows. This type of structure is widely used, but there are occasional problems with subsequent movement of the slab due to poor original soil compaction or under-slab water erosion of the soil. To reduce this, some legs may be added to the slab, by drilling through and pushing them through the holes, to help prevent lateral movement.

One further problem with the above approach is that in cutting out a relatively low mass section of the floor, we may be creating a structure which will be more prone to local vibrational effects, that is, the lower mass will tend to move more when it is walked on or when a piece of equipment on it produces vibration. A different approach for the slab on grade floor is to cut as before but this time completely remove the cut section. A pit may then be dug and a large mass pedestal poured so that it is not in contact with the edge of the pit or the rest of the floor (Figure 7.4). This type of vibration isolation is often found in buildings housing sensitive electron-beam lithography equipment.

Cutting out isolated floor sections usually does not work very well when a basement is beneath, for obvious reasons. Here a different type of structure is required, as shown in Figure 7.5. Here the isolated section of floor is supported by its own columns and foundations. These columns pass through the basement floor but do not make any contact with it to prevent vibrations entering from basement equipment. The gaps in both floors are again filled with a vibration-absorbing material such as polystyrene or foam rubber. A critical point to remember here is that this structure should be designed so that its resonant frequency is as far away as possible from any external or internal sources.

To end our discussion on floors, we should mention raised floor structures as used in floor return cleanrooms and test facilities. From a vibration control point of view, these must be chosen with great care. Standard computer room

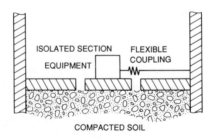

Figure 7-3 Slab-on-grade vibration isolation.

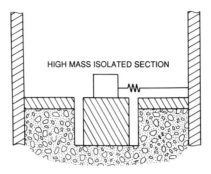

Figure 7-4 High-mass block, slab-on-grade vibration isolation.

raised flooring tends to have lightweight sections, made from laminated material, supported on adjustable jacks, the idea being that sections can be lifted easily for cable maintenance. Generally speaking, this flooring was not designed with vibration in mind, as it will tend to be displaced fairly readily due to its inherent flexibility. Cleanroom flooring should always be more substantial than this, with higher mass rigid sections being supported by short thick jacks. Even with this type of flooring, vibration-sensitive equipment should rest on isolated raised stands to prevent floor-carried vibrations from interfering with it.

Plant Layout

As discussed previously, much of the internal vibration originates from equipment in the building. Much of the building plant is heavy and thus may rest directly on structural elements in the upper floors or directly on the basement

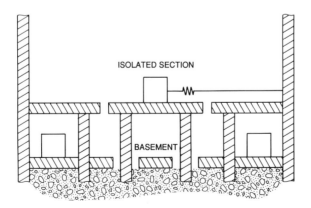

Figure 7-5 Vibration isolation with basement.

floor. However, much of the heavy equipment does not actually have to be in the cleanroom building at all.

The Satellite Building Concept (SBC) is a facility layout which has separate office/canteen, maintenance/stores, and plant buildings. Heavy vibration-producing plant equipment such as air compressors, chillers, and so forth is placed in the plant building and may readily be vibrationally isolated from the cleanroom building using flexible couplings and ground isolation trenches. Other service operations such as the ultrapure water production facility may also be placed in this building.

Unfortunately, not all vibration-producing plant machinery can be segregated in this fashion for practical reasons. This is discussed in the next section.

INTERNAL VIBRATION ENGINEERING

Plant Isolation

While much of the vibration-producing plant may readily be placed in a remote building, other vibration sources such as vacuum pumps may not. These have to be in close proximity to the equipment they serve, as long pipes take a long time to pump out and have a poor vacuum conductance. Therefore, to reduce the injection of vibration into the environment from this equipment, we must isolate it from the structure as much as possible.

In most cases, vibration producers can be mounted on or suspended from antivibration mounts. These generally contain an absorbing element such as a rubber pad or spring; vibrational kinetic energy is turned into stored potential energy in the absorbers and may be dissipated as heat. The vibrating equipment is allowed to float on these absorbers so that the displacement of the mass of the equipment is not transferred to the building structure. The type and stiffness of these mounts depends on the mass of the machine and the amplitude and frequency of the vibrations it produces. Noise producers, including air ducts, can be insulated or lagged with an absorbing material (for example, fiberglass), care being taken to allow cooling to take place where necessary.

Sensitive Equipment

In the case of particularly sensitive equipment, for example, high resolution lithography apparatus, it is always best to bring as much vibration isolation and control as possible to the instrument. This never removes the need for environmental vibration control, but it does reduce the ambient requirements.

At a very simple level, the same techniques which are used for plant isolation may be used for equipment. Some researchers have found car tires to be a great equipment isolator, although we would not suggest bringing these into the cleanroom. Frequently, the equipment is placed on massive granite or similar dense material blocks which are themselves resting on absorbers. This approach is found in many optical systems, including inspection systems, as the increased mass makes the entire system less susceptible to movement by vibration.

The mounts described above are called *passive* elements, as the vibration absorption is inherent in the material itself. They are usually chosen to damp out vibrations in a particular frequency range but may actually be quite useless at other frequencies. *Active* mounts are also used for particularly critical applications. These generally use compressed air as the absorbing element. This is an extremely good absorber as it will compress and flow very easily, preventing any coupling over a wide range of frequencies. Active mounts are so called because the air pressure has to be maintained at a constant optimal value by a control system, even under changing load conditions, to keep the supported surface level. These systems typically have their own small compressor as well as the control system elements (controller, valves, and so forth) and the air pistons or sacks. Unfortunately, all of this hardware makes them rather expensive.

Vibration Practices

A familiar concept, often stated in this book, may also be applied to vibration control: even the largest investment in vibration control can be rendered ineffective if vibration practices are poor. Fortunately, we have only a few key points to keep in mind, and these are discussed briefly here.

People should always be isolated as much as possible from vibration-sensitive equipment and vice-versa. It is a good idea to structurally separate "people floors" from "equipment floors" whenever possible to prevent vibrations caused by walking getting to the equipment. Wheeled pedestrian vehicles, such as cleanup and dispensing carts, should always have soft tires.

As is often the case, there is no substitute for education of the direct labor force and any person who is likely to be near the equipment while it is operating. Running, "hard" walking, jumping off ladders, and unnecessary movement in general should be avoided as much as possible. Finally, never lean, with any part of the body, on or near equipment and inspection/measurement stations. It is surprising how many operators rest their arms on alignment tools during exposure. This should always be discouraged.

SUMMARY

As we have seen in this chapter, vibration can be as devastating to certain operations as the contamination we discussed previously. There is a multitude of external and internal sources with vibration arising from the earth or from the equipment in and around the facility. Fortunately, we can decouple these vibrations from our sensitive areas by good structural design and by the use of vibration blocking and damping systems on equipment.

8

Cleanrooms, Codes, and Legislation

Due to the nature of the processes carried out by many high-technology industries, specifically with regard to the hazards inherent in these processes, some control is necessary to protect the work force, the community, and the environment. This control appears in the form of codes and legislation. We will discuss them in this chapter.

PRINCIPAL CODES

Three regional building codes and their companion fire codes regulate the construction and fire protection of buildings in the United States. The *Uniform Codes* cover the western United States roughly from the Mississippi River westward. The *Southern Codes* cover the southeast, and the *Basic Codes* cover the northeast. These are model codes, that is, they are drafted by building safety officials, fire chiefs, and related professionals as consensus documents. The codes are then published by the parent organization, such as the International Conference of Building Officials and The Western Fire Chiefs Association, who are responsible for development of the Uniform Codes. At this point, the codes have no authority. They must be adopted by state law or local ordinance in order to give the local fire department or building safety official the authority to enforce them.

All model codes have specific requirements for cleanrooms. A fourth organization has authorized a committee to draft regulation for the protection of cleanrooms. The National Fire Protection Association (N.F.P.A.) has started a committee of cleanroom users, insurance carriers, government, and cleanroom suppliers to draft a standard for the N.F.P.A. This standard will apply to government agencies and will also selectively regulate some cities as they select N.F.P.A. standards to supplement building and fire code issues during the adoption process. N.F.P.A. standards also have some international implications because a large part of the world looks to this organization as the leader in the United States in terms of fire protection standards.

Building Codes

The building codes follow the same general course throughout, that is, they call for separation structures between cleanrooms and other occupancies in the building to be *fire rated* (hold back fire and smoke for a given minimum time), restrict volumes of hazardous materials in storage, and segregate exit corridors from those corridors where hazardous materials are transported. Specifically, the requirements are as follows:

Separation. One-hour fire-rated separation construction is required to separate the cleanroom area from the remainder of the building.

Floors. Noncombustible floors are required. Openings through floors may be unprotected when interconnected levels are used for mechanical equipment only, that is, the interconnected levels of the structure may not house "people" operations.

Ventilation. One cubic foot per minute per square foot of floor space general ventilation is required for cleanroom areas. This may include the recirculated air of the laminar flow system. Exhaust ventilation duct systems may not connect to another duct system outside the cleanroom area within the building. Smoke detectors are required to be installed in the recirculating air stream and shall transmit their alarm to a constantly attended location. The general ventilation system shall be equipped with a remote switch capable of shutting down the ventilation system. This switch shall be located outside the cleanroom area.

Electrical. Hazardous location wiring is not required if the ventilation rate is 4 cubic feet per minute per square foot as an average and when the rate is 3 CFM/sq. ft. at any location within the cleanroom area.

Exits. Handling of hazardous material within an exit corridor is prohibited. Exit corridors shall not be interrupted by intervening rooms. A minimum of two exits are required where the occupant load exceeds 30. The total width of exits in feet shall be not less than the total occupant load divided by 50. The width shall be divided approximately equally among the separate exits. If only two exits are required, they shall be placed a distance apart equal to not less than one half of the length of the maximum overall diagonal dimension of the building or area served measured in a straight line between exits. The maximum distance of travel from any point to an exit shall not exceed 100 feet. Every corridor serving an occupant load of ten or more persons shall not be less than 44 inches in width (two "exit units"). Exit corridors shall be separated from cleanroom areas by one-hour fire-rated construction. There is an exception in the rules to cover existing buildings that exempts them from the "separation of corridors" provision if the building was built to a previous code. The exception falls out when the building is modified and the exemptions are accompanied by a number of restrictions such as local alarm requirements, sprinkler protection, and restrictions on cart design and cart procedures.

Service corridors. This is the way building and fire officials envision hazardous materials will be transported to the use points and storage locations within the cleanrooms. They are required to be separated from exit corridors for people by a one-hour fire-resistive barrier. Mechanical ventilation is required at six air changes per hour or one CFM/sq. ft. of floor area, whichever is greater. Maximum travel distance to an exit from a service corridor may not exceed 75 feet. Dead ends may not exceed 4 feet and there may be not less than two exits. Not more than half the required exits may be into the cleanroom.

Storage of hazardous material. Amounts of hazardous materials that exceed a certain level as prescribed by a table in the codes are required to be in internal storerooms specifically designed for hazardous material. The storage room may not exceed 6,000 square feet and must be separated from all other areas by not less than two-hour fire-resistive construction if the room is 300 square feet or larger. If the room is less than 300 square feet, the construction may be one-hour rated. If hazardous materials are dispensed within the storeroom, the area is limited to 1,000 square feet. All hazardous material storage rooms are required to have at least one exterior wall. This wall shall be not less than 30 feet from the property line. Explosion venting may be required depending on the character and volumes of storage. Hazardous materials storage rooms are to be classified as Class I, division I, electrically, whether or not dispensing takes place.

Piping and tubing. Pipes and tubing used to transport hazardous materials to use locations shall be metallic unless the material being transported is incompatible with metals. Systems supplying gaseous hazardous materials shall be welded throughout, except for connections, valves, and fittings, to the systems which are in ventilated enclosures. All hazardous materials' piping or tubing in service corridors shall be open to view. Hazardous material piping and tubing may be installed within the space defined by the walls of exit corridors and the floor or roof above, or in concealed spaces above other occupancies under the following conditions:

- Automatic sprinklers are installed in the space, unless 6 inches is the largest dimension of the space.

- Ventilation of six air changes per hour is provided to the space. The space may not be used to convey air from any other area.
- A catch pan is installed to catch any drippage from liquids.
- Separation from the exit corridor is required to be one-hour fire rated.
- Manual or automatic shut-off valves are required at branch lines into the cleanroom and at entries into exit corridors.
- Class I, division II, electrical wiring is required within the enclosure.

All of these requirements do not apply if the material is transported in a pipe within a pipe made from ferrous material.

Fire Codes

The three major fire codes in the United States have some common ground rules. This commonality lies in the fact that after the Uniform Fire Code Article 51 was accepted, the semiconductor industry championed the idea of spreading the requirements to the two other code bodies. The idea was to provide common ground rules for the industry which has factories in all three jurisdictions. Consequently, the fire codes for the other two areas resemble Article 51 of the Uniform Fire Code quite closely. All three codes impose restrictions on the quantities of hazardous materials which are allowed to be in a cleanroom, and all three have a table similar to the one shown in Table 8.1. In calculating these quantities, the fire codes generally except material in use at a workstation or within piping. They also do not count hazardous materials which are in reservoirs and filters or in approved safety containers within systems which are provided with local exhaust ventilation.

All of the fire codes concentrate heavily on "excess flow control" provisions. Hazardous materials in piping systems present a potentially severe hazard to firefighters during an emergency, and there is concern that a firefighter entering an area to fight a fire will find himself literally up to his hips in hazardous material leaking from a pipe which has failed during the emergency. The requirement is for gases or flammable liquids in piping systems to be provided with a fail-safe system which will shut off flow of gas or liquid due to rupture in the pressurized piping.

When gases are used or dispensed and the physiological warning properties for the gas are at a higher level than the accepted permissible exposure level for the gas, a continuous gas monitoring system shall be provided to detect the presence of a short-term hazard condition. Ammonia, for instance, has good warning properties. Nearly anyone can smell ammonia long before it reaches a hazardous concentration. The same thing can be said for a number of compounds including, for instance, hydrogen chloride and silicon tetrachloride. There are, however, a number of materials used in cleanrooms with relatively poor warning properties. People could be exposed well above per-

Table 8.1   Maximum Quantities of HPM Permitted in Each FAB Area

Flammable Liquids	
Class 1-A	90 gal
Class 1-B	180 gal
Class 1-C	270 gal
Combination Flammable Liquids	360 gal
Combustible Liquids	
Class II	360 gal
Class III-A	750 gal
Flammable Gases	9,000 cu ft @ 1 Atm & 70° F
Liquified Flammable Gases	180 cu ft.
Flammable Solids	1,500 lbs
Corrosive Liquids	150 gal
Oxidizing Gases	18,000 cu ft
Oxidizing Liquids	150 gal
Oxidizing Solids	1,500 lbs
Organic Peroxides	30 lbs
Highly Toxic Material and Poison Gas	Included in the aggregate for flammables as noted above

missible exposure levels long before warning takes place in the form of smell, fumes, tears, or other response. These conditions generally require continuous gas monitoring equipment to provide early warning of engineering controls not working properly or indication that something has gone wrong within the system.

The hazardous materials restrictions apply not only to the cleanroom in general but are also applied to the individual workstation. Maximum quantities of hazardous materials per workstation generally follow the amounts shown in Table 8.2. The fire codes uniformly require local exhaust ventilation to capture and exhaust fumes and vapors at the workstation. Further, they all require that the exhaust system be safeguarded from fire or explosion, by restricting the duct to those materials which are compatible in the gas stream. Thus, the regulations all state that two or more operations shall not be connected to the same exhaust system when either one or the combination of the substances removed may constitute a fire, explosion, or chemical reaction hazard within the exhaust duct. The codes also require that exhaust ducts which penetrate occupancy separations shall be in a shaft of equal fire resistance to the separation being penetrated. Ducts may not penetrate area separation walls, and fire dampers are not permitted in exhaust ducts. Ducts used for exhaust purposes are required to be internally sprinklered if the duct diameter is equal to or greater than 10 inches and the duct is inside the building and the duct conveys gases or vapors in the flammable range. All three versions of the fire code have the same exception to this rule, that is,

Table 8.2 Maximum Quantities of HPM
per Workstation

Flammables and Toxics	Gases	3 cylinders
Combined	Liquids	15 gal
	Solids	5 lbs
Corrosives	Gases	3 cylinders
	Liquids	25 gal
	Solids	20 lbs
Oxidizers	Gases	3 cylinders
	Liquids	12 gal
	Solids	20 lbs

ducts which are below ceiling level that are less than 12 feet long or listed in the fire codes may be used without sprinklers. All three fire codes also require that the exhaust system be connected to some form of emergency power capable of operating the system at one-half of its rated capacity. Electrically, they require that unless the exhaust system can provide dilution to produce a nonflammable atmosphere on a continuous basis, all workstations involved must be Class 1, division 2, if they use flammable gases or liquids.

Workstations using liquid hazardous material are required to have a compatible drain with the work surface sloped toward that drain. Within the cleanroom, hazardous materials may be stored, but only in approved storage cabinets. Approval is generally defined as "acceptable to the authority having jurisdiction," in other words, the Fire Chief. Metal cabinets are approved if they meet a detailed standard in the codes which requires local exhaust ventilation, 2 inches of liquid spill storage space in the bottom of the cabinet, self-closing doors with latches, and sprinklers for cabinets containing flammable hazardous materials. The fire codes also require that chemical storage cabinets provide separation of hazardous materials by class to provide compatibility.

In existing buildings the chief is authorized to permit transporting of hazardous materials either by hand (with restrictions) or by cart in the building's exit corridors. In order to qualify for this permission, the user must transport in approved containers, via approved cart, only in sprinklered corridors with an emergency alarm or telephone every 150 feet along the corridor. Procedurally, the cart may not be parked in the corridor and the number of carts in the corridors is also limited. Service corridors are required for all newly constructed cleanrooms for the transportation of hazardous material from storage to point of use. The service corridor is required to be a minimum of 5 feet wide, or cart width plus 33 inches. Manual alarm or signaling devices are required, and none of the fire codes will permit dispensing of hazardous material within a service corridor.

All three codes require an emergency plan, defined as the maintenance of plans showing the amount and type of hazardous material stored, shut-off valve locations, telephones or alarm station locations, as well as locations of all emergency exits. The codes also require that there be an emergency response team organized on each operating shift to respond to emergencies and assist the fire department in an emergency. The emergency response team is required to be drilled on at least a quarterly basis with drills and training sessions recorded for review by the fire department.

Other Code Requirements

As was stated earlier, the National Fire Protection Association has authorized a committee to prepare a standard for the N.F.P.A. on fire protection for cleanrooms. This standard will primarily impact those bodies which reference N.F.P.A. standards for code compliance. Most notable in this group is the federal government in the United States. U.S. government installations are exempt from compliance with local building and fire codes and instead opt for compliance with national standards as generated by the N.F.P.A. There are a number of cities and states which also adopt N.F.P.A. standards either directly or write them into local codes by reference.

The International Association of Fire Chiefs generally views the N.F.P.A. standards as the preferred way to regulate fire issues. The other major international fire protection authority that is widely relied on is the F.O.C., which is the Fire Officers Committee of Great Britain. F.O.C. standards are used extensively in most of the areas of the world where Great Britain was a colonial power. F.O.C. standards are seen in Hong Kong, Australia, Malaysia, India, and Singapore. Insurance carriers which provide risk protection for cleanrooms are concentrated in the United States and generally wish to follow N.F.P.A. rules throughout the world. Thus, an international company or a multinational company is usually faced with a set of regulations which seeks to make a marriage between F.O.C. and N.F.P.A.

FIRE PROTECTION

Fire Systems

Fire protection in the modern cleanroom is a critical issue. Over and above various fire code requirements, there are insurance requirements and those requirements which are self-imposed by a good loss prevention program. Fire insurance carriers are frequently overwhelmed by the concentration of values as seen in the cleanroom. It is not unusual to have single pieces of equipment in the cleanroom with values well in excess of half a million dollars, and several of these pieces of equipment may well occupy the same general area of the cleanroom. The costs of equipment added to the cost of the building plus the value of the product being produced in the area can easily represent tens of millions of dollars of value in a single area. These values, and related business interruption costs, are enough to make the underwriter nervous about the risk he is accepting for insurance purposes.

Protection programs to prevent losses are key to retaining favorable insurance treatment in terms of rates as well as maintaining a good relationship with the fire regulatory authorities. The protection program begins with management programs. Management should establish programs which demonstrate a willingness and a determination to reduce the probability of loss. If a loss occurs, those same programs should be immediately responsive to reduce the impact of that loss both on the property and the productive capacity of the organization. Programs should include:

1. An organization of trained personnel to respond to fire and other loss potential emergencies. This emergency organization needs to have an intimate understanding of the physical plant and its operations and must be present during all operating shifts.
2. Trained employees to inspect and maintain loss control equipment such as water supplies, sprinkler equipment, ventilation systems, and alarm systems.
3. Procedures for restoring all impairments to loss control equipment on a priority basis.
4. Procedures to effectively control all phases of new construction and remodeling. These procedures should include routine as well as extraordinary maintenance operations that involve the use of hazardous materials or open flames.
5. Procedures that call for the duplication or separate storage of critical records, design data, or other unique materials without which the operation could not restart or would be impaired. Computer tapes are a good example of this type of material. Tapes should either be duplicated and stored in another location or should be stored separately from the use point to prevent total loss of the operating capacity.

Buildings used for cleanroom operations should be of fire-resistive or noncombustible construction. Combustible construction materials should be avoided where possible. High property damage and business interruption values may warrant the establishment of multiple cleanroom areas in separate buildings or separated by minimum 3-hour fire walls with Class A fire doors on communicating openings. To meet the intent of this concept, ventilation systems, fume exhaust systems, and other utilities should be independent from adjacent buildings or areas to the extent possible.

Individual process operations within the cleanroom should be subdivided to the greatest extent possible using minimum 1-hour noncombustible walls to reduce potential damage from smoke, corrosive vapors, fire, or water in one area from contaminating adjacent areas. Interior partitions should be constructed of noncombustible materials such as steel, concrete, or masonry. Covering, if used, should have an interior finish having a flame spread rating not over 25 without evidence of continued progressive combustion, and fuel contributed and smoke developed ratings of not over 50 when tested in accordance with the "Method of Test of Surface Burning Characteristics of Building Materials" (ASTM E-84, UL-723, NFPA-255).

Heating, ventilating, and air-conditioning (HVAC) ducts, connectors, and appurtenances should be constructed of noncombustible materials such as steel or aluminum or of Class 0 or Class 1 materials as tested in accordance with the "Standard for Factory Made Air Duct Materials and Air Duct Connectors," (UL181). Duct coverings, linings, tapes, and core materials in panels and adhesives should have a flame spread rating not over 25 without evidence of continued progressive combustion, and fuel contributed and smoke developed ratings not over 50 when tested in accordance with ASTM-E84 as above. Air filters should be Class 1 when tested in accordance with the "Standard for Air Filter Units," (UL-900). High Efficiency Particulate Air (HEPA) filter units should be listed as acceptable as tested in accordance with the "Standard for High Efficiency Particulate Air Filter Units" (UL-586). Smoke and heat venting that uses the HVAC system to full exhaust capacity and is actuated by smoke detection equipment in the return air ducts should be provided in production buildings. The recirculation air fans should be stopped automatically during this mode of operation. Separate air handling systems should be provided for each of the distinct operations conducted in the cleanroom environment. The air handling system in cleanroom areas should be arranged for 100 percent supply and exhaust with no recirculation.

Automatic sprinklers should be installed throughout all buildings and areas including the interstitial spaces. Process areas should be designed on the basis of ordinary hazard, Group 3 occupancy. Storage areas should be designed in accordance with appropriate portions of NFPA 231 (bulk storage) or NFPA 231C (rack storage). Automatic total flooding Halon protection should be provided in production areas containing high-valued equipment such as electron beam lithography machines, step-and-repeat camera enclosures, and beneath raised floors of computer installations in test areas. One-inch hose connections, each equipped with 75 feet of 1-1/2-inch woven jack-lined fire hose with adjustable spray nozzle, should be provided in selected areas of high combustible loading such as storage areas. Hose connections should be arranged to remain in service if sprinkler protection for that specific area is shut off. At least one 2-A-rated fire extinguisher should be provided every 3000 square feet and other B-rated units provided in areas containing electronics equipment. Additional 2-1/2- to 3-pound carbon dioxide or Halon fire extinguishers should be located where small quantities are used at wet benches, with these units accessible to people at the work area. Dry chemical type extinguishers should not be allowed in any areas containing circuit masks, wafers, or electronic equipment. A number of companies are currently providing built-in fire extinguishers for wet bench applications. These are triggered by a detector of some type and automatically discharge the extinguisher into the wet bench in event of fire.

Ducts from heat-producing equipment such as epitaxial reactors and diffusion furnaces should be of metal construction and vented separately to the outside of the building. This same treatment should be applied to exhaust ducts which are expected to handle pyrophoric (spontaneously flammable) gases. Ducts used for other exhaust systems should be flame-retardant fiberglass-reinforced plastic having a flame spread of not over 25 without evidence of continued progressive combustion, and a smoke-developed and fuel-contributed rating no higher than 50 when tested in accordance with the surface burning standard (ASTM E-84, and so forth). If the duct is of metal construction with polyvinyl chloride (PVC) or epoxy interior coating of four mils maximum thickness, this same test standard should be applied. Automatic sprinkler protection using corrosion resistant heads with separate control valves should be installed within all ducts in which combustible residues may accumulate and in all ducts exceeding 8 inches in diameter of fiberglass-reinforced plastic or metal with interior coating exceeding 4 mils. It is desirable in plastic ducts to install a sprinkler in each vertical drop. Sprinklers in the ducts should have a separate water flow alarm. Sprinklers should be installed within and over fume scrubbers of plastic construction as well as in plastic exhaust ductwork downstream of the scrubber. Fiberglass exhaust ducts which carry insurance approval without internal sprinkler systems are available.

A two-source water supply, either of which will be of adequate capacity to meet the anticipated sprinkler demand plus a minimum of 500 gpm for hose streams should be provided. One acceptable approach would be connection to an adequate city water supply and one 1500-gpm fire pump taking suction from a 250,000-gallon tank. There are a variety of options in meeting the water volume requirements, and these should be explored with the insurance carrier and the water authority in the area. One facility asked the city, who was interested in building a 1-million-gallon water tank on their property, to increase the tank size and reserve the bottom 300,000 gallons of water for fire protection at the facility. The city was delighted with the idea, and the insurance carrier bought into the idea as well. Yard mains to protect the exterior of the building should be a minimum of 8 inch, completely looping the facility. This loop should supply all internal fire protection systems. Multiple control valves should be provided to assure maximum flexibility and reliability of the overall water distribution system. Each sprinkler system should be supplied by a separate connection to the yard main controlled by an exterior valve. Where high values suggest the desirability of increased reliability, consideration should be given to providing an interior cross-connection with a normally open control valve between two sprinkler systems. Fire hydrants are normally required at approximately 300-foot intervals on the looped yard main.

Recirculation of Contaminants

The cleanroom may be likened to a living organism in that it has a respiratory system. It has, however, a very unusual respiratory system, in that it recirculates large volumes of air in addition to exhaling some. In this respiratory system, recirculated air is forced through a series of filters by large volume fans. The air is also passed through coils to condition temperature. If humidity control is required (and it usually is), the air is passed through a humidity control unit. It is then forced through HEPA absolute filters either using a ducted system or a charged plenum system. This air finally enters the cleanroom and passes through either top-to-bottom or side-to-side, depending on the design. This done, the air reenters the filtration system. If it were not for losses, this would be a very efficient system.

Air losses do occur through a variety of avenues. The cleanroom must be made positive in terms of pressure in relation to the spaces around it. Each time a door is operated, there are losses. Even if the doors are kept closed, the air escapes through door cracks, utility openings, window and wall cracks, and via the local exhaust system required to prevent exposure to chemical substances in the work area. A more or less typical cleanroom area must inhale approximately 25 percent of its air as make-up air to replace these losses.

A large source of contaminants in the cleanroom is that 25 percent of the air that is drawn into the building to make up for losses. Contamination levels seen in that air will depend on a number of variables, tied to the basic location of the cleanroom building itself and what contaminants are put into that incoming air via our own exhaust discharges. Location near a freeway or major highway will result in the air intakes pulling dust, carbon monoxide, hydrocarbons, rubber, and brake lining material into the air-conditioning system. Location in a desert area such as Arizona results in high concentrations of fugitive dust, particularly in the summer storm season. Salt might be expected in seacoast areas or where salting of roads occurs in winter climates. Other contaminants in the airstream will depend on who your industrial neighbors are and how carefully they treat their exhaust air or other emissions.

Strangely enough, most contaminants that enter the building via the make-up air system are of our own making. For instance, we may place the air intakes directly over the receiving dock and be surprised when diesel fumes enter the cleanroom area and people start reporting a variety of respiratory symptoms. We also may not do a good job of locating exhaust discharge points in relation to where we pull air back into the building. Prevailing winds, stack heights, stack velocities, temperature inversion, rooftop design, building profiles, and exhaust system scrubbing efficiencies all play an important role in reentry of contaminants. Smoke testing of exhaust discharge stacks has shown occasions when the exhaust air from a 50,000-CFM system travels only 20 feet up after leaving the end of the stack. It then rolls over onto the roof and reenters the building via the air intakes. These contaminants then load up the filter system, if they are filterable. This, in turn, causes increased costs as new filters must be placed into the system. The problem usually comes to light when particle counts climb to unacceptable levels in the cleanroom. The area is usually shut down for a thorough cleaning while the filters are changed.

If the contaminants are not filterable, they pass through the filter system and enter the cleanroom. They contaminate any product left exposed to the air or if the contaminant has an odor or irritating properties, it starts to make people sick. Either way, production stops because of contamination, and we are in the middle of a *yield bust* (a sudden, usually unexplained, drop in product yield) or the operators are too sick or frightened to work in the area. Panic calls are made in the sickness situation to the Safety Department to try to discover the source of the problem. By the time the safety professional can get to the area and into a bunny suit, the odor is gone, and he is left to try to reconstruct what happened from witnesses. These reports are usually less than helpful. There are many operators who describe every odor as "natural gas." Others cannot report facts at all; they offer an opinion as to the source of the odor. Those opinions are often familiar: "It was cooking odors from the cafeteria," or "It was definitely phosphine." Still others smell nothing at all, but they share the symptoms of the other employees. The safety professional usually is forced to leave the area with the problem unresolved. A cycle is odor—complaint—investigation—nonresolution—odor, and so forth. This goes on to reach a crisis point, and people refuse to work out of fear.

The yield bust phenomenon is even more spectacular. The quality control people in the probe area advise engineering that one or more of the probe parameters has gone wildly out of specification. The engineers immediately descend upon the area and accuse everything and everyone in sight of being the source of contamination. For example, "The DI water is dirty" may be the claim of the day. The facilities engineering department changes all of the filters and polishing resins in the system. They shut down the entire plant for a day while they decontaminate the DI water system. Alternatively, "Raw materials are 'bad'." Vendors are raked over the coals to prove that materials are up to or exceeding specifications. "Gases are bad"—test all gases. "Chemicals are bad"—test all chemicals. "Equipment is not being profiled, maintained, and so forth." The entire factory is on edge trying to find and solve the problem and get the fabrication area back into operation.

The answer to the contamination problem may well lie in good planning during the design of the building itself. That is an obvious statement. Of course, during the design of the factory, we took into consideration all of the necessary parameters to assure that vented air would not reenter the facility through the air intakes. Most who have been in the business for a time understand that the design of a cleanroom is in a state of almost constant flux. If a company starts design work on a new cleanroom, executes those plans, and then starts the design for a second cleanroom, the plans will not be the same. The two plans may be similar, but not the same. Strong ideas and personalities enter into the design sequence. One engineering manager does not like the idea of short walls providing bays for air to return to the filter system. He wants the appearance of an open area, and insists that the laminar flow system must return its air through an underfloor system. Newer fabrication areas are being built with full basement-type return air plenums housing the first filters, air-conditioning coils and supply fans mounted on springs to isolate them from the frame of the building. An older building in need of remodeling is a good example of poor or total lack of planning for the reentry of contaminants. A building built in the early 1960s, before clean-rooms became widespread, requires a major renovation to make it suitable for a modern cleanroom. A 1960s building was not designed to handle the huge volumes of both exhaust and intake air necessary to support the modern cleanroom.

There are methods of discovering sources of reentry contaminants after the building is in place. Computer modeling has some use in this area. Computer programs have been written to profile buildings and predict, with some considerable accuracy, potential sources of contamination. The program takes a number of variables concerning the building and atmospheric conditions into account, and applies some guesses and gas dynamics rules to the puzzle, and quite frequently shows the source of the problem. If the conditions are relatively constant, the models will work quite well. If this is not the case, computer modeling becomes largely guesswork and not much better than the field engineer's speculation.

Another method is scale modeling. It is not as fast as a computer model, nor is it less expensive. If time is a factor, scale modeling is not the answer. A 1/40 scale model of the building or building complex is constructed. The model includes terrain details including elevation changes within 1500 feet of the facility being studied. All air intake locations are constructed in great detail so that the amount of air capable of being drawn in the sample is proportional to actual flows. Exhaust stacks are also detailed with propor-tionate flows. The entire model is placed in a wind tunnel. It can thus be rotated 360 degrees and subjected to a variety of wind speeds, directions, temperatures, humidities, and so forth. Testing is quite simple. Carbon mon-oxide in a known concentration is fed out through the exhaust stacks and sampled at all air inlets under whatever atmospheric conditions the client wishes. In this way, programming one exhaust stack at a time, it is possible to determine the fate of exhaust air relative to the air intakes, one stack at a time or in combination. Smoke testing in the wind tunnel helps to show air current profiles around the building and demonstrates how contaminants travel over and around the structure. With the model it is possible to increase stack heights, velocities, or any number of variables. In this way design theories are tested to determine impact on the system. The models usually cost about $50,000 to build. The total cost of the study will depend on how extensively one wants to study the model in how many variable situations.

Smoke Removal

It is 6:30 A.M. and the foreman is in the wafer fabrication area getting things ready for Monday's day shift to start work at 7:00 A.M. He is turning on equipment, allowing ovens to heat up and stabilize, making sure chemicals are available, wafers are ready to start, DI water is at the proper readings, cascade rinses are full, and all of the other details necessary to start the shift have been attended to. He turns on the power to an acid wet bench with a heated acid bath inside and moves on. The acid bath is a sink made from polypropylene. In the bottom of the bath is a resistance heater to heat the bath to 100° C. The heaters are protected mechanically by a standoff, that is, a false bottom. The mechanism is controlled by an electrical controller and sensed by a thermocouple. Liquid level is sensed by an electromagnetic switch which floats in the acid. If the acid level falls, the switch opens a contact and interrupts heater power. Today, the acid bath is empty and the electromagnetic switch fails. It has been corroded in the up or full position. Since the foreman is busy about his start-up activities, he does not check the bath visually, but relies on the sensor to do the job. The seeds of disaster are sown.

At 6:50 A.M., he is near enough to the wet bench to see it fully enveloped in flames. The empty sink has been set on fire by the heaters, and the fire has already spread to the polypropylene deck, exhaust duct, and canopy of the equipment. Three sprinkler heads have fused, one in the exhaust duct, and two in the ceiling below the laminar flow filters. He trips the electrical power and proceeds to aid in firefighting with first-aid fire extinguishers. By 7:10 A.M. the fire is extinguished and the water is turned off in the sprinkler system. The smoke from this fire has contaminated the entire laminar flow HEPA filter system for a 20,000-square-foot area. Acid vapors from the combustion products of burning plastic combined with smoke have damaged all of the optics on all of the alignment tools in the wafer area. All exposed product is contaminated with smoke particulate, acid, and dirty water from the sprinkler system. This loss resulted in a claim to the insurance carrier for $5 million and, more importantly, six months of downtime while the wafer area was being rebuilt.

Smoke, or more accurately, the products of combustion, are more devastating to the wafer cleanroom than the damage caused by the fire. The major keys to prevention of losses from this kind of an accidental fire are two-fold: (1) limit the amount of combustible material in the area of heated surfaces; (2) get the smoke out of the building before it has a chance to recirculate.

There is a great deal of pressure from insurance carriers to deal with the second option. The amount of damage seen in this single small fire provides graphic evidence of the reason for concern. The laminar flow HEPA filters from the ceiling involved in this fire had to be completely replaced with new units, and the entire area had to be recertified for cleanliness before production could even be thought of starting again.

The new articles in the fire and building codes require us to place smoke detectors in the return air system. The ability to detect smoke at an early stage is already there if we have complied with this requirement. Now it is necessary to take a signal from this detection and alarm system and do something else with it besides transmitting the alarm to the central station. Suppose we could trigger other events with that same signal through a secondary panel. We could open a relief vent and take 100 percent of our air from outside of the plant. We could trip open another vent in the return air system and let the laminar flow system fan take all of the smoke to the outside of the building. Roof vents are established passive smoke relief systems. Their effectiveness in the cleanroom setting is still speculative as they have not been applied to this use with any regularity. The designers of cleanrooms are so concerned about keeping the cleanroom "tight" in terms of unfiltered air, that roof vents may be difficult to adapt to cleanroom designs.

Another, more basic, consideration deals with prevention of fire in the first place. Obviously, if a fire does not, or cannot, start, the need for smoke removal becomes academic. If the etch station used in our example had been constructed out of a noncombustible material, the fire would not have started. Some locations are using quartz etch tanks or ceramic or other material that will not burn. There is still the risk that the tank will be empty when electrical power is energized to the heaters. A gas tube thermocouple attached to the heaters will not sense overtemperature as there is no liquid in the tank to act as a heat sink. In this case, the heaters will overheat and the heater element will fail. However, since there is no fuel for a fire, there is no fire and no smoke to be removed. Since most wet benches are built by independent industry suppliers, basic design tends to be for the general market (with some customization for large customers). This frequently dictates that the equipment be manufactured from low-cost materials. Scrimping on cost of materials keeps overall prices competitive and allows the small manufacturer to stay in business. Polypropylene is a relatively inexpensive material, considering its resistance to chemical attack, ease of fabrication, stability, and a host of other needs. These considerations are important. The perfect wet bench in terms of its ergonomic usability, ability to capture and control hazards, control temperature, and other variables is still no good in the cleanroom if it is not clean too. Polyvinyl chloride (PVC) is a good substitute for polypropylene. Unlike polypropylene, PVC will not burn; it will simply melt and deform. A fire in a wet bench made of this material is still possible but the fire is not likely to spread.

There is a current movement in the industry to emphasize sprinkler protection in chemical hoods and wet benches and to attempt to detect fires early through the use of localized fire detectors. A battery-operated detector on the market sells for around $200. It mounts to the outer wall of the wet bench and "sees" the interior of the wet bench. The eye of the system has the capability to see for 180 degrees, so it can effectively see the entire interior of the wet bench. It is expected to detect a fire in its early stage and transmit the alarm to a central panel as well as ring a local alarm. The unit has an additional output terminal which will allow the equipment engineer to use the signal to interrupt power, discharge a built-in fire protection unit, or whatever else is wanted.

Sprinklers in wet benches and chemical hoods are a sound fire protection measure. However, one must keep in mind the basic purpose of a sprinkler system—it does not actually prevent fires. The purpose of a sprinkler system is to react to a fire situation and control the fire at the source, or at least, prevent spread to adjacent rooms, buildings, or equipment. When the sprinkler system actuates, there is invariably a great deal of water damage, and equipment in the cleanroom is particularly sensitive to this type of damage. Installing a sprinkler head or two into a wet bench or chemical hood is not particularly difficult. Sprinklers in exhaust ducts exposed to the corrosive vapors from acid baths typically last eighteen months to two years with very little extra effort to protect them from corrosion. In the specifications for installation, one merely requires the lead-coated head to be dipped into beeswax twice and then to be wrapped in a plastic bag. This arrangement has a minimal effect on the operating temperature of the head while effectively protecting it from corrosion.

LEGISLATION

To protect the work force, the community, and the environment, a large number of regulations must be followed when dealing with hazardous materials and operations. The codes discussed above mainly cover construction and plant aspects with regard to the storage and use of HPMs, but there is also considerable legislation pertaining to the manufacture, use, and disposal of process effluents, which include unreacted substances and by-products. Much of this legislation is aimed at protecting the work force, the community, and the environment.

The reasons for the protection of the work force are obvious when the toxicological properties of HPMs are considered (see Chapter 12). The reasons for the protection of the environment become obvious when the popular press continually carries articles about the depletion of the ozone layer, acid rain, the greenhouse effect, and so forth. The need for protection of the community is perhaps not so obvious until we perform some simple calculations concerning the destructive power of HPMs. For example, if a standard cylinder (200 cubic feet) of pure arsine was accidently rapidly vented, for example, it falls over and the valve is sheared off, it would create a volume of lethal gas (at around 250 ppm) of 800,000 cubic feet. The volume of extremely dangerous levels would be much larger. A large plume of deadly gas could envelop highly populated areas near to the plant (homes, offices, schools, hospitals, and so on).

The first protective legislative acts to appear earlier in this century were those pertaining to protection of the worker from the poor safety standards of the employer. These acts are well known and have led to the type of regulations enforced by OSHA. There have now been cases where criminal charges have been brought against senior company officials for concealing information on hazards from the work force (for example, Cook County, IL, company president, plant superintendent, and plant foreman indicted for murder of an employee). A more current and advanced version of work force protection legislation is the "Right to Know" or Hazard Communication law. The determination of the hazard properties of the material is the responsibility of the manufacturer of the material. This information is conveyed to the user by way of the Material Safety Data Sheet (MSDS), which must be supplied by the manufacturer or seller and updated when necessary. It is the responsibility of the management of a company to relate this information to the work force by documented training sessions and open access to MSDS files. In addition, all chemicals should be properly labelled with the type of the material and hazard properties. The labels must follow the chemical through to disposal. If the chemical is transferred to another container, it must be labelled in exactly the same way. Laboratories and universities are exempt from this legislation but pilot plants are not.

In addition we now have the Toxic Substances Control Act (ToSCA), which is a complex piece of legislation, belonging to no one agency or political group, designed to control the production and use of HPMs. The most significant elements of the act are contained in sections 4, 5, and 8. Section 5 requires that producers or importers of new chemicals give a 90-day premanufacturing notice (PMN) to the EPA. New chemicals are those which do not appear on the chemical inventory in section 8. (Note that articles which contain and release hazardous chemicals are considered to be chemicals under the terms of the act.) The information to be given on this notice is defined as "reasonably ascertainable" and includes chemical name, structure, use, by-products, possible employee exposure, disposal, and any other pertinent facts. If information is limited, the act allows controlled and limited use of the substance by way of a significant new use rule (SNUR). However, section 4 makes it abundantly clear that manufacturers and processors must bear the responsibility of the costs of testing. Section 8 is really the database part of the act and contains production use data, inventory of chemicals, known adverse reactions in humans, available health and safety studies, and details of substantial risk reporting.

In the early seventies we saw the first serious environmental acts in the form of the Clean Air and Clean Water acts, passed by the federal government and enforced by the Environmental Protection Agency (EPA). These acts were designed to stop blatant dumping of toxic wastes into rivers, lakes, wells, and the air. The acts put maximum limits on how much of a particular chemical could be dumped in a set time period. They were fairly difficult to enforce due to a lack of a serious monitoring requirement, and fines were not very harsh. Since then, more potent environmental acts have been passed, including the Resource Conservation and Recovery Act (RCRA). This was originally designed by the EPA, but due to public pressure Congress put hammers in a revision to give the act more bite. RCRA is meant to take care of the types of wastes which are not currently neutralized, including substances like chlorinated solvents, which cannot easily be burned due to the toxic by-products. These substances were, for many years, collected in underground tanks and then pumped out into drums which were then legally dumped in specially designated landfills (or illegally dumped in the countryside).

RCRA brings in "cradle to grave" responsibility and liability with no statute of limitations. The law states that waste producers are still responsible through disposal even though another company may handle this task. Since landfilling is environmentally unsatisfactory, this will cease altogether in the mid-1990s (unless the EPA can show a totally safe way of doing this), forcing waste producers to enter into recycling programs. RCRA also requires permits for generators (90-day maximum storage limit) and disposers (minimum 90-day waste storage limit) to ensure that storage (including temporary holding) and disposal is carried out properly.

On the subject of temporary storage, underground storage tanks for liquid (usually solvent) wastes have been used for many years, but the startling fact is that almost all of them will eventually leak or are currently leaking. This has allowed disastrous pollution of soils and groundwater over the past forty years. Groundwater is generally carried horizontally in porous strata regions called aquifers. If leaking contamination soaks into these aquifers from the saturated soil, the water becomes contaminated and the contamination can be carried underground for many miles to be taken up by wells and put into the domestic water supply. Cleanup of these accidents is difficult and expensive (typically $12M per tank) as the soil has to be removed and disposed of as hazardous waste, and the water has to be pumped from the aquifer, filtered, and returned clean. This situation has prompted the Leaking Underground Storage Tank (LUST) legislation which is EPA-enforced and is very stringent. For instance, no new tanks may now be installed unless they are protected under the provisions of the act. Protection includes double containment of the tank and any underground pipework and/or cathodic protection. There is a notification component of this act, and all tank owners have to notify the appropriate state agency of tank and content details (including old and unused tanks). The law also states that owner/operators must:

Have methods for detecting releases

Keep records of the methods

Take corrective action when leaks occur

Report leaks and actions taken

Provide for proper closure of tanks

Provide evidence of financial responsibility for cleanup

Community protection is covered by the Superfund Amendments and Reauthorization Act (SARA) Title III, also known as the Emergency Planning and Community Right-to-Know Act (1986). It is intended to ensure that every community has a good emergency plan in the event of a serious accident. There are four main components to the law:

Emergency planning

Emergency notification

Community right-to-know

Emissions reporting

For emergency planning, Local Emergency Planning Committees (LEPCs) are responsible for the response plan. State Emergency Response Commissions are set up to ensure that these plans are in place. The LEPC and SERC must be immediately informed of any releases of hazardous materials. The community is given access to the MSDSs by the company. All MSDSs are also filed with the LEPC, SERC, and local fire department. The amounts and locations of hazardous materials must be reported to the LEPC and fire department, and this information must be available to the public through the LEPC. Routine and accidental releases must be reported annually to the EPA and state government. A database on this information will be maintained by the EPA.

There is more to come. On an international scale, it is likely that certain chemicals may be banned from use altogether. One such group is the chlorinated fluorocarbons (CFCs), which are apparently destroying the earth's ozone layer. Materials such as Freon 11, 12, 113, 114, and 115 are widely used as nonflammable cleaning solvents and refrigerants, and are extremely chemically stable. Unfortunately, this stability allows them to travel as far as the Earth's ozone layer, where they may react with the ozone in the presence of ultraviolet light, breaking up the O_3 and thereby removing the ultraviolet blocking power of the layer. The internationally respected Montreal Protocol calls for 1986 production levels of these materials to be maintained from July of 1989 (effectively a 20 percent cut at this point), winding down to a 50 percent cut in production by 1998. This is a tall order, as the worldwide use of Freon 113 was estimated at 360 billion lbs per year in 1988. However, as an incentive for the reduction of CFC manufacture, the EPA fines for overproduction in the United States could run as high as $25,000 per kilo.

Since we have built up a dependence on these materials, it is obvious that replacements are necessary. Chemical producers are working on less stable hydrogen-containing substances, which are unfortunately more flammable and more toxic, but they will be kinder to the environment. For instance, whereas 113 has a lifetime of 100 years and an ozone depletion potential (ODP) of 0.8, newer hydrogen-containing variants such as 141 and 123 have lifetimes of 10 and 2 years and ODPs of 0.1 and 0.08 respectively. To circumvent the flammability problems, manufacturers are experimenting with solvent mixtures, for example, Freon 123 is not a particularly good solvent but if present in quantities greater than 30 percent by volume in Freon 121b will considerably reduce the flammability of this latter substance.

SUMMARY

Due to the nature of the materials and processes used in high-technology industries, large volumes of codes and legislation have been developed to reduce or control the associated hazards. These are designed to ensure consistency of design and construction and to protect the workforce, the community, and the environment from the potentially detrimental effects of the materials we frequently use in our production facilities. Even though small producers and research laboratories can be exempt from much of the legislation, the spirit of the guidelines should still be followed in these cases.

9

Ultrapure Water

THE NEED FOR ULTRAPURE WATER

Semiconductor processing has always included operations which involve wet chemicals. In general, the first action in a typical silicon-integrated circuit fabrication process is a cleaning step using a strong oxidizing agent (usually a mixture of sulfuric acid and hydrogen peroxide) to remove organic contaminants, followed by a dip in hydrofluoric acid to remove native and chemically grown oxides. Later, hydrofluoric acid may be used to pattern or remove silicon dioxide, hot phosphoric acid used to remove silicon nitride, mixtures of phosphoric and acetic acids to etch aluminum, and so on. Wet processing is still utilized to a great extent in the semiconductor industry even though plasma etching and stripping are now widely used. It is not only the product which undergoes this wet processing. Equipment components are frequently wet cleaned. In other non-semiconductor industries, wet chemicals may also be used for cleaning purposes, for example, to prepare surfaces by etching before deposition of materials.

One of the unfortunate aspects of such cleaning and etching operations is that the reaction must be stopped in a uniform fashion across the surface of the components. This usually means that rapid dilution by immersion in a water tank or complete rinsing with a water spray is necessary. In addition, water soluble reaction product residues must be removed from the surface in the same manner before processing can continue, as these will act as sources of contamination. Once the component has been water rinsed, it may be dried in a stream of nitrogen or similar inert gas. It is therefore obvious that we require not just any water but water which has been engineered for this purpose, particularly with regard to its contaminant content, so that we do not add contamination by these processes. Not only must the water be extremely pure, it must also be produced in large quantities for materials processing on an industrial scale. As an example, a finished wafer may require as much as 100 gallons of this ultrapure water during its creation, for both substrate and equipment cleaning.

As we can see, ultrapure water is a critical component in the processing of contamination-sensitive materials and structures. In this chapter, we will discuss the types of contamination found in raw water and how we may use a variety of techniques to provide us with the necessary water purity.

CONSTITUENTS OF RAW WATER

Sources of Water

We use water for a large number of applications. Water is supplied to homes for domestic purposes, and many industries use large quantities of water in a variety of manufacturing processes. A typical large urban development will tend to take its water from as many sources as possible so that demand may still be met even when one source dries up.

Cities in or near mountain regions tend to take water from natural lakes or man-made reservoirs fed by rivers and streams. The water in these feeders may come from melting snow and ice in high regions or from a wide catchment area, which may cover thousands of square miles, which acts to channel rain into streams and rivers. Many cities will also use wells as a source of water. Layers of porous rock bounded by less permeable materials (for example, clays) will act as underground channels or aquifers. Water may be pumped from wells driven through the overlying strata into these aquifers. When the water table drops, the supply may be replenished to some degree by pumping water back into the aquifers or by allowing water to percolate into the ground from natural or specially constructed pools. In the case of artesian wells, the natural internal hydrostatic pressure is great enough to force the water to the surface without pumping, but over-use will reduce this natural pressure over time. In arid regions near the coast, desalination could be used to provide relatively salt-free water, but this process is extremely costly. In many areas, city water is recycled, that is, waste water is collected and treated to provide clean water, especially where demand is high and supply is limited. This may not be done indefinitely as certain materials cannot be removed completely.

The local water authority will often treat even fresh water. A typical treatment would involve coarse filtration to remove floating and suspended material using, for example, a gravel bed, followed by chlorination (the addition of chlorine compounds in small quantities) to kill harmful living organisms. Soluble fluorine compounds such as sodium fluoride are also added in some regions as they have been shown to reduce tooth decay. If the feedwater is particularly alkaline, acidic chemicals could also be added to reduce the pH.

It should be noted that the characteristics of water will depend greatly upon its origin and any subsequent treatment; characteristics of the water supply can vary throughout the year as one source is substituted for another. It is important that we know these characteristics so that we can take appropriate action to purify the water for contamination-sensitive applications. At the most elementary level, we may characterize the water supply using a few simple terms. These are: *total dissolved solids* (TDS) which is merely a measure of dissolved material in units of weight per volume of water; *pH*, which is a measure of the amount of hydrogen (H^+) or hydroxyl (OH^-) ions in the water (acidity or alkalinity respectively); *turbidity*, which is a measure of the lack of clarity of the water; and *suspended material*, the amount of undissolved solid or gelatinous particulate material carried by the water. To these terms we may also add *color*, which, as one may guess, is an indication of color added by suspended or dissolved material.

Water from lakes or rivers can have a TDS of between 150 and 1500 mg/l, depending on the solubility of material the water has passed over. Between 150 and 750 mg/l may be considered soft, whereas the higher range is hard water. It will almost certainly be turbid (murky), containing a large amount of suspended material. This suspended material will typically be mostly of plant origin and hence the water will also have color. In contrast, water from well sources will have a TDS of less than 750 mg/l and relatively low turbidity, as the porous rock it has passed through will act as a filter. Since there will also be fewer plants underground, the amount of suspended solids is less. However, due to the nature of the water-carrying rocks, in most cases the water tends to be very alkaline (a pH of 9 is not uncommon). Water taken from high saline sources such as the sea or estuaries has a TDS greater than 1500 mg/l and once again turbidity, suspended matter, and color are high. Since many water-using plants will get their water from the pretreated city supply, we should also include this in our discussion. The TDS is typically below 750 mg/l and turbidity, alkalinity, and suspended matter are low, which is assential because this is drinking water. We will now look more closely at the main contamination groups in raw water, inorganic and organic materials and micro-organisms.

Inorganic Materials

We have discussed the basic nature of inorganic substances in Chapter 2. The material in this category we typically find in water is diverse in make-up and in form. The three forms are: (1) dissolved material, where the compound has been broken apart and reduced to ions by the water, (2) suspended material, where dissolution has not (or not yet) taken place and the solid particles are carried by the currents in the fluid, and (3) colloidal material, where an immiscible liquid or gel is suspended in the water. Examples of various ions are given below.

Cations	Anions
Na^+	F^-
K^+	Cl^-
Mg^{++}	NO_3^-
Ca^{++}	HPO_4^{--}
Fe^{+++}	$CO^{--}{}_3$
	SO_3^{--}
	PO_4^{---}

Note that sodium chloride (NaCl) will break up into singly-charged positive sodium ions and singly-charged negative chlorine ions, and calcium carbonate ($CaCO_3$) will form doubly-charged positive calcium ions and doubly-charged negative carbonate ions. Ions increase the electrical conductivity of the water, so we may determine the concentration of dissolved materials by measuring the conductivity or resistivity (1/conductivity). A conversion chart for conductivity, resistivity, and dissolved solids is shown in Table 9.1. Note that the highest resistivity at room temperature is approximately 18.24 Megohm-cm, as this is effectively a measure of the naturally occurring hydrogen and hydroxyl ions in the water. It is also important to note that only dissolved material will contribute to conduction and suspended material will not.

One of the major contaminants, due to its abundance in the earth's crust, is silicon dioxide. This is one of the few materials which can actually take on the three forms discussed above. In addition, silicon dioxide is also found in *diatoms*, the skeletons of rather unique micro-organisms, which shows that

Table 9.1 Resistivity (1/conductivity) to Dissolved Solids Conversion

Conductivity (micromhos-cm)	Resistivity (ohms-cm)	Dissolved Solids (ppm)
10,000	100	5,000
1,000	1,000	500
100	10,000	50
10	100,000	5.00
1.00	1,000,000	0.500
0.100	10,000,000	0.0500
0.056	18,000,000	0.0277

(All at 25° C)

inorganic materials can sometimes have an organic source. There will also be metals, metal oxides, sulfates, and carbonates as listed below.

Dissolved SiO_2	Al_2O_3
Colloidal SiO_2	ZnO
Solid SiO_2	TiO_2
$CaSO_4$	Diatoms (SiO_2)
$MgSO_4$	$CaCO_3$
Solid Fe_2O_3	$CaHCO_3$
Manganese salts	$MgCO_3$

Table 9.2 gives an average water analysis showing typical relative amounts of the major inorganics present. Note that the major compounds are sulfates and bicarbonates, and a prominent metal is the highly soluble sodium, which is rather unfortunate considering that sodium is one of the greatest enemies of semiconductor devices. The numbers in this analysis are typical averages, but the actual amounts will vary from region to region. The sources of these compounds and elements are basically whatever the water comes in contact with, which includes rocks, soils, canals, and pipes. Longer contact times and higher temperatures mean a higher TDS content; soils with a large small-particulate content mean a higher suspended material level. One additional source of inorganic contamination in water is rainwater itself. For example, in regions where fossil fuels are extensively burned, high levels of sulfur dioxide occur. This dissolves in rain to form sulfuric acid (acid rain).

Table 9.2   Typical Analysis of Raw Water

Constituent	Amount (mg/l)
Silica	13.3
Iron	<0.01
Manganese	<0.05
Calcium	60.0
Magnesium	19.4
Sodium	88.2
Copper	<0.01
Potassium	4.1
Zinc	1.03
Bicarbonate	144
Sulfate	195
Chloride	71.9
Fluoride	0.96
Nitrate	0.46
Orthophosphate	0.97
Total Dissolved Solid	570
Resistivity	0.001 megohm-cm

Organic Materials

The sources of organic contaminants in water can be both natural and industrial (the latter includes agricultural sources). The natural materials are generally of plant and animal origin (we will discuss micro-organisms in more detail in the next section), for example, dead or decaying cells, cellular contents, microbial by-products, and so forth, or naturally formed oils, fats, or organic acids. The remains of bacteria are called *pyrogens*, and are chemically very diverse; for example, the contents of a typical animal cell can range from simple salts to complex proteins. Natural oils and fats can be of plant or animal origin or from oil-bearing strata in the ground (oil originates from decayed plants). The rotting process on plant and animal remains is essentially driven by bacterial action, and organic acids (as well as noxious gases like methane) are generally by-products of this.

Industrial and agricultural sources of organic contaminants have become increasingly more significant in recent years, so much so that a plethora of legislation has appeared in an attempt to control the pollution of lakes, streams, and groundwater. There is typically a large variety of oils (for example, motor oil) and other long-chain hydrocarbons which originally leak from automobiles and storage tanks and are carried to water sources by rainwater or in aquifers. Leaking storage tanks are a large and somewhat sinister source of other water contaminants. Much of the highly toxic chlorinated hydrocarbons found in the water of cities which are hosts to semiconductor or aerospace plants have come from leaking underground tanks. Detergents from industrial and domestic sources can also be present in significant quantities. If materials are not biodegradable, that is, if they cannot be broken down by natural action, they will build up in the ecosystem. The large amounts of chemical fertilizers, pesticides, and herbicides used in modern agriculture will also find their way into the water supply. Below is a short list of organic water contaminants.

Oils   Pesticides
Grease   Amines
Alcohols   Nitriles
Humic acid   Tannic acid
Detergents   Benzene, phenol, and microbial
Chlorinated hydrocarbons   by-products
Colloidal organics

Micro-organisms

In a chemical sense, micro-organisms are organic. However, they also exhibit one further important characteristic; they are alive. Being alive gives them a unique ability, the ability to replicate themselves utilizing available nutrients. These nutrients and examples of potential sources are shown in Table 9.3.

Table 9.3 Nutrients for Bacterial Growth and Examples of Sources in a Water Delivery System

Nutrient	Origin
Carbon	Organic Acids
	Pipe Plasticizers
	Lubricants
	Micro-organisms
	Human Debris
Nitrogen	Nitrogen Compounds (Nitrates)
	Microbial By-products
Trace Metals/Salts	Process Piping
	Human Debris
Phosphorus	Phosphorus Compounds (Phosphates)
	Microbial By-products
Sulfur	Sulfur Compounds (Sulfates)

Note that major sources of nutrients are the piping systems which carry the water, and therefore there is a wealth of sustenance for growth and replication even in clean water. We will return to sources of nutrients in ultrapure water systems later.

There is an incredible variety of micro-organisms to be found in water. These can be grouped into three basic categories: bacteria (microscopic animals), algae (microscopic plant life), and yeasts/molds. As may be seen by Table 9.3, they can utilize a large range of materials for life, but what is even more astounding is the variety of adverse conditions under which they may survive. Some will grow in the light, some in the dark, some will grow where there are only trace quantities of nutrient materials (these are called oligotrophs, and they spend most of their time harvesting and concentrating nutrients), some will even grow in extremely acidic water. They also have a wide range of characteristics and forms. Examples of micro-organisms and certain characteristics are listed below. Not all of these may be found in a particular water sample (which is just as well as some are very unpleasant to humans).

Algae, blue-green	Slime formers
Basidio-mycetes	Spore formers
Diatoms	Sulfuric acid producers
Iron-depositing molds	Yeasts
Nitric acid producers	

Micro-organisms, pyrogens, and many other organic materials will not actually contribute much to electrical conductivity and cannot be considered to be in the category of dissolved solids. However, the amount of organic carbon or carbon which has not been oxidized (oxidized carbon is CO, CO_2, and CO_3) can be represented by TOC, total organic (or oxidizable) carbon, measured in units of weight per volume of water. This factor is extremely important in characterizing water, and TOC (in mg/l) is frequently quoted along with TDS (also in mg/l). As one may expect, there is a direct relationship between TOC and bacterial counts.

PURIFICATION TECHNIQUES

Water Purification

We have seen in the previous sections that the water which arrives at our facilities is anything but clean. We must now ask ourselves a number of very important questions. Which purification methods are available, just how clean can we make the water, and what levels and types of contaminants can we tolerate? The latter two questions are actually very closely related, as in some cases we will have to take what we get. We will try to answer these questions in this section. A further question we have to ask is, how will we know when the water has been purified to whatever standards we are attempting to attain? To this end we will discuss detection and measurement techniques later in the section.

Let us first turn our attention to the methods of water purification which are available to us. Perhaps the oldest and best known method is distillation. In this method, water is heated in a vessel, usually to boiling point, so that it evaporates. The water vapor is then condensed and collected in a cooled pipe. The contaminants, both dissolved and particulates, are thus left behind in the vessel. Some of the more volatile contaminants may also evaporate, but this effect can be reduced by cascading evaporation vessels so that the output from one is used as the feedwater for another. This method is favored by pharmacists and biochemists, as the boiling of water kills any microorganisms present. However, it has severe drawbacks for large-scale ultrapure water production. The biggest problem is that it is extremely energy-inefficient, as we have to continuously heat feedwater only to remove the heat energy in the condensation process. Also the hot concentrate has to be removed from the vessel to reduce the build-up of contaminants.

Another method of cleaning water is by absorption. Here we use materials which will absorb contaminants from the water. The absorbent materials will, of course, become saturated after some time with the contaminants, but they may be regenerated or discarded at this point. Activated (porous) charcoal is a good example of an absorber, as organic molecules tend to stick to it. The drawbacks of this method are that large amounts of material are required for large water throughputs, and that the absorber may also be a source of particulate contamination. Absorption tends to be used as a first line of defense and not as a singular purification technique in industrial pure water engineering.

A number of other techniques are efficient in water contamination removal. Electrodialysis, in which ions are enticed across a permeable membrane and hence out of the water by an electric field, is such a technique. This tends to be more of a laboratory technique, as it is rather slow and can only be used for low flow rates. However, the use of membranes is common to many types of purification. A membrane, or more specifically a semipermeable membrane, is actually a filter of sorts. Membranes generally are polymeric materials which are discontinuous on the molecular scale; that is, although the linked polymer chains are continuous, there are holes left between them due to the inherent structure of the polymer. Thus semipermeable membranes can be used for the removal of solid materials, including very small particles (<0.01 micron diameter). What is perhaps more remarkable is that the pores in some materials are actually small enough to prevent the passage of large molecules and ions. We shall return to this point later, as these abilities are a key part of such water purification techniques as membrane filtration, reverse osmosis, and ultrafiltration. These techniques, especially the first two, are very widely used in industrial water purification and therefore deserve to be discussed in detail in sections of their own (following this one). It should be noted here, however, that throughput in membrane processes can be very low.

One of the best techniques for the removal of ionic contamination is a chemical technique known as *deionization*. This is also a widely used water purification technique and will be discussed in the next section.

Deionization

The process of deionization uses insoluble solids containing cations or anions. These solids are capable of undergoing a reversible exchange of these fixed ions with mobile ions in the water to be purified. The deionizer solids, sometimes called resins, are actually polymers with added acidic or basic ions. The polymers are typically styrene copolymers or divinyl-benzene (both for cation exchange resins), or phenol-formaldehyde or epoxy-polyamine (for anion exchange resins). They are formed into beads, approximately 0.5 mm in diameter, and packed into (fiberglass) cylinders, approximately 2 feet in diameter and 5 feet tall. The beads provide a very large effective surface area so that water flowing through the cylinder comes in contact with as much of the active (surface) area of the polymer as possible.

The surface of the cation resins is covered with hydrogen (H^+) ions. Cations in solution are exchanged with these hydrogen ions. For example, if the dissolved mineral is sodium chloride (NaCl), the ions will be Na^+ and Cl^-. The Na^+ mobile cations will displace the H^+ fixed ions and become attached to the surface of the resin, leaving the H^+ ions in solution. This is true of most chlorides and sulfates. If the dissolved mineral is $NaHCO_3$ (or other bicarbonates), the exchange reaction will produce trapped Na^+ and carbon dioxide (CO_2). The liberated H^+ will combine with the remaining hydroxyl (OH^-) ions to produce more H_2O.

The cation exchange process is generally followed by an anion exchange. The surface of the anion resins has hydroxyl ions attached. The anions in the solution are exchanged with these OH^- ions. For instance, the Cl^- or SO_4^{--} left in solution from the cation exchange step becomes attached to the surface, and the newly liberated OH^- joins with the previously liberated H^+ to make more water. Therefore, since the harmful ions have now been removed from solution, and since the by-product of the above reactions is primarily water, we have succeeded in purifying the water.

Much of the industrial success of this technique lies with the ease of regeneration of these exchange resins. This is typically done in-house or off-site (by the company that rents the deionizers). By placing the cation resins in a dilute acid, or the anion resins in a dilute base, the adsorbed ions may be replaced with hydrogen or hydroxyl ions respectively, thereby allowing the resins to be put back to work. However, since no process is absolutely ideal, we should mention some of the problems. The resins can become fouled with bacteria and algal slimes if the feedwater is not properly treated or if water is allowed to stagnate in the resin tanks for long periods. The resins may also become poisoned by certain ions. It is thought that potassium can bond to certain resins and will not be displaced during regeneration. The effectiveness of the resins becomes reduced with further exposures until little exchange takes place, at which point the resins must be discarded (new resins are expensive).

There are some variations on the deionization theme. Certain polymers, known as organics scavengers, have the ability to remove organic molecules. They are frequently used prior to cation-anion deionization. Also, anion resins are frequently called strong base resins due to their use of hydroxyl ions. An alternative is a weak base resin which can adsorb entire acid molecules. This is useful after cation exchange as the products of sulfate and chloride exchange (that is, the sulfate and chloride ions plus the liberated hydrogen ions) are sulfuric acid (H_2SO_4) and hydrochloric acid (HCl) respectively. Unfortunately, the action of these weak base resins is not strong enough to pick up the weaker-charged carbon dioxide and silica groups. These resins are cheaper and easier to regenerate (with sodium carbonate) than strong base resins and thus are preferred when silica is not a problem. They are not used very frequently in the semiconductor industry. As a final deionization or polishing step, cation and anion resins are placed in the same tank so that this one unit is a complete deionizer. They are stratified within the tank and may be regenerated by first floating off the lighter anion resins and treating them separately.

As a rule of thumb, if the inlet water TDS is below 150 mg/l, or if the water throughput is less than 0.5 gpm (gallons per minute), it is economically viable to use deionization to produce semiconductor grade DI (deionized) water, also called UPDI (ultrapure deionized) water. For higher TDS or higher flow rates, the cost of regeneration would be prohibitive and a predeionization-dissolved contaminant reduction would be desirable. One such technique is reverse osmosis, described in the next section. If the treated feedwater to the deionizers has around 10 ppm dissolved minerals, they should be able to run for over 100 hours continuously without regeneration. This translates to 30,000 gallons at 5 gpm throughput.

Reverse Osmosis

Reverse osmosis (RO) is a process in which a continuous membrane is used as a molecular filter. The technique is capable of removing up to 95 percent of dissolved minerals, 95 to 97 percent of dissolved organics, and 98 percent of biological and colloidal material. It is therefore not suitable by itself for the production of ultrapure water as it does not have the ion removal efficiency of deionization resins (but it has found its way into homes in a simplified form for the cleaning of drinking water). However, RO is extremely energy-efficient and is the ideal predeionization technique.

The principle of reverse osmosis is actually quite simple. If a concentrated solution is separated from a dilute solution by a semipermeable membrane, (that is, one which will allow water to flow through but not particulates, dissolved, or colloidal matter), the solvent (water) will flow from the dilute side of the membrane to the concentrate side by osmosis. This occurs because nature dislikes concentration differences and will do her best to even up the balance. Since the dissolved material cannot pass through the membrane but the water can, the water flows into the concentrated solution to reduce the concentration per unit volume. This process will continue until the concentrations of the solutions on both sides of the membrane are equal, that is, equilibrium is reached. The force driving this movement of material is called the *osmotic pressure*. The more astute reader will realize that the above-described process is actually not producing clean water and thus is not exactly what we want. However, if we apply a pressure greater than the osmotic pressure to the concentrate side of the membrane, the water will flow from the contaminated side to the pure side; this purification process is thus called reverse osmosis.

In industrial RO, the pressure is 400 to 600 psi, generated by suitable pumps at the feedwater side of the membranes. A number of membrane configurations and materials are used. The most widely used membrane form is a spiral-wound configuration, as shown in Figure 9.1. These units are typically 4 inches in diameter and 40 inches long. The feedwater (brine) is forced into a channel on the feed side of the membrane spiral where it may pass through the membrane due to the elevated pressure. The water which

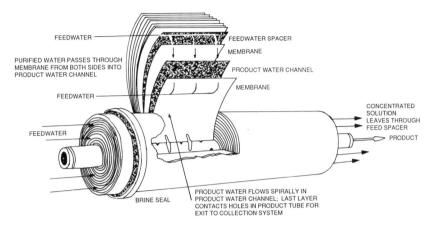

Figure 9-1   Spiral-wound RO element.

does not pass through the membrane (the concentrate) is drained at the opposite end of the RO unit. The product water spirals into the center of the unit and is collected by a holed tube and taken off to the next stage (which may be another RO membrane in a dual pass system). The beauty of this system is that it is a continuous process, capable of product water flowrates of up to 10 gpm. An alternative configuration is the hollow fiber unit in which the membranes are formed into long tubes and bundled as shown in Figure 9.2. These tubes can be made to be very compact while still maintaining a large membrane area. The feedwater passes from the outside of the tubes and is collected from the insides. The concentrate leaves from the other end of the unit as before.

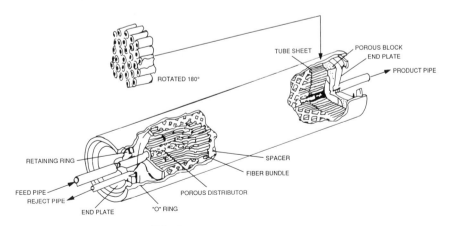

Figure 9-2   Hollow fiber RO element.

Only certain materials are suitable for RO membranes, as they have to have the correct molecular structure to be porous to water molecules but little else. The two most common materials in use are cellulose acetate, normally in spiral-wound form, and polyamide (essentially nylon), which is used in both spiral-wound and hollow fiber form. Other materials which are gaining popularity are polysulfone and hollow fiber Aramid (a Dupont tradename). Different materials have different properties and thus care should be taken in selecting the best membrane material for a particular application. For instance, cellulose acetate tends to hydrolyze in alkaline feedwater and thus the recommended pH range is 4–7.5. Therefore, if the feedwater is alkaline (and it frequently is), acid should be added prior to RO to bring the pH down. The working temperature range of membranes made from this material is also relatively small, around 35–95° F. This means that the RO units should be operated indoors with some degree of environmental temperature control. Finally, cellulose acetate is also the favorite food of certain types of bacteria, so chlorine (0.5 ppm) should be added to the feedwater to ensure that any incoming organisms are killed. Fortunately, cellulose acetate has a reasonable tolerance to chlorine, not as much as polysulfone, but considerably more than polyamide or Aramid, which have practically no chlorine tolerance at all. Since city water inevitably has chlorine added, sodium bisulfite can be used to remove the free chlorine. The non-chlorine tolerant materials may then be disinfected with certain other biocidal agents (discussed later), but in the case of polyamide, there are no known bacteria which can digest it anyway. Polyamide also has a wider pH range (4–11), is more able to withstand alkaline feedwaters, and has a wider operating temperature range (36–122° F for spiral-wound thin film composites, 32–95° F for hollow fiber types).

RO units suffer from other problems, particularly fouling with a variety of materials. Fouling is caused by hydrated oxides of iron, nickel, copper, and so forth, calcium precipitates, colloids (particularly aluminum silicates and organic chemicals), and bacterial and algal slimes. Various chemicals are available which can be added to the feedwater to reduce this problem. Unfortunately, the problems described above, and to some extent the solutions, eventually lead to the demise of the membranes, and therefore they may not be expected to remain efficient beyond five years of service.

Filtration

A number of configurations and materials are used in filters for ultrapure water, but their basic function is to remove particulates, including bacteria. High-performance solvent cast membrane filters and (small pore) ultrafilters, like reverse osmosis, can remove large molecules. However, the smaller the pore size, the lower the throughput, as it is more difficult for the water to pass through, so ultrafilters tend to be employed at the point of use only.

Filters are usually "staged" in ultrapure water production systems so that the largest pore filters come first to remove the big material from the water to avoid loading the high-performance units. Less demanding filters, for example, those which filter particles greater than 5 microns in diameter, are typically a pleated, pressed-fiber material, whereas higher efficiency filters are frequently cellulose esters, nylon, acrylic, polypropylene, or various other polymeric materials. Membrane filters such as these are typically rated at 99.999 percent to 99.9999 percent at 0.1 micron, but the more advanced materials are capable of removing 0.01-micron or even 0.002-micron particles for ultrafilters. Not only should the materials be able to trap particles, but they should also be extremely chemically inert. To this end, expanded PTFE materials are now being used, but they are expensive.

Water filters work in much the same way as any other filter, but the fluid dynamic aspects are naturally different for water as it is more dense and viscous than air. There are two types of filter ratings commonly used. *Nominal* filters are rated to remove particles which are smaller than the actual pore size of the material. The smallest particles become trapped by colliding with and sticking to the filter element. Unfortunately, these filters are not particularly reliable as the retention of smaller particles largely depends on process conditions (for example, temperature, flowrate, pressure changes), and retention will also drop off sharply when the filter becomes saturated. This type of filter should thus only be used for noncritical applications, for example, prefiltering. *Absolute* filters are considerably more reliable as the pore size is smaller than the smallest particle size to be removed. Retention is thus independent of process conditions, and saturation causes an increase in pressure across the filter without a loss of retention.

Measurement Techniques

In order to determine exactly what is in our raw water and to find out how clean our water has been made, we must perform certain measurements. Different measurement techniques can be used to determine various aspects of water purity. There are two main groups into which we may partition water measurements, namely, those which can easily be performed in-house and those which usually have to be performed in an analytical laboratory. This may be a risky split, as some of the so-called laboratory techniques of a few years ago may now be performed by relatively untrained operators rather than by highly trained technicians due to advances in equipment automation technology. The first four groups of techniques described below fall into the in-house category.

The simplest technique we can use is resistivity measurement. Two inert electrodes are placed a fixed distance apart in the water, and the resistance of the water is determined by passing a current between them. We have already given the conversion chart for resistivity to TDS in Table 9.2, but we should remind the reader that this technique should never be used alone to determine water quality as it is only an indication of the ionic content. Bacteria, colloids, and so forth are not included in this measurement and thus it only tells us how well our RO or, more specifically, the deionizers are working. (This is an important point as many people use resistivity measurement as proof of how great their water is, even though it could be green with algae.)

One of the most effective methods of determining how many bacteria are in the water is to take a known sample volume and spread it on a culture plate containing a nutrient agar (gourmet food for bacteria). The plate is then placed in an incubator for one or two days to allow the bacterial colonies to grow. Once they grow it is very easy to see and count them. Since each colony is usually derived from a single bacterium, we now know how many bacteria were in our initial sample volume. Actually, it is not quite as simple as that, as some species of bacteria will form a colony in a matter of hours whereas others may take weeks to produce a colony large enough to be seen. This technique is sensitive to around 1 cell per ml of water, which is generally adequate, but the delay is not very convenient, especially if an instant response is required. There are other, more immediate techniques; for instance, the bacteria can be collected in a membrane filter and then counted under a microscope. This is a rather tedious technique, but if performed properly can be sensitive to 0.01 cells per ml. A further technique, which has a sensitivity somewhere between the previous two, is to pass the water containing the bacteria through ultraviolet light. As with many organic materials, the bacteria will tend to fluoresce under this type of illumination and may thus be seen and counted, manually or automatically. The main drawback of these latter two techniques is that they can be sensitive to interference, that is, objects other than bacteria can be counted by mistake.

Since we are obviously as concerned with particles in the water as in the air (perhaps even more so as the surface tension forces created by the water will hold particles to surfaces very tightly), we will now consider how we can measure particle counts in liquids. This is not as easy as it is in air, as although we can use light-scattering techniques, with appropriate corrections for the refractive index of the liquid, bubbles will be detected as particles. Perhaps the best way to circumvent this problem is to degassify the liquid prior to counting by passing it through a low pressure region so that any dissolved gas is desorbed and bubbles will not form. Once again, a direct count may be performed by allowing a known volume of liquid to evaporate on a plate and counting the particles under a microscope (compare witness plates in air counting). The advantage of this technique is that the particles are retained for analysis. A related technique is to trap particles in an appropriate filter and perform a direct count. High frequency ultrasound is also a contending technology for particle counts in water as bubbles have a different density than solid particles and thus will reflect sound waves differently.

The fourth group of measurement techniques is used to determine TOC. The best in-house technique is infrared spectroscopy, which is sensitive down to 10 parts per billion (ppb). When molecules are excited by infrared irradiation, they will absorb certain wavelengths, and this absorption can be readily detected. The wavelengths absorbed are specific to the bond types present, and thus carbon-carbon bonds can be detected; hence total organic carbon levels may be determined. Another more specialized measurement, which tends to be more of a laboratory technique, is flame ionization. Here, the test material is burned and the elements present can be determined by their ionization potentials.

Hardness, heavy metal contamination, and silica may be determined by colorimetry (to less than 1 ppm) or atomic absorption (to a few ppb). The latter is a specialized technique, performed in laboratories. Fortunately, there are many laboratories which can perform these measurements (for an appropriate fee), and thus water samples may be sent out on a regular basis for analysis. Silica may now be readily measured in-house by using modern automated colorimetry systems.

RO/DI SYSTEMS

Typical System Configuration

In this section we will discuss a typical RO/DI system configuration as used for the production of ultrapure water. Figure 9.3 illustrates this configuration, showing the main components.

The first part of the system is the pretreatment section. As the name implies, this is where the incoming water is prepared for ultrapurification. The water is first passed through a sand and carbon bed. The filtering effect of the sand removes most of the larger suspended solids, and the carbon adsorbs much of the oils and other organic substances. The tanks containing the sand and carbon may be backflushed periodically to remove the trapped materials, effectively regenerating these units. The next step in the pretreatment is the addition of various chemicals. An important addition at this stage is a biocidal agent which kills living organisms in the water. Depending on the RO membrane materials used (see page 000), chlorine or iodine may be added. The solubility of iodine is not particularly high, but if it is dispersed adequately in the water, contact with bacteria is still highly likely. The addition of biocidal chemicals, usually by diaphragm pump, should be tightly controlled as too much biocide could destroy the RO membranes or damage the DI resins. Another chemical which could be added at this stage is acid, usually sulfuric, to balance the pH (reduce the alkalinity where necessary). Once again, an automatic control system with a pH sensor in the downstream flow is necessary to prevent too much acid reaching the membranes. A scale inhibitor may also be added here to reduce the build-up of scale on the membranes, and in regions where the iron content of the water is high, an iron inhibitor would also be of use. Finally, a filter is used, usually better than 5 micron, to remove

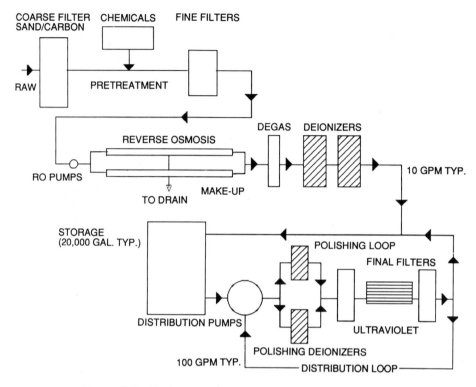

Figure 9-3   Production of pure water—typical arrangement.

any remaining coarse material, including that which has been added by the prior pretreatment.

The next section of the system is the RO units. The RO high-pressure pumps supply water to the membranes through stainless steel pipework. The parameters which should be measured here are the rejection ratio (an indication of how much water is sent to drain as concentrate compared to the total supply volume), the resistivity of the product water, and the supply pressure. Once again, an automatic controller is recommended here, linked to the pretreatment sensors, to protect the membranes and assess their performance. If acid addition is used in pretreatment, any bicarbonates present will form dissolved carbon dioxide in the water, and this should be removed. A typical degassification method is to allow the water to fall down a tower in which there is an updraft of HEPA-filtered air. Care must be taken to ensure that the air is clean, otherwise recontamination of the water will occur. The water is now ready for deionization, as described on page 170. The above section of the purification system is generally referred to as the make-up, as it is only run to top-up the storage tank, which immediately follows the DI units. It is therefore used to replenish ultrapure water which is dumped to drain. This dumpage is not as much as one may first think, as much of the water which is sent into the production facility is recirculated back to the tank. We will discuss this aspect more fully later in this chapter.

The storage tank, typically 20,000 gallon capacity, is the reservoir for the ultrapure water system. The distribution pumps will remove water from this tank at a rate of 100 gpm (typical) and will pressurize the supply system to 100 psi. Since the water in the tank will tend to lie around for a while, it will become a little contaminated (contamination from system materials is discussed later). It must therefore be polished before it can be sent into the production facility. This is achieved by passing the water through mixed bed deionizers. Since we have not yet performed any fine particle filtering to this point, the polished water should now be filtered with membrane filters rated at around 0.5 to 0.1 micron absolute. This will also remove any particles which arise from the polishing operation. Another problem with the temporary storage in the tank is that bacteria will begin to grow, admittedly in small quantities, but nevertheless enough to be concerned about. Since we cannot add biocide at this late stage in the purification, we may use a physical agent rather than a chemical. The ideal noncontaminating physical biocide is high intensity ultraviolet light. The lamps are arranged around a quartz tube so that the water passing through is illuminated uniformly. Since the uv kills the bacteria but does not actually remove them, we must now use 0.3- to 0.1- micron absolute (or better) filters to catch the pyrogens. The water is now ready to be sent into the facility.

This is not the end of the story, however. The water is handled in a particular manner within the facility, and most of it is recirculated back to the storage tank to be repolished and sent back into the facility. We will discuss what happens in the facility, concentrating on the point of use, in the next section.

Point of Use

The ultrapure water is now ready to be used in washing, chemical dilution, and rinsing applications. It is generally piped to wet benches and automated wet processing equipment. At the wet benches, the water is brought to faucets and spray guns. The operation of these is obvious but their design can lead to problems, as we will see later. An often-used device is the cascade or weir, where water is allowed to fill a tank and spill over into another tank at a lower level. Materials may be dipped into these tanks (downstream first, then upstream) for rinsing and cleaning purposes. Nitrogen is often bubbled through these tanks to provide agitation to help the cleaning process. In the automated spray etch equipment, the water is dispensed as a spray to rinse substrates after a cleaning or etching step.

To counteract contamination of the water which could arise in the system after the final filters, point-of-use filtration/purification is often used. At its most basic level, this involves the insertion of a fine particulate filter in the water line leading to the discharge point. Membrane or ultrafilters are much better as point of use elements as they can be used to remove more than particulates, especially in the case of ultrafilters which can remove bacteria and pyrogens. The low throughput of these ultrafilters is not as much of a problem at the point of use as demand is fairly low. There are also small capacity self-contained RO units which could be used at the point of use with a mixed bed tank and filter set to purify pretreated water in low-volume applications where the installation of a full RO/DI system would not be economical.

System Materials

Unfortunately, we cannot use just any material to carry our ultrapure water around our facility for one simple reason; the water is so pure that it is an excellent solvent and will dissolve many materials, thereby becoming recontaminated. We are therefore restricted to using materials which are as chemically inert as possible.

The least expensive (and unfortunately the least inert) piping material is polyvinyl chloride (PVC). To add flexibility to this robust material, plasticizer is added, and this tends to leach into the water. Fortunately, this effect drops off fairly quickly as the inner surface of the pipe becomes depleted of the plasticizer with time. The materials used to join these pipes contain solvents which also leach into the water with time, as will the colorants used to make the pipe look more attractive. In addition to these problems, since the PVC pipes are extruded, the inner surface is longitudinally striated, and this can create a place for bacterial growth. Also, if too much jointing cement is used, the internal bead that forms will allow bacterial growth. Fortunately, components are now available in expanded polypropylene, which is becoming comparable in cost to PVC but is more chemically inert even though it contains plasticizers (colorants are generally avoided). It is also possible to thermally weld this material, thus avoiding the problems associated with cements, but welding increases the cost of system installation considerably.

The "creme de la creme" of system materials are the fluorine-containing perfluoroalkoxy (PFA) and polyvinylidene fluoride (PVDF). Most fluorinated polymers are incredibly chemically inert, much more so than chlorinated materials due to the large bond strength of fluorine to carbon. PFA contains polytetrafluoroethylene (PTFE), perhaps the most inert polymer available in industrial quantities. This clear material has virtually ideal characteristics, being resistant to ultrapure water and even corrosive chemicals. It exhibits extremely low leaching and is very smooth with a low coefficient of friction so bacterial cling and growth are highly unlikely. PVDF has similar qualities, having a wide temperature range and high impact strength. It is also resistant to uv light and ozone, both being used as biocides. Once again, they may be orbitally butt-welded to provide practically seamless joints. An obvious question is, therefore, "Why doesn't everyone have these materials in their systems?" The short answer is that PFA components are forty to fifty times the cost of PVC, and PVDF components are around thirty to forty times more expensive. A possible fluoropolymer contender is plain PTFE (or Telflon, to give it its Dupont trade name), which in unmachined form is about ten times the cost of PVC. One reason PTFE has not been used alone (that is, not as a mixture with something else) is that it is difficult to work with due to its extremely inert nature. A material is not very useful if it is difficult to make elbows, valves, and so forth out of it. However, these problems are being overcome, and it is likely that we will see more of this material in RO/DI systems in the future.

Of course, our main concern is the water going into the facility, which should be supplied by the purest method possible. The recirculated water, which will be repolished before being put back to work, can be returned in a less expensive material. The storage tank is generally made of fiberglass, as are the DI resin containers, for strength and economy. Large components, such as the storage tank, need only be coated with inert polymer if fiberglass alone is deemed to be insufficient.

OPERATIONAL CONSIDERATIONS

Recirculation

The recirculation of water through the facility and back to the polishing section is extremely vital to maintain purity while keeping the production costs as low as possible. Put simply, the water must be kept flowing and not be allowed to stagnate at any time in the distribution system. If the water was brought to the point of use and only flowed when a valve was turned on, during the stagnation period the water would become highly contaminated by leaching and bacterial growth. This contaminated water would then flow out onto the sensitive product after the valve was opened, until the pipework was flushed of the contamination. This could take many hours if the bacterial colonies have taken a hold on the surface of the pipes. As a rule of thumb, the water should be kept flowing at around 5 feet per second so that bacteria cannot take hold on the surfaces. Smooth, low-friction surfaces on the inside of the pipes and within system components will ensure that bacteria will not be able to hang on and hence will not be able to populate the pipework. Any bacteria, particulates, pyrogens, and dissolved materials which do find their way into the water will be returned by the recirculating water to the polishing section to be removed.

The need to keep the water moving has some far-reaching implications in system design. We cannot have any "dead legs" or stubs in the system, which poses problems in design for flexibility. This also means that faucets and spray guns are a bad idea, as these are temporary dead legs. If standard units are used, the water should be run for several minutes; a 15-minute flush is usually fairly adequate. Note, however, that this wastes both time and water. A nonclosing valve may also be used so that there is a constant trickle of water when the faucet/spray is off. However, this solution is not ideal because the flow would have to be fairly large to prevent the problems described above. The best solution is to use a unit which has a send and a return line so that flowing water is brought right up to a valve at the discharge point and, when the valve is off, is returned to the recirculation channels.

We do not have these problems with cascades because they have a constant flow of water through them. However, there are other considerations. If the water which flows through the cascade is unused, that is, no objects are placed in the tubs for rinsing, it will not pick up a large amount of contamination and may thus be recirculated. However, if a batch of wafers which has just been cleaned in sulfuric acid is placed in the downstream tub for initial rinse and then placed in the upstream tub for fine rinsing, the water leaving the cascade will be contaminated with sulfates. Clearly, this water should not be returned into the system—or should it? Initially, the level of contamination will be extremely high, and this water can be sent to drain. However, as the flowing water rinses the wafers, the ionic content of the water will be reduced. It is possible to have a resistivity sensor in the drain water which switches from drain to recirculation at around 1 to 4 Mohm-cm (depending on preference). Many systems like this have been in use in the semiconductor industry for some time. There are still problems with this approach which make it unpopular with designers of the highest quality ultrapure water systems. The most obvious is that we should never deliberately introduce any contamination into the system if we are really serious about keeping it ultrapure. It is arguable that the loss of water by dumping all contaminated water is worth it, as the polishers will last longer. Another argument for dumping all contaminated water is that certain ions will poison the deionization resins; hence great care should be taken in choosing what we put into the system.

Changeovers

One practical aspect of ultrapure water system operation is that it should be capable of being run continuously. However, since deionizer resins will require regeneration, filters will block up and require replacement, pumps will need service, and so on, the system has to be taken apart periodically. This means that a certain amount of parallelism is required in an RO/DI system. For instance, the DI resins should be arranged as dual banks, with only one bank being active at any one time. The same goes for other elements such as the RO pumps and membranes, distribution pumps, polishing resins, and final filters. Both sides are run alternately during normal operation, but the burden of water production falls upon one side when the other is being maintained. This is an expensive approach but nevertheless absolutely necessary in a production context. Changeovers may be initiated by the technical staff or by an automatic controller, based on resistivity measurements, volume of water processed, maintenance schedules, and so forth.

Changing from one bank to another has its pitfalls, however. Since only one side is active at any one time, water will stagnate in the other and will therefore become contaminated. When a changeover is performed, this contaminated water will be injected into the system. This low-quality water should thus be flushed to drain before the resting section is connected into the system. Note that even the water in the deionizer resin tanks, although free of ionic contamination, can still have particles and even bacteria in it. Even ball valves, which are frequently used in large diameter pipes for shut-off, can cause problems, as the bore through the center of the ball will trap water when the valve is put to the off position. This cylinder of not-so-clean water will be shot into the system when the valve is turned to the on position. Once again, the contaminated water could be dumped to drain, but this will complicate the system. In any case, changeover practices are important, for example, if the ultrapure water maintenance crew is about to initiate a changeover sequence, the personnel in the production area should be forewarned so that appropriate defensive action may be taken. This makes good sense and is a further example of how the different areas within a clean production facility have to interact.

We have yet another problem in the context of changeovers and maintenance: when we break the system, how do we know that we have put it back together again properly? This question is not as frivolous as it sounds— many companies have very large problems in this respect (and some of them do not know it). Elements like deionizer tanks are usually little problem; if the connections to the tank are loose, water sprays out, and you know a problem exists. However, it is more difficult to see internal seals like those of RO membranes and especially filters. Normally filters are cylindrical units containing a central pipe which seals to the casing unit and the rest of the system by means of double O-rings. If this O-ring seal leaks, dirty upstream water will be forced under pressure through the seal and into the post-filter area. Hence, the filter is not really doing its job. We may check seal integrity by two means; by pressure testing (assuming the sealed pressure drop is known) and *bubble pointing.* This latter method is very sensitive and is akin to looking for a leak in an inner tube. The filter casing is filled with water, and the filter unit is pressurized with nitrogen gas. If a stream of bubbles is observed coming from the seal, a leak is present.

Biocidal Treatments

In addition to supplying a biocidal agent during pretreatment, operation of most ultrapure water systems requires a system sanitization every four to six months. This ensures that any bacteria which have gained a hold at some imperfection in the pipework will be destroyed. This action usually requires a shutdown, the length of which depends on the biocide used (that is, how long it takes to kill the bacteria and how long it takes to be removed from the system).

As mentioned previously, chlorine and iodine are effective biocides. If used at dosages of around 100 mg/l (Cl) or 200 mg/l (I) for contact times of around 2 hours, they will kill virtually 100 percent of all living material. Unfortunately, they will not decompose in the system, and chlorine may attack some system components so they must be flushed out as rapidly as possible once their job is done. Chlorine dioxide acts like chlorine but is somewhat kinder to the system. Ammonium compounds have also been used as biocides, but the dosage levels are high (up to 1000 mg/l) for contact times of up to 3 hours. Organic materials such as formaldehyde (around 1 percent by volume) can be used but tend to be reserved for individual components, such as polyamide RO membranes. One of the most popular system biocides is hydrogen peroxide (H_2O_2). Its main drawback is that very large concentrations are required (typically >10 percent by volume, which is expensive as electronic grade chemical is required) for 2 to 3 hours. The great advantage is that hydrogen peroxide will decompose into water after some time and hence biocide removal is not a great problem. The most effective biocidal agent is ozone (O_3). A mere 10 to 50 mg/l for less than 1 hour is effective. Although ozone is highly toxic, it will also decompose (especially if a catalyst such as steel is used at the system vent).

Operational Problems—Examples

We will end this chapter by supplying the reader with two real-life examples of what can go wrong in ultrapure water systems as a warning to those who may become too smug about their system operation. Figure 9.4 presents some data reported by a large semiconductor company. This data shows the relationship between the RO pump current (really an indication of when the membranes are being pressurized) and particle counts per liter of water. As

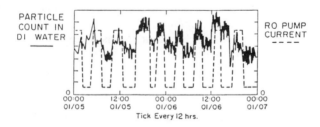

Figure 9-4   DI water supply particle count and RO pump current correlation of DI suuply particles and RO pump current (from *Jnl. Env. Sci.* Nov./Dec., 1986).

we may see, there is a direct correlation between these factors, indicating that whenever the RO section was functioning, it was somehow introducing particles into the system. First thoughts may bring us to the conclusion that the membranes are either no longer an effective barrier to the upstream particles or that the membranes themselves are falling apart. The real answer was actually simpler than that. A section of plumbing had slipped and was resting on the pump motor. Every time the motor came on, the vibration sent shock waves along the pipe, which released particles from the surface of the pipe material. This was a good indication that the pipe really was not free of particles in spite of having been in service for several years. (Incidentally, deliberately vibrating pipework during cleaning radically increases the cleaning efficiency.) It was also an indication that someone didn't do his or her job on facility maintenance.

The second example is more shocking. A European study was performed to assess which final filters did the best job of removing particles of 0.5 micron and greater. A number of production facilities were tested, and the results were astounding. The particle counts for similarly rated filters (for example, 0.2-micron nominal nylon) ranged in different facilities from a few tens of particles in the 0.5 to 1.0 size range to over one million! Clearly this is not the fault of the filter material and is much more likely to relate to problems of maintenance and improper filter-changing procedures. Buying expensive elements does not guarantee that you will get ultrapure water.

SUMMARY

To summarize, we have seen that water which arrives at our facility can contain a great variety of contaminants depending on its origin and subsequent treatments. Using purification techniques such as deionization for the removal of dissolved salts, reverse osmosis for the reduction of all water-carried contamination, and filtration for the removal of fine particles and bacteria, we can clean up the water so that it may be used in contamination-controlled environments. It is important to control the organic carbon content of the water, as there is a link between carbon levels and bacterial counts. In addition, the water has to be supplied to the cleanroom using noncontaminating materials, and it must be kept flowing at all times to avoid the effects of stagnation.

10

Production Materials

It is obvious that we must bring a number of things into our controlled environment besides clean conditioned air, equipment, and ultrapure water. If we are to do any production work, or research involving the production of contamination-sensitive items, we must bring the materials used to create our items into the cleanroom. Naturally, unless we can have a robot-run facility, we will also have to allow people into our pristine area. This allowance is a problematic one and one which we shall discuss in Chapter 11. We will treat the bringing in of production materials in this chapter. Please note that many of these materials are not only contamination-sensitive but are also extremely dangerous. We will discuss the hazards of our trade in considerably more detail in Chapter 12, but we cannot decouple the facets of contamination control and safety, and so we will slant our discussions of production materials in both directions in the following sections.

GASES

A large number of different gases are used within the high technology industries in production processes. Some are relatively innocuous (nitrogen, for example), presenting a hazard only in unusual circumstances, whereas some are among the most toxic substances known to man (for example, arsine). Some gases will cause little problem in terms of contamination in the event of a small leak (oxygen, for example) whereas some can cause a great deal of damage due to their corrosive properties (for example, hydrogen chloride). In general, gases may be stored and used in bulk form or packaged in smaller quantities in metal cylinders. We will discuss bulk storage and use in the following section.

Bulk Gas Storage and Distribution

A number of gases are used in such vast volumes that it makes economic sense to use a bulk storage and delivery system. The most common of these in the semiconductor industry are nitrogen and oxygen, although other materials such as argon, hydrogen, and carbon dioxide may also be used in bulk form. They are typically piped around the facility, being routed to where they are required in much the same way as other services, such as power and water, for example. Gases delivered in this way are usually called *house gases*. The best way to handle large volumes of gas is to cool them to below their boiling points so that they become liquid. For instance, nitrogen liquefies at $-196°$ C and occupies approximately $1/100$ of its gaseous volume at room temperature and atmospheric pressure. Gases may be kept liquid by using a vacuum insulated vessel and high pressure; such storage tanks are commonplace in facilities. The science of materials at low temperatures is called *cryogenics*.

The liquefied gases may be delivered by insulated road tanker. If a permanent tank is located at the facility, the liquid is pumped from the tanker by way of an insulated hose. It is important to keep a flow of an inert gas such as nitrogen through the delivery hoses in between fills to prevent contamination of the inside of the hoses. This precaution is frequently ignored but will prevent a great deal of moisture and many particulates from getting into the tank. Alternatively, a skid-mounted tank may be delivered complete with liquid and connected into the gas delivery system. In some locations, a gaseous nitrogen pipeline is run from the gas production plant to the facilities which use it. This system is not as common as the use of cryogenic tanks due to the large expense of installation and tends to be found only at the largest facilities. The advantage of a permanent pipeline delivery system is that it eliminates road transport and reduces on-site storage requirements.

In the case of cryogenic tanks, the inner tank material is typically steel, the better tanks having stainless steel to reduce contamination from the impurities in the metal. Some liquefied gases are good solvents so impurities can leach out of low-quality metals, perhaps to find their way into the gas. It is therefore very important to ensure that all system materials which handle the cryogenic liquids have been treated to remove as much particulate material as possible and any organic films. It is especially important to remove moisture, which can adhere to metal surfaces, pass into the liquid, and subsequently evaporate to appear in the gas stream.

We may use our liquefied gases in two ways: (1) pipe them into the facility by way of insulated pipes or carry them in in liquid form in smaller portable vessels and use them as cryogenic coolants in such applications as vacuum system cold traps (nitrogen only), or (2) heat them to boiling point and pipe them into the facility in gaseous form to use them in processes. Since the boiling points of our cryogenic materials are very low, passing them through a large-finned heat exchanger at ambient temperature will transform the liquid back into a gas. It may then be put into the distribution system at delivery pressure (typically 100 psi). There is one big advantage of taking gas from a liquid source. Most of the contaminants, especially particles, will remain in the liquid and will not pass into the gas phase unless the boiling action is particularly turbulent. Once again, it is critical that the pipework used to carry the gases is installed with contamination control in mind. This installation is not easy.

The first problem we have with pipework is the material itself. For many years, house nitrogen was carried by copper pipes. Unfortunately, studies of copper pipes have shown copper to be a poor choice if control of particulates is of prime concern. Figure 10.1 shows particle data from a standard copper nitrogen line along with the typical emissions from a stainless steel line. The stainless steel line is at least an order of magnitude better than the copper line. This is because during manufacture the copper pipe will become oxidized, and the surface deposits will tend to flake off in use, especially if subjected to shock or vibration. There will be fewer surface deposits in the case of the stainless steel line; however, there still will be some, as stainless steel is not without its problems. Stainless steel pipes are formed by an extrusion process at high temperature. This process leaves the steel highly contaminated with oxides of iron and other added impurities, and other compounds such as

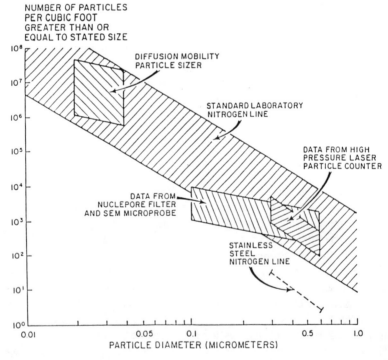

Figure 10-1 Particle concentration vs. size in a gaseous nitrogen line (data provided by the BOC Group, Inc., Murray Hill, NJ, from a paper by J. M. Davidson and F. K. Kies, Submicron Particles Analysis in VLSI Gases, given at Osaka Semiconductor Conference, June 28, 1985).

sulfides. The material is cleaned with a chemical etchant to remove these heavy deposits both inside and out, but there will still be impurity precipitates (small regions of different composition) within and at the surface of the steel. These precipitates can remain until the steel is stressed by bending or other trauma at which time they become detached. In addition, the surface of the cleaned steel will still be rough, the surface acting as a trap and ultimately a source of particles and absorbed gases and vapors. We have much the same kinds of problems with machined components such as valves and regulators which must be included in the system to control the gas flow. The act of machining the surface by milling or drilling creates a rough finish and will expose the subsurface precipitates.

Is there any way to reduce these problems? Fortunately, there is, but the methods tend to be expensive and so only the highest quality delivery systems are treated. A particularly effective method of reducing the roughness is to use an electrochemical etch technique called *electropolishing*. In this method, the machined and precleaned object to be treated is made an electrode in an electrochemical bath. When a current is applied, the surface is etched. Any features which protrude from the surface are more rapidly etched due to the increased local electric field around them, hence the surface becomes highly polished. Unfortunately, electropolishing will not remove the surface precipitates and some further cleaning may be necessary, for example, agitating the part by vibrating it to shake the particles free and flushing them out with an inert gas. An alternative technique is *chempolishing,* in which the surface is etched using a chemical etchant. This is better than electropolishing in that it can be used to simultaneously etch out exposed surface precipitates, but the resulting finish is not as smooth. Incidentally, any moving parts in the system such as valves should be made of materials which will not shed particles. This is also not a simple requirement, as a rubbing contact between materials will tend to degrade their surfaces and hence produce particles. However, this problem has been reduced considerably by the use of highly stable polymeric valve components and coatings.

Our next problem with the gas delivery system lies with the installation techniques. It is extremely critical to ensure that as little contamination as possible enters the system during installation. Delivery systems are difficult to clean once they have been assembled due to the complexity of the pipework and system components. We will return to the question of cleaning later in this section. When the pipes and components arrive on site, they are (or should be) cleaned and capped with an impermeable plastic. As soon as the caps are removed, contamination will enter. This is especially significant for pipes and tubes due to their large internal surface area. To reduce contamination, the free ends of the pipes can be connected to an inert gas supply to prevent the ingress of contamination while the joint is made at the other end. Nitrogen or argon can be used, depending on the method of joining employed. In the case of a small component such as a line valve, this should also be flushed with gas while being connected. If small components or assemblies have to be stored uncapped, they should be placed in a box fed with nitrogen. Extreme care must be taken when cutting the pipework not to leave a rough edge or any loose metal filings. A well-sharpened pipe cutter and finishing tool must always be used (never a saw).

We can join the pipework in a number of ways. The most permanent method is to weld the steel tubes together. Specially designed orbital welders are used, which rotate around the tube using an electric arc and a continuously fed welding rod. The welding area, as well as the inside of the tube, is bathed in argon to prevent oxidation of the metal, which would lead to a poor quality weld with a large contamination potential. The permanent nature of the weld makes this technique ideal for joints in the pipework but is not advisable for system components such as valves as these may have to be replaced at some time, especially if they have to undergo many operating cycles. In this case we may use one of a number of fittings. The simplest to use is the so-called *compression fitting*. A shaped metal ring is placed around the pipe and the ring and pipe are pushed into a tapered fitting. A nut is tightened onto this fitting, which compresses the ring onto the pipe and simultaneously into the tapered fitting. The resulting dual seal of ring-to-pipe and ring-to-fitting is very gas-tight. The only problem we have is that if we break the fitting for any reason, it is inadvisable to reuse it, especially if multiple breaks have to be made. The subsequent metal-to-metal seals will never be quite as good as the original as the materials have already been deformed to fit the original seal. We therefore should cut away the old seal region and use a new compression fitting. The alternatives to compression fittings are the VCO and VCR fittings. In these, one side of the fitting is welded to the pipe and the other side is attached to the component (which may actually be another pipe). In the VCO case, the seal between the two halves is by a nonmetallic O-ring, whereas for the VCR it is a soft metal washer. When these fittings are broken, the old seals may be discarded and replaced.

Cylinder Gas Use

Gases which are used in smaller quantities will be delivered to the facility in cylinders. These cylindrical containers are usually made of steel, although stainless steel, Teflon-lined steel, and aluminum cylinders are becoming more popular due to their reduced contamination properties. It should be kept in mind that the conventional carbon steel cylinders were never actually designed for low contamination applications. Cylinders are made by heating a thick sheet of steel and bending it around a form (mandrel) while hot to make the cylindrical shape. The top and bottom are added and the seams welded. The surface of ordinary carbon steel is thus left with a great deal of surface particulate matter. The somewhat porous surface so formed is also able to absorb water and organics. This is illustrated in Figure 10.2, which shows

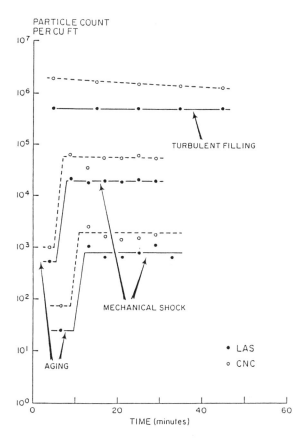

Figure 10-2 Particle concentration of cylinder gas sampled by orifice type pressure reduction device under different test conditions (from *Proc. Inst. Env. Sci.* 1987 by H. Y. Wen and G. Kasper).

how turbulent filling or mechanical shock can release these particles into the gas flow. In addition, Figure 10.3 demonstrates how gas (in this case arsine) can become contaminated by water from the cylinder itself over a period of time. Water will enter the cylinder during wet cleaning between fills and will become absorbed into the steel surface. The cylinder can be dehydrated by heating or the surface can be made water resistant by polymer coatings.

The cylinders come complete with a valve which is either mechanically (screw) or pneumatically operated. A specially shaped fitting, the CGA (Compressed Gas Association) fitting, is included, the shape being unique to the class of material within the cylinder to prevent the accidental connection of a toxic material to an inert gas line, for example. The cylinder will never be connected directly to the supply line as the cylinder pressure can range from a few psi (gage) to 5000 psig, depending on the gas and cylinder type.

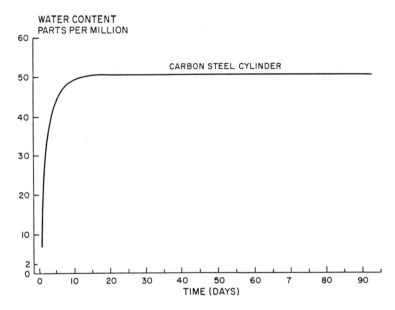

Figure 10-3 Water content of arsine gas as a function of time in a carbon steel cylinder (from *Microelectronic Manufacturing and Testing,* February 1988).

Therefore a pressure regulator will be used to both reduce the pressure to a safe line value (20 psig is typical) and to maintain this pressure even as the cylinder pressure drops as it becomes depleted. The regulator is frequently built into a gas panel, which can also include additional control and isolation valves, excess flow preventors which shut off the supply in the event of an accidental line break, and purging systems which flush process gases out of the supply system, including the gas panel components, to a vent to facilitate cylinder changes and maintenance. The cylinders and gas panels are placed in reinforced, vented steel gas cabinets which minimize the risk of personnel exposure in the event of a cylinder/panel leak. The same types of fittings discussed previously can be used for cylinder gas delivery systems although it is usual to employ VCR fittings for particularly hazardous materials. Coaxial pipes (a pipe within a pipe) are also frequently used for this classification of material. The gas is carried in the inner pipe, and any leaks from this line go into the outer pipe which is monitored, thus reducing the risk of personnel exposure.

There are essentially two types of valves used in delivery systems. Packed valves are used for noncritical applications. In this case, the actuating stem of the valve passes through a rotating seal (for example, O-rings) which isolates the inside of the valve from the environment. Packed valves are somewhat prone to leakage, especially if they are poorly maintained or used very frequently, and hence can result in the escape of the gas or the ingress of contamination. Critical applications, particularly the delivery of hazardous gases, will use packless valves. There are two forms of this valve type; the bellows valve and the diaphragm valve. In both cases there is no rotating seal around the actuator shaft. Instead, the contact between the inner "wetted" parts of the valve and the outer components is via a sealed metal bellows or flexible metal diaphragm. The wetted volume is thus fully enclosed at all times, and leaks are much less likely.

Monitoring and Cleaning Processes Gases

Process gases may be contaminated with a variety of substances. It is the responsibility of the gas manufacturers to clean up the contaminants prior to delivery and also to supply some kind of analysis so that you know what is present (as-delivered gas purity is usually expressed as a percentage, for example, 99.999 percent pure argon). Unfortunately, the gas manufacturer can make no guarantees that you will keep the gas clean. As we can see from the previous sections, there is ample opportunity for particles, water vapor, and perhaps some organic materials to enter the gas from the distribution system materials or from poor installation/maintenance practices. The point of this section is to consider how we can monitor and clean process gases.

As far as particles are concerned, we can use a light-scattering technique to detect and count them. Our only problem here is that the gases will inevitably be at elevated pressure so the sensing chamber will have to be able to withstand this. If a high-pressure particle counter is not available, a filter technique may be used in which we run the gas at a test point through a particle-trapping filter. The filter is removed and inspected under a microscope to determine the particle content of the gas stream. Water vapor is more difficult to measure and requires special analytical equipment, especially if we require sub-ppm capability (which we usually do). This usually means the use of a mass spectroscopy system. Fortunately, such systems are becoming highly automated and therefore may be used by relatively lightly trained personnel. Once we know how bad our problem is, we can begin to rectify it.

The most obvious particle reduction technique is filtration. For process gases this is not a simple affair, as the filter will have to withstand high pressures. Some gases are corrosive, so not just any filter material will do. A good high-pressure, high-stability filter is a sintered metal pad. In this filter type, metal particles are fused together to create a filter matrix with a low pressure drop (a few psi at worst). Filters should definitely be placed as close to the point of use as possible. Water vapor, once again, causes more problems. Various systems are available based on the molecular sieve or selective absorption principle, but we have to be careful that the absorbing material does not add to the particle content or otherwise contaminate the gas stream.

However, prevention is considerably better than cure in the case of process gases.

The removal of particles and absorbed water vapor and organics can be extremely difficult, especially in complete operating delivery systems. The usual way of cleaning pipework is to flush with copious quantities of nitrogen. This may remove some of the construction debris but is unlikely to do a good job of cleaning the pipe surfaces. One of the most effective ways of removing the absorbed vapors is to bake out all the metal components at temperatures in excess of 100° C while flushing with nitrogen or other inert gas. This combined approach results in the desorption of much of the water and organics, but care must be taken not to overheat system components such as valves which may contain heat-sensitive nonmetallic parts (check with the manufacturer about safe temperatures). One effective way of removing particles is to vibrate the pipework with a simple 60-Hz buzzer while flowing nitrogen through. The vibration effectively shakes the particles loose. The buzzer should be moved along the system from source end to delivery points, allowing enough time at each location to maximize particle removal. A particle counter at an outlet point will tell you if particle emission has slowed down significantly, indicating that the unit should be moved. Another technology we may apply to the cleaning of internal pipe surfaces is the dry-ice snow we met in Chapter 4. When used in conjunction with vibration, dry-ice snow has been shown to be very effective at particle removal, especially the removal of severe debris which entered during construction.

Safety Aspects of Process Gas Systems

We will discuss in detail the safety issues pertinent to controlled environments in Chapter 12. However, due to the nature of process gases and their potential for harm, we will briefly discuss some of the safety aspects of gas handling in this section.

Any discussion of the safe handling of chemical compounds, whether gas, solid, or liquid, must begin with an exploration of the idea of acceptable risk. Once we have established the level of risk one is willing to accept, either personally or on behalf of others, it will flavor the judgment of all design decisions for the safety of equipment, process, or procedure. The first school of thought in this regard is based on the belief that a little exposure is like being a little pregnant—either you are exposed or you are not. This school of thought regarding risk acceptance postulates that all exposures, no matter how small, have a harmful effect on the body. Our ability to locate the site of harm and quantify the damage is the only variable in the equation. If this is a valid assumption, all exposures or potential exposures must be controlled absolutely. Zero exposure is the only acceptable design criteria, and if controls fail human beings will suffer impairment. The second school of thought postulates that there is a safe level of exposure to substances; a level below which no harmful effect occurs. This is the concept upon which Threshold Limit Values (TLV) and Permissible Exposure Levels (PEL) are based. It holds that for each substance there are three possible dose (exposure) levels of interest in a toxicological sense: harmful effect, therapeutic effect, and no observable effect. These two schools of thought regarding exposure are of primary importance, for they determine what levels of potential exposure we will design control systems to. They become the basis for the protections specification.

Most safety professionals subscribe to the second school of thought, that there is a safe exposure level, below which no damage occurs to humans. They understand that a safe level is sometimes a moving target based upon current knowledge of toxic substances and human reaction to them. Most also believe that one does not choose TLV or PEL as the design criteria for controlling exposures in the workplace. Since the PEL is a legal limit in the United States, enforced by the Occupational Safety and Health Administration (OSHA), once we have crossed this barrier we risk legal enforcement action to force us back into compliance with the imposed limit. The prudent safety authority in industry will pick some number less than the PEL as an internal engineering trigger or design criteria so that corrective action gets started long before there is risk of citation by legal authority.

Suppose we choose one-half the TLV or PEL, whichever is lower, as the design limit for controlling exposure. This becomes a trigger for starting engineering action to correct the potential problem while it is still potential and not worthy of citation. This conservative approach assumes that TLV and PEL are not necessarily the same number and not necessarily set for the same purpose. The rationalization document for the TLVs published by The American Conference of Governmental Industrial Hygienists makes it apparent that toxic response is not the sole criteria for establishing exposure limits. The exposure limit for hydrogen chloride is based principally upon its irritation properties rather than its ability to produce injury. This is still a valid exposure design target. An employee exposed to HCl at some level above TLV will experience irritation. The result will be mistakes, complaints, and disruption of the production operation, all of which are unacceptable to the mission of the operation. It is part of what the safety professional is trying to safeguard—the productive capacity of the operation.

The purpose of this treatise is not to to discuss toxicology. Nor is it the intent to try to condemn or defend either school of thought regarding safe handling of gases. Full coverage of both of these concepts would require a great deal more space than is available. Toxicology is a living science, and we are learning more with the passage of time. The purpose of this discussion is to point out that the safety engineer has the responsibility to provide guidance to the designer of equipment, systems, and buildings regarding what is acceptable. In short, it answers in part the question, "What do I design to?" Gases commonly used in the cleanroom now were considered to be little more than scientific curiosities, war gases or by-product toxic agents, until the semiconductor industry started to use them in the 1960s. Arsenic, boron, phosphorus, and antimony were all used in solid or liquid form until then. With the availability of highly purified sources of these materials as hydride gases, the technology changed to accept and use them in closed or semiclosed systems. The gases bring with them not only highly purified sources of the base metals, but also some rather unique toxic and physical hazard properties. Designing a system to handle these gases safely requires an understanding of their properties, both physical and toxicological.

The hydride gases have a very low TLV, typically less than one part per million (see Chapter 12). They cause death at relatively low levels of exposure in short exposure times. Odor properties are poor indicators of exposure since odor threshold is usually above the TLV. Low-level exposures can result in serious injury, and since most of the hydride gases are metals of metabolism, even these low levels of exposure may result in long-term injury to body systems. Two variables in connection with these gases are of interest to the safety engineer when evaluating hazard potential: concentration and volume. Phosphine, for instance, is used in concentrations varying from a few parts per million in a carrier gas to pure phosphine. Volumes run from lecture-size cylinders to 330-cubic-foot cylinders. The ultimate hazard potential becomes the product of volume and concentration of the material in the system.

Of immediate interest to the safety organization is the potentially exposed employee in the cleanroom. A second level might be employees in areas surrounding the cleanroom who are conducting related operations. A third level of interest might be people in nearby buildings on the same property. Finally, we must be concerned about neighboring areas surrounding the cleanroom area building and the property. Each of these classes of people must be considered in the design of the control system, if we are to do the complete job. The employee in the cleanroom is of primary interest. His or hers is the most frequent potential exposure, and if an accident occurs, this will be the area of greatest concentration. The employee works with the material on a daily basis and must make the correct decisions and perform correct actions regularly or exposure will occur.

A veritable mountain of work has been done in the last few years on toxic gas cabinets for the containment of toxic gas cylinders in use. These cabinets, made commercially by a number of suppliers, are generally well designed and facilitate containment of small leaks which may accidentally occur in the system. Most suppliers are willing to install piping internal to the gas cabinet to individual design requirements. The cabinets all come with a connection for local exhaust ventilation, a sprinkler head ready for connection to a water supply, and a sturdy metal outer shell with access limited to a port. Piping systems generally include dedicated pressure-reducing regulators, check valves, separate nitrogen lines (from a cylinder source) for purging purposes, excess flow provisions, exhaust failure alarms and whatever else you may want to design into the system. It is the user's responsibility to devise an operating procedure for the use of the cabinet. Some users allow only engineering or supervisory employees to change cylinders in the cabinet. Others will allow only maintenance technicians to touch the cabinets. Each user has realized that changing a cylinder of gas whose concentration is immediately dangerous to life and health is a highly critical operation to be conducted only by well-qualified and trained individuals. Safety personnel have the responsibility to assure that hazards are recognized by whichever employee group is selected to change cylinders. Safety requires training programs and written procedural controls. It is prudent to establish a "buddy" system during changing operations. It is also advisable to work from a written procedure or checklist, which should be followed each time. It is particularly critical to have and follow written procedures in areas where cylinders are changed infrequently, such as research areas.

Control of exposure also requires exposure monitoring. The newer cabinets make provisions to permit monitoring at the cabinet. Monitoring equipment currently available will allow the user to monitor for the hydride gases at extremely low levels. It is possible to detect and measure concentrations as low as 10 percent of the current exposure limits with confidence. The user who applies area monitoring systems to the use of toxic gases in the production area or research laboratory is still faced with some decisions if a safe system is to be maintained. Alarm levels and shut-down levels must be specified. Standard industry practice is to sound a local alarm at one-half threshold limit value and shut the system down automatically at TLV. Shutdown is accomplished by a high pressure shut-off valve triggered by the alarm system.

Location of gas cabinets has been a source of argument between processing engineers and safety engineers for some time. The process engineer wants the gas cabinet close to the equipment served to eliminate long piping runs with their potential problems, and the safety engineer wants the gas outside the building. A number of fire situations have forced more and more processing engineers to opt for outdoor storage or storage in attached buildings which can be fire isolated from the contamination-sensitive cleanroom.

Catastrophic release situations must also be addressed. Exposure limits are generally designed for working age adults. They do not generally apply to the elderly, infirm, or children. If cylinder failure occurs, the neighboring areas surrounding the cleanroom will undoubtedly be exposed at some level. Degree of exposure will depend on cylinder size, concentration, dilution by exhaust, treatment systems, and atmospheric conditions. Decisions about number of cylinders, concentration, location on the property, and control systems must take all of these requirements into consideration.

Hazardous Gas Detection

A variety of gases are frequently used in the cleanroom. Some of the gases have good warning properties (ammonia, chlorine), and some do not (arsine, phosphine). Gases with good warning properties advertise a leak very quickly, as they have a characteristic odor and produce symptoms at a low level of exposure. Those without good warning properties may have no odor at all or the odor threshold is greater than the exposure limit. With these gases a person could be exposed at potentially dangerous levels and not be aware of the exposure until damage was done. This latter class of gases includes arsine, phosphine, diborane, hydrogen selenide, hydrogen sulfide, and carbon monoxide. All of these gases present a hazard to the health of living beings or cause danger because of reactivity, instability, or flammability.

Most of the semiconductor industry uses gas detection systems in conjunction with processes utilizing these toxic gases. Manufacturers generally mount sensors to monitor gas cabinets, process equipment, and vacuum pumps using the gas. A number of vendors supply toxic gas monitoring systems to the industry. They use a number of different monitoring approaches to obtain monitoring data that is of value to the user. Some of the systems use infrared analyzers, others use gas chromatography or chemiluminescence or reaction with chemicals impregnated on a paper tape. Rather than theorize on the capabilities of each system, let us look at one company's specification for monitoring.

1. A toxic gas monitoring system shall be used in conjunction with all processes utilizing toxic gases with low warning properties. Sensors shall monitor gas cabinets, process equipment, and any vacuum pumps using the gas.
2. System failure of the toxic gas monitoring equipment shall cause an alarm to sound and shut off the gas supply within the gas cabinet.
3. Each monitoring device will be the state-of-the-art at the time of installation and must be capable of monitoring below and alarming at or below the TLV.
4. The system shall alarm at 10 percent of the TLV of the gas monitored. At 50 percent of the TLV the system shall alarm and automatically shut off the gas supply at the gas cabinet. Gas cabinet alarms may alarm at 50 percent of the TLV and shut down at TLV levels. All monitoring equipment shall be subject to the approval of the Safety Manager.
5. If no suitable monitoring device is available for a particular toxic gas then it will be used by Special Approval Only. Special Approval Only means authorization by the Plant Manager, the Director of Research and Engineering, and the Safety Manager of the facility, acting as a hazard review committee.

Most of the detection systems currently available can deliver detection limits within the 10 percent factor required by this specification, but it is done at the expense of time. For instance, an eight-point monitoring system will cycle through the monitoring points in a very short time if it is required to monitor at TLV or $2 \times$ TLV. This gives the user a frequent look at what is going on at each monitoring point. If the user wants to monitor each point at $0.1 \times$

TLV, then the analytical time lengthens the cycle time as each point takes longer. Microprocessors have helped this problem to some degree. They will allow nearly continuous monitoring of all points. If something is seen, the microprocessor cycles through the points individually until it finds the one in an alarm state. The technology is moving very rapidly, and each user must make some basic decisions about what is important to that individual operation in terms of the philosophy of monitoring.

WET CHEMICALS

Acids, bases, solvents, metals (such as beryllium, lead, and so forth) all require the same level of concern as the hydride gases. They require segregation in storage, delivery, use, and disposal. They are not generally used in closed systems as is the case with the gases, so they require the additional burden to the employee of personal protective equipment on top of whatever clean-room clothing is required. Liquid and solid chemicals must be considered in terms of accident potential, as with the gases.

Chemical controls, both engineering and administrative, must be used to address all of the requirements. Engineering controls will be related to local exhaust, storage conditions, protective clothing, and disposal controls. The administrative controls will deal with reasonable quantity limits in the storage cabinet, room, building, or on site. They will also require review of orders for chemicals by safety and environmental specialists to assure that Environmental Protection Agency regulations, OSHA regulations, and the various codes and ordinances are satisfied. Once the controls are in place and protective clothing is specified and available, the key to safety is training. People must be trained to recognize the hazards of handling liquid chemicals and gases. A water-clear liquid must be recognized in the chemical handling area as potentially a hazardous chemical. Employees must learn to differentiate between classes of hazards. Corrosivity or oxidizing potential of acids, flammability of solvents, internal organ effects potential of solvents and many of the gases are all important basic information which will impact an employee's judgment in deciding whether or not to take a chance.

Bulk Chemical Storage and Distribution

Economies of scale can be achieved in storage and distribution of liquid chemicals, such as nitric and hydrofluoric acids, by bulk storage and distribution systems. This approach offers the advantage of allowing sampling of large lots for incoming inspection and reduced handling of individual containers. The source may be large bulk tanks located externally to the fabrication area, or may be smaller *day tank* (a small tank used to hold chemicals for immediate use) type tanks located in an attached building. In either case the bulk storage is external to the fabrication area, which effectively reduces the volume of chemicals stored in the area and the handling associated with those chemicals. Storage tank provisions provide protection against overfilling, accidental spill, fire, explosion, and other emergencies.

Piping systems of a material appropriate to the chemical being carried are used to deliver the material to the use point in the wafer area. Unfortunately, many pipe materials (for example, polypropylene) will discolor in the presence of strong chemicals, indicating a loss or the alteration of a chemical component. This may be a sign that the material being carried is becoming contaminated by the pipework to some degree. Fluoropolymers such as PTFE are much less likely to discolor, and this suggests that their inert nature is better in terms of contamination control in process chemicals. The piping systems generally run in the underfloor area but have also been designed into the space above the HEPA filter system in the ceiling. In the ceiling, the piping runs are concealed from routine, regular inspections and are generally provided with a trough or drip tray to catch any leakage which may occur. Better still is the use of coaxial piping systems with interspace leak monitoring.

The piping systems deliver the chemicals directly to the use equipment. Direct delivery has the effect of reducing storage space requirements within the wafer area as well as reducing the risk of contamination from the handling of chemical containers into and out of the area. Dispensing is accomplished via dead man type switches at the equipment which require the operator to depress a switch the entire time that dispensing takes place. This keeps the operator in attendance at the operation to supervise the task and provides someone to take action in event something goes awry.

In practice, the newest wafer-processing cleanrooms use both the bulk tank and day tank approach. The bulk tanks are used for single chemicals, and the day tank version is used for mixed chemicals such as combinations of acids for etching, cleaning, and so forth.

Bottled Chemical Use

In spite of bulk distribution advantages, there are some chemicals which simply do not lend themselves to bulk dispensing. Examples of these are photoresist, spin-on dopants, and other similar viscous materials. Also, special low-use chemical mixtures will be dispensed from individual containers. Thus, handling of these containers is still necessary in the cleanroom environment. These containers are made of a material which is compatible with whatever they are carrying, for example, glass for strong oxidizers and plastic for hydrofluoric acid mixtures. Safety coatings are used for glass bottles to reduce breakage hazards. These plastic coatings are applied externally, will help to prevent the glass from shattering, and also contain leakage.

Individual chemical containers are usually received at a building or dock dedicated to the handling of chemicals. They are received in, for example,

Department of Transport approved shipping containers, which are usually specially designed impact-absorbing cardboard cartons. Arrangements can be made with the supplier to ship the individual bottles in sealed double plastic bags so that they may be cleaned before they leave the shipper's location and arrive ready to be transported into the cleanroom. It is routine to specify two plastic bags for each bottle of chemicals. The outer bag is stripped off just as the chemical container is placed into the chemical storage cabinet at the delivery point, and the second bag is removed as the material is taken into the cleanroom.

Chemical storage cabinets must be placed strategically about the cleanroom so that travel distances to use points are kept to a minimum. The design of the cabinet is important for a number of reasons related to safety.

1. The cabinet must be limited in capacity. This is a code requirement for flammable liquids and makes good sense from a risk management standpoint.
2. Cabinet design is usually dictated by code requirements regarding strength, material, and fire resistance.
3. Shelving should be restricted in height so that employees are not required to handle hazardous materials over their own heads. It is not uncommon for caps to loosen in transport, producing a drip/spill hazard.
4. The upper shelves should be reserved for empty containers. This assures that overhead lifting is limited to the lighter containers and protects against spills by requiring the user to assure a tight cap just before placing the bottle on the shelf.
5. Appropriate spill materials, such as absorbers and neutralizers, and first-aid/emergency equipment must be convenient to the storage location. If an employee with chemicals splashed onto his face has to go through doors to another room to find an eyewash, the injury severity will be increased.
6. Provisions for transporting the material from the cabinet to the use point must be made. This may be a wheeled cart or a plastic or rubber bucket to contain the chemical in transport.

It will be necessary to train employees in recognition of different chemical types. If employees are expected to store acids in an acid cabinet and solvents in a solvent cabinet, they must understand the difference. That understanding must be specific enough to assure that the correct decisions are made each time. Chemical compatibility is a concept that all chemical users must understand if they are to handle more than one class of hazardous material during work. Employees must also be trained in emergency response and recognition of when things have gone wrong. The emergency response training must include not only what to do if the employee is personally exposed, but how to assist others in an exposure situation. It must prepare the employee to handle a spill or a cracked, leaking, or unlabeled container. A wrong decision in any of these situations invariably results in an accident and subsequent injury.

Monitoring and Cleaning Process Chemicals

As with process gases, we must monitor process chemicals to determine their cleanliness and take action to remove particulates and other forms of contamination whenever necessary. This, once again, is much more difficult than it sounds as many process chemicals are extremely corrosive, flammable, toxic, reactive, that is, generally unpleasant. We therefore have to rely on the manufacturers to supply us with the cleanest chemicals possible. In the early 1980s, the quality of chemicals supplied to the high-technology industry by United States' suppliers was considerably lower than that of similar Japanese materials in terms of numbers of particulates and trace metals. The situation in the United States now is very much better, and quality is much more likely to be met and guaranteed. However, we should not be complacent, as there are still likely to be some particles in chemicals (anywhere in the world), possibly arising from packaging materials or delivery systems. Table 10.1 illustrates measured numbers of particles of various sizes which were found in process chemicals. A number of sources and actions were seen to be responsible for this, including mechanical shock on the containers. Figure 10.4 shows how particles in process chemicals can easily end up on semiconductor wafers (or any other surface, for that matter).

Wet chemical particle counters for harsh liquids are available but are not very common outside of chemical bottling plants. These counters are typically like water units, but the cell and pipework is made of a chemically inert material such as PTFE (for acids) or quartz (for solvents). A PTFE filter membrane may also be used to collect particles for counting and to clean the liquid of particles. An alternative scheme for particle reduction in liquids involves bubbling a relatively inert gas such as nitrogen or argon through the liquid. Particles stick firmly to the surface tension regions surrounding the bubbles and the contaminated foam may be collected at the top of the container. The technique is not as efficient as membrane filtration (60 percent removal at 0.5 micron is typical), but could be used to pretreat chemicals prior to filtration to increase filter life.

Complex chemical analysis, such as chromatography, may be performed under laboratory conditions to assess other forms of contamination, but no simple in-house methods are yet available. Perhaps the system described in Chapter 3 or a similar method may be the answer to this dilemma in the near future. There is no simple way of removing nonparticulate materials such as dissolved trace metals or organics from process chemicals. Once again, we have to put our trust in the suppliers, so it is probably a good idea to

Table 10.1 Measured Particulate Contamination in Chemicals

Chemical	Number of Particles/liter					
	≥0.5 μm	≥0.6 μm	≥0.7 μm	≥1 μm	≥2 μm	≥5 μm
H_2O_2	80K			2K		80
NH_4F	360K	240K	92K	16.8K		1.6K
H_3PO_4	284K	84K	8.9K	560		280
10:1 NH_4F:HF	340K	190K	48K	10.6K	2.3K	600
HF	28K					
20:1 NH_4F:HF	620–9200K					
10:1 NH_4F:HF	280K					
H_2SO_4	13M					
Al etch	2.3M					
Acetone	230–9200K					
Isopropyl Alcohol	54–1800K					
Photoresist Stripper	15M					
Photoresist Developer	26–700K					
HMDS	18K					
NH_4OH	1.8M					
H_2O_2 (unfiltered/filtered)	2M/3.2K					
NH_4OH (unfiltered/filtered)	19M/3.3K					
HCl (unfiltered/filtered)	620K/290					
H_2SO_4 (unfiltered/filtered)	12M/200K					
100:1 H_2O:FH (unfiltered/filtered)	60K/10					
Chemical/Handling	**0.5–1**	**1–2**	**2–5**	**5–10**	**10–20**	**20–25**
Methanol	1.4K	60	40	10	0	20
20 Cap Tightenings						
Bakelite cap	7.2K	880	260			
Polyproplyene (heavy)	60K	4.2K	1.0K	60		
Mechanical Shock	6.8K	850	390	120	10	
Mechanical Vibration	1.9K	210	120	70	10	

From C. M. Osburn and R. P. Donovan, The Effects of Contamination on Semiconductor Manufacturing Yield. Originally published in the March/April 1988 issue of *The Journal of Environmental Sciences*. Reprinted with permission.

maintain a good, trustful, and long-lasting relationship so that quality expectations may always be met.

The other aspect of wet chemical monitoring is mainly safety related. The wafer processing area generally requires monitoring equipment to detect and measure chemical upset situations. Whether the cause is a leak in a process gas line, in the bulk chemical distribution pipe, a spill on the floor, or a leaky

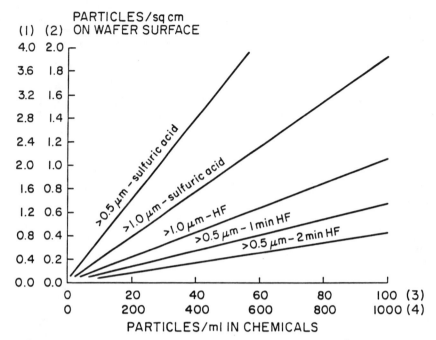

(1) Sulfuric and hydrofluoric acid — >0.5 μm particles/cm² on wafer
(2) Sulfuric and hydrofluoric acid — >1.0 μm particles/cm² on wafer
(3) Sulfuric and hydrofluoric acid — particles/ml >1.0 μm
(4) Sulfuric and hydrofluoric acid— particles in bath > 0.5μm

Figure 10-4 Effect of particulates in chemicals on the deposition and defect density on semiconductor wafers. (1) Sulfuric and hydrofluoric acid—>0.5 μm particles/cm² on wafer. (2) Sulfuric and hydrofluoric acid—>1.0 μm particles/cm² on wafer. (3) Sulfuric and hydrofluoric acid—particles/ml >1.0 μm. (4) Sulfuric and hydrofluoric acid—particles in bath > 0.5 μm (from *Semiconductor International,* July 1988, p. 44).

drain, chemical contaminants in the cleanroom pose a risk both to the people occupying the space and to the process itself. Very sophisticated monitoring equipment is available to do this monitoring on a real-time basis. Monitoring points are selected commensurate with the exposure risk and are connected to a central monitoring unit. The unit usually monitors several points, either simultaneously or sequentially, and compares the values monitored against a known standard. When the alarm set point is reached it takes whatever action steps are programmed into it. Specialized units can monitor acid and solvent vapors. Gas chromatography units have been adapted to the monitoring situation to search the area for literally any organic material (for example, vapors from a solvent spill), identify it, compare to a standard, and record the results or sound an alarm and take other action, usually through a microprocessor control system.

Monitoring and alarm systems, even with automatic actions taken to resolve the immediate situation, are only half the job. People still are required to react to the emergency. Someone must determine the cause of the problem and repair the damage. People must be removed to safety until the exposure level is acceptable again. Employees with acute exposures because of the accident must be cared for medically. Precautions must be taken to assure that the emergency does not extend beyond the original area to other spaces within the cleanroom or other occupancies in the building or on the property. These actions are normally accomplished by an Emergency Response Team. Its mission is first to safeguard people and then to control the emergency. To discharge this mission, the team must be well trained and schooled in every conceivable emergency situation. This training is discussed further in Chapter 12.

Spill Control

A very old safety axiom states, "Each time a material is handled there is an additional opportunity for accident." Certainly this is true with regard to the handling of liquid chemicals which are capable of being spilled. Spills in the cleanroom can be minor or major depending on a number of variables. The chemical, size of spill, location of spill, and air circulation system all play a part in the degree of danger. If one gallon of concentrated sulfuric acid is spilled onto the floor inside the cleanroom and the vapors are recirculated by the air-conditioning system, the problem can be major. The air-conditioning system can broadcast the vapors throughout the area very rapidly, causing exposure to people and damage to equipment.

One response to this potential problem is to zone the air-conditioning so that only part of the cleanroom is contaminated by the spill. Another is to provide a false floor so that spilled chemicals fall through to a subfloor area which is always awash with a flow of water to the chemical drain system. Both of these precautions reduce exposure potential to both people and equipment, but they do little to prevent spills in the first place. Prevention would require that the container be capable of being dropped or bumped without breaking. This could be assured by coating the container in a plastic coating or providing a rubber bucket to carry the glass container.

Some basic procedural rules will also help:

- At most, two containers at a time may be carried by hand from storage to use point.
- Containers at the use point shall be kept inside use equipment, never on the floor or on top of equipment.

- Containers shall be examined for leak potential before being removed from chemical storage cabinets. Caps shall be examined for tightness or signs of leaks.

In the event of spills, speed is essential. Neutralizer or absorbent material (as appropriate to the chemical spilled) must be available quickly in the area. The area must be isolated quickly, both as far as people are concerned and from an air-conditioning standpoint. The material must be thoroughly neutralized and removed, then the area must be washed of all residue to prevent residual contamination. Another approach that is being used more frequently in preventing spills is the use of piping systems. A bulk supply is located outside the cleanroom and is piped to the equipment within specialized piping systems. The operator depresses a "press to dispense" button, and only while it is held is the chemical allowed to flow into the container. These systems require a systems safety approach to respond to potential safety problems. How will we assure that the operator cannot use the wrong chemical, or overfill the container? Moving hazardous materials by piping systems has been done successfully in the petrochemical industry for years. The concept has merit, but on the scale usually encountered in the semiconductor industry not a great deal is known, and so it is approached with some caution.

HANDLING OTHER MATERIALS

A variety of materials we are likely to encounter in our controlled environments do not fit into the gases or wet chemicals categories. These are solids, and may be production materials or even parts of equipment which require special handling attention. Once again, we will examine the problems from the viewpoints of microcontamination and safety.

Microcontamination Control for Solid Surfaces

We have discussed in Chapter 4 how we may clean surfaces using a variety of techniques. Many of these techniques may be used to clean solid production materials as well as equipment components. However, in many cases production materials are considerably more sensitive to residual contamination from cleaning techniques than are other surfaces (equipment, workbenches, and so forth). For instance, we may clean a workbench using a cleanroom wiper which is made of a non-linting material. However, even the small amount of *extractables* (chemical contamination which may be released from the wiper material) left after cleaning may prove problematic to a sensitive material such as a semiconductor substrate. It is therefore very difficult to remove particles from sensitive surfaces without further contaminating them. One widely used method is to use rotating nylon brushes with ultrapure water to remove particulate material. The danger here is not so much from residual contamination but from mechanical damage from dragging hard particles across the surfaces. We therefore are encouraged to examine alternative methods.

The dry-ice snow technique described in Chapter 4 has potential for substrate cleaning. Electrets, formed into brushes, may also be a viable alternative. Unfortunately, electret brushes still rely on physical contact between the brush and the surface to break the particulates free in the first place; this could lead to redeposition. However, the idea of using an electrostatic field alone to pull dust off a surface is very attractive, as it reduces the chance of recontamination from the cleaning element. If we put a high negative DC voltage probe near a particle-covered surface, the particles will be charged to a potential opposite to that of the probe by a process called *charge induction*. This should cause an attractive force between the particle and the probe that might, in theory, remove some of the particles. However, this isn't all of the story. The negative probe will generate negative ions that will tend to charge the particles to a negative potential. This produces a repulsive force between the probe and the particle. The net result is that a DC-charged probe does not remove particles from a surface.

The situation is somewhat different if we consider the use of AC or chopped DC voltages. We recall that particles may be held onto a surface by organic layers. The particle/glue system can be modeled as a simple mass spring system, the mass being the particle and the spring being the organic layer. If this is the case we might expect that driving the mass spring system at its resonant frequency would feed energy into the system to the point that the spring might break (for a discussion of resonant frequency, see Chapter 7). If the spring breaks, the particles will jump off the surface and might well be collected by a local vacuum probe. Theory predicts that the resonant frequency of a particle/spring system will go up as the particle size decreases; for example, experiments performed with high voltage AC at 60 Hz and 15.7 kHz revealed that the lower frequency causes the larger (100–200 micron) particles to break free, whereas at 15.7 kHz we see the smaller (50–100 microns) begin to move. This technique is not yet in commercial use but may be an excellent way of selectively removing particles from a surface (for example, an integrated circuit).

There are alternatives to the removal of particles after they have reached the surface in question. We could attempt to prevent them from reaching their destination by protecting the surface in some manner. For instance, we could coat the surface of the materials with an inert, easily stripped coating (for example, Hydragel; see Chapter 11). Any particles which land on the coating will come off with the coating during stripping when it is time for the materials to be used. Unfortunately, we do not know much about the residual chemical contamination after stripping, and this should be a factor in the selection of a coating material. An innovative technology that has been in-

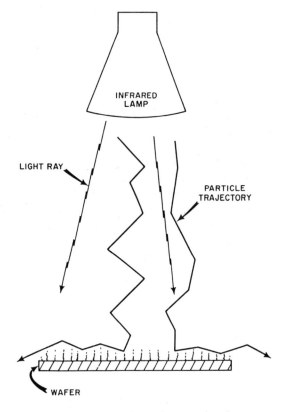

Figure 10-5   Thermophoresis for reduction of wafer contamination.

vestigated involves keeping a substrate at a temperature above the ambient to preclude the deposition of particulates and organic vapors. The system is shown in Figure 10.5. When a material is above the ambient temperature, the excited air molecules immediately above the surface will not allow particles or vapors to reach the surface. This effect is known as *thermophoresis*. Numerical data on the results of exposing wafers at various temperatures to ammonium chloride smoke particles or glycerol (glycerine) vapors are shown in Figures 10.6 and 10.7. As may be seen, a surface temperature of 60° C is adequate to prevent most particles from the ammonium chloride test smoke or glycerol from reaching the surface.

Safety Aspects of Other Materials

The hazard natures of many solid production materials are discussed in Chapter 12. However, we will discuss some of the other hazards related to the handling of solid process materials here.

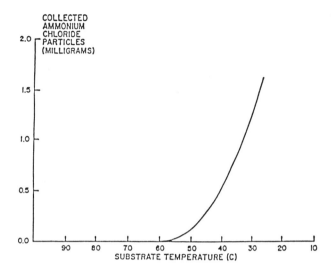

Figure 10-6 Effect of surface temperature on the deposition of ammonium chloride smoke particulates. Substrate exposure time 5 minutes.

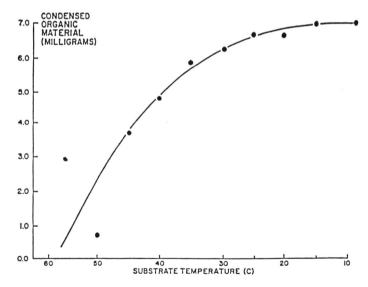

Figure 10-7 Effect of substrate temperature on the deposition of vapors from liquid glycerol at 240° C. Substrate exposure time 30 minutes.

Solid materials can become hazardous when they are transformed into a state where they may be readily taken in by the body. For instance, when substrates or layers are ground into dust during a sawing, polishing, or cleaning operation, this dust may be breathed in or injested. Also, solids may have dangerous volatile components which become gases when the solids are hot (for example, heavily phosphorus-doped hot wafers will outgas phosphorus).

SUMMARY

In order to bring gases, wet chemicals, or solid production materials into the cleanroom, we must handle them with care to avoid microcontamination and safety problems. Gas cylinders and pipework can be a considerable source of contamination and must therefore be lined or specially cleaned/decontaminated. Emphasis must be placed on the proper design of gas delivery systems to avoid contamination and leaks. Detection and monitoring must be an integral part of the design for hazardous materials. Leak prevention and spill control are also critical for wet chemicals.

11

Personnel and Contamination

Everyone concerned with contamination control dreams of the day when automation will remove the need for employees to actually handle the wafers. This wish arises from the fact that humans are such a major factor in cleanroom contamination. Not only do they generate a large number and variety of contaminants, but they also tend to be in close proximity to the product at many stages during the production process. The introduction of advanced manufacturing technologies, as discussed in Chapter 13, will ultimately allow the separation of employees from the product in an operating facility. However, we must recognize that many of the techniques described are still in their infancy and are likely to remain beyond the reach of most companies (especially the smaller organizations) for some time. Therefore, since we are faced with the continued marriage of people and product, we must use appropriate contamination control equipment and practices for personnel to reduce the human factor in cleanroom contamination.

In the following sections we discuss certain aspects of personnel and contamination, including the key aspects of cleanroom apparel and work practices in controlled environments.

APPAREL

Garment Materials

The purpose of cleanroom apparel is to protect the environment from the human wearer (and not vice-versa as in the case of safety equipment). The first aspect we should consider is how to choose a suitable material for this apparel. Materials used in standard street clothing suffer from two major disadvantages; they tend to be contamination generators and, on the microscopic scale, they are "open" structures. Openness leads to warmth in the bulkier fabrics and to comfort (by allowing vapor transmission) but it allows the passage of particulates through the material.

On the question of the thread used, natural materials such as cotton are unsuitable as cotton threads are spun from short fibers (the spinning process merely twists these plant fibers together, and they are essentially held by friction). Over-tensioning the thread will tend to break and dislodge the fibers and they may then be introduced into the environment by gentle abrasion (the larger lengths appear as lint). In addition, natural fibers are inevitably contaminated with other natural contaminants such as fine dusts and pollen which are not always completely removed by laundering. Therefore, synthetic or man-made materials are preferred as the threads are formed from twisted monofilaments. The filaments are actually long lengths of polymeric materials, such as nylon or polyester, extracted from a chemical reaction vessel. The threads made with these filaments are extremely strong, relatively contaminant-free, and may be easily woven into cloth. The strength of the thread allows a very tight weave and consequently small interstices in the resulting cloth. The material is therefore a more effective filter in reducing the transport of contamination from the surface of the skin to the environment.

The form of the weave may vary. The two most common weaves are taffeta and herringbone. The difference in these weaves is evident in the micrographs of Figure 11.1. In the taffeta weave, each horizontal thread passes under one vertical thread before passing over another. In herringbone, each horizontal thread passes under and over double vertical threads, one of which passes over the next parallel horizontal thread, the other passing under. This leads to a characteristic pattern in the cloth. Herringbone is favored in many cases for strength, although when the cloth is new, taffeta tends to trap particles slightly better. A popular woven material for cleanroom suits is Dacron. It is a low-cost synthetic which is hard wearing and capable of being laundered a few times without loss of integrity. This durability is a very important factor as the materials will become contaminated with use and must be cleaned (cleaning is discussed later in this chapter). Unfortunately, whereas pore size may be as little as 0.5 micron immediately after weaving, the stress on the fabric during wear can open the pores to 30 microns (herringbone), or as much as 100 microns (taffeta). Therefore, although the threads may

Figure 11-1   Scanning electron micrographs of various materials used in cleanroom clothing. (a) Polyester taffeta (100×). (b) Polyester taffeta (500×). (c) Polyester herringbone (100×). (d) Polyester herringbone (500×). (e) Spun-bonded polyolefin (100×). (f) GORE-TEX™ expanded PTFE laminate (100×). (g) GORE-TEX™ expanded PTFE laminate (500×). (h) GORE-TEX™ expanded PTFE laminate (5000×) (micrographs courtesy of W. L. Gore and Associates.™ Trademark of W. L. Gore & Associates, Inc.).

(a)

(b)

(c)

(d)

(e)

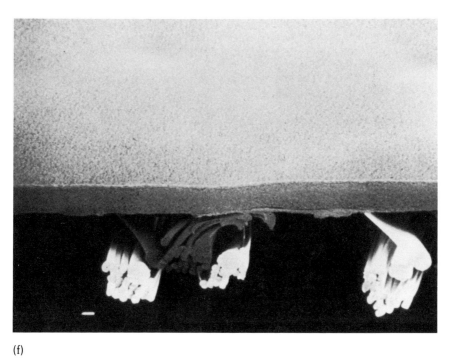

(f)

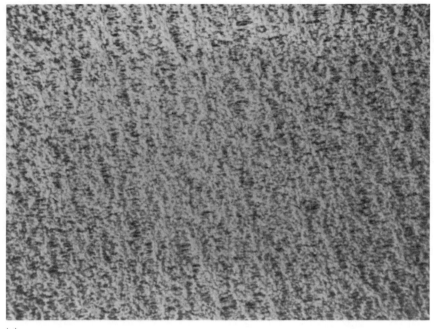

(g)

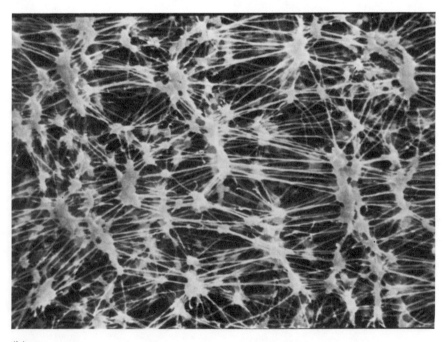

(h)

still be intact, the cloth can become unsuitable for cleanroom use over time. Also, continued exposure to acid vapors (as often happens in semiconductor manufacturing) results in the denaturing of the synthetic fibers, making them brittle and prone to breaking.

One may come to the conclusion at this stage that a woven cloth, even one woven from synthetic thread materials, may not be the ideal choice for cleanroom applications. Such materials are in fact very widely used in controlled environments due to their desirable properties, as discussed above. However, other low-cost, non-woven alternatives do exist. One such material is spunbonded polyolefin, also called Tyvek (a Dupont registered trademark). The fibers are once again synthetic and extremely strong. They are formed into cloth by a compression process to create a multilayered structure of randomly oriented fibers. The cloth has no appreciable open regions, typically 0.5 microns maximum even after wear, and is therefore an extremely good particle barrier. For the same reason, it is also a good barrier to moisture and hence it can be extremely uncomfortable for the wearer during extended use, but can provide good short-term protection from chemical spills. Unfortunately, the fibers will tend to break after continued wear, especially if the material is constantly abraded, and may be dislodged from the surface. This effect may manifest itself in as little as two hours of normal wear and hence the suit has a short cleanroom lifetime (a few days maximum). In addition, because the fibers are not held together in a woven structure, repeated laundering will degrade the cloth; hence soiled Tyvek should be discarded. On the other hand, since the material is formed in much the same way as paper, it is extremely low in cost (considerably lower than polyester) and may thus be used in a disposable mode. Future versions of Tyvek cleanroom suit materials will be coated in materials such as polyamide to reduce many of the problems discussed above.

The conclusion we may now come to is that the ideal cleanroom material would be some combination of woven and nonwoven materials. Such a marriage exists in expanded PTFE laminate materials such as GORE-TEX (a W. L. Gore and Associates registered trademark), where a taffeta woven material supplies the structural integrity and a semiporous PTFE backing prevents particles crossing but allows a high degree of vapor transmission (the backing layer acts as a membrane filter). The resulting material is strong, comfortable, has a long life, may be repeatedly laundered, and may be used in even the most demanding controlled environments. Some concern was originally voiced over how well this material would stand up to laundering (delamination of the pad from the polyester was a worry), and if it would trap chemical contaminants, but so far these fears have been proven to be unfounded. Unfortunately, it is the least elastic of the materials discussed, and hence the suit design must accommodate this. The reason why it is not universally used in cleanroom apparel is that it is prohibitively expensive. However, it goes without saying that the best cleanrooms deserve the best materials. A microscopic comparison of Tyvek, polyester, and expanded PTFE laminate is provided in Figure 11.1, which shows scanning electron micrographs of these materials. Note the apparently large interstices in the polyester cases as compared to the more opaque Tyvek and GORE-TEX (all materials are unworn). The 5000X magnification of the expanded PTFE, showing the extremely open microstructure, helps to explain why this material traps particles while allowing moisture to pass.

We have mentioned comfort frequently in the above discussion of materials. Much of the comfort factor lies with moisture transmission. This has recently been quantified in a measurement called MVTR (moisture vapor transmission rate). MVTR is the amount in grams of water which can pass through 1 m² of the material in one day. The higher the MVTR, the larger the moisture transmission, and the more comfortable the material is to wear. The MVTR for Tyvek is 7,000 g/m²/day, whereas for polyester herringbone and taffeta it is around 20,000 g/m²/day. Because of the unique nature of the expanded PTFE laminate, it has a very high MVTR at 32,000 g/m²/day.

Suit Styles

A good synthetic material alone does not constitute a cleanroom suit. The two-dimensional cloth must be converted into a particular style of three-dimensional suit by first cutting out the appropriate shapes and then joining them together. Here we are faced with two problems: (1) the act of cutting the cloth leaves frayed ends in woven or non-woven materials and these can shed lint when abraded, and (2) the strongest method of joining cloth sections is by sewing, which opens up holes in the material allowing free passage of particles. Therefore, the cloth edges in cleanroom apparel are carefully seamed to ensure that the edges are tightly wrapped in the fold. Unfortunately, sewing will always puncture the cloth, but the effect may be reduced by using a fine synthetic thread (put in with a thin, sharp needle). Double stitching is usually employed to ensure that the joints/seams are strong and do not leak. However, long-term use will result in the needle holes becoming wider due to the lateral forces on the cloth by the threads, and older suits will therefore leak at sewn areas. Tyvek tends to be particularly bad in this respect as the strong fibers tend to move readily with the thread and thus wide holes open up. One further method of reducing the problems of sewing is to chemically bond or heat seal the seams after sewing. These are expensive processes, but we retain the strength of a sewn seam while eliminating the particle transmission problems.

As far as suit styles are concerned, we have a number to choose from, depending on the application and cost considerations. In a less demanding environment, for example, class 10,000, a round-necked long lab coat or smock made of polyester or Tyvek would typically be used along with an elasticated bouffant cap and overshoes of the same material (the soles of the shoes would be made of textured plastic for strength and to prevent slipping on vinyl floors). The suit must have a tight-fitting zipper, preferably nylon to reduce particulate generation, and the sleeves should be elasticated or be able to be closed tightly with "popper" fasteners. Poppers would also be used at the neck to reduce leakage here. In the class 1,000 case, booties could be used instead of overshoes, with the smock being long enough to cover the tops of the leg sections, or a jumpsuit could replace the smock. Personnel who work in close proximity to the product should also wear gloves and perhaps even a facemask. These items of clothing are discussed in greater detail in later sections. For class 100, a more complete covering is required, for example, a full Tyvek or polyester jumpsuit or bunny suit is generally mandatory, with booties and a hood that covers the neck as well as the head. Once again, gloves and facemask may be used, in addition to eye covering such as goggles when the employees are close to the product. At class 10, we have something of a break point. A jumpsuit (sealing with poppers at the sleeves, legs, and neck), booties, hood with integral facemask or separate mask, gloves, and goggles are all mandatory. Additional anticontamination techniques may also be used, such as hairnets to prevent hair contamination during gowning and moisturizing creams to reduce skin emissions. The major difference is that only the highest quality polyester or expanded PTFE laminate materials are considered to be suitable for class 10. Suit types for class 1 are discussed later.

If a small number of suits is used in a particular medium class facility, for example, in a small class 100 research lab or a class 10,000 low-volume assembly area, then disposable Tyvek apparel would probably be the most cost-effective option. However, large-scale operations would almost certainly demand launderable Dacron or a similar material. The economics of the decision is simple: woven synthetic suits will cost more to purchase than nonwoven alternatives and there will also be laundering and some repair costs. The purchase and laundering costs per suit will decrease with larger quantities of suits. If these costs over the lifetime of the suit are less than what it would cost to purchase Tyvek suits over this time, then Tyvek is obviously not the most cost-effective option. A common mistake in these calculations is to overestimate the life of the woven suit. Chemical damage, tears, expansion of the weave, and thread holes all reduce the average suit life. Although most suit manufacturers will probably disagree, two years is probably a good figure to use in the case of many class 100 suits.

Gloves and Skin Coatings

We begin our discussion of gloves and skin coatings by first reminding ourselves what the problem actually is. Hands generate contamination in the form of skin flakes, as does the rest of the body. However, contamination from hands is much more significant, as they tend to be the parts of the body closest to the product and equipment. Studies have indicated significant differences between employees in terms of the number of skin flakes shed per unit time. There seems to be no general information at this time on why some employees shed more than others, however, tests have indicated that hand washing can increase skin scale shedding by factors of 1.5 to 2. Each time the hands are washed, things get worse, possibly because of a loss of natural oils from the surface of the skin.

Merely covering the hands with a suitable cleanroom suit material is not the solution to the contamination problem as this would effectively remove the operator's sense of touch. Therefore we have to use a good noncontaminating barrier which allows free finger and hand movement and as much feeling as possible. The options currently available are thin gloves and skin coatings, the latter being applied as a liquid or cream.

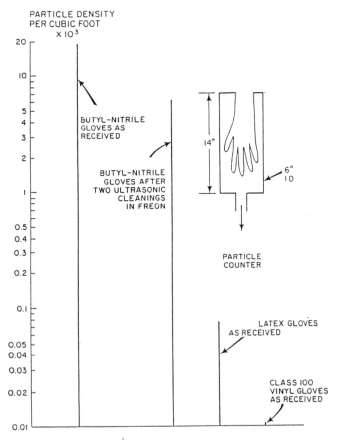

Figure 11-2 Experimental results particle generation by various gloves. Particle size range 0.2 to 12 microns. All measurements were done in a HEPA-filtered environment.

The question of gloves is worth a book in itself in that the gloves are closer to the wafers than anything else worn by the employee. In Figure 11.2 we show some data on new gloves of various types used in one company. This data includes chemical (safety) gloves, as these will also be worn during many operations, usually on top of anticontamination gloves. As may be seen, the chemical gloves seem to be particularly bad.

In Figures 11.3 and 11.4 we present some data on particles shed by human hands with and without gloves. Clearly gloves are better. However, the latex gloves used in these experiments start to shed particles after several hours of use. This is very evident if we compare the data in Figure 11.3 with that in

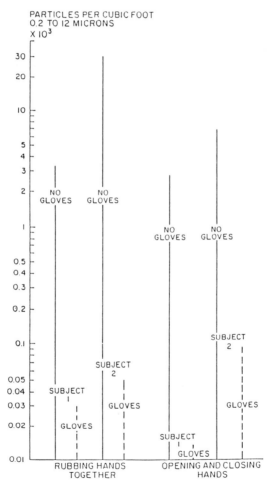

Figure 11-3 Experimental data effect of cleanroom gloves on particle generation by human hands (new latex gloves).

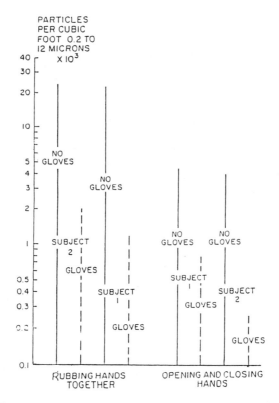

Figure 11-4   Experimental data effect of cleanroom gloves on particle generation by human hands (latex gloves, 2 hours of use).

Figure 11.4. We understand that in one major semiconductor company the employees in a critical operation change their gloves every 45 minutes. This may be worth doing, particularly with a valuable product. Gloves are not very expensive. However, we must admit that it is difficult to say how long gloves should be used, how they should be tested, or even what materials should be purchased for a given application. There are a number of standard tests for gloves, but none of them really models the environment that clean gloves are exposed to. Our best advice is to test various gloves for a particular application by designing a test relevant to that application, for example, if the gloves are used near solvents, expose test samples to solvent vapors and look for degradation.

We noted earlier that there was concern about employee-generated particles, particularly skin flakes. This has generated a lot of interest in skin creams or lotions that might be used to reduce skin flaking. The application of these materials in the United States has been limited by the fear of medically related lawsuits, but there is some use of this technology in Europe. In Figure 11.5

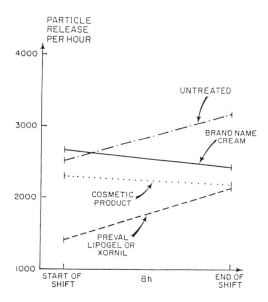

Figure 11-5 Effect of skin treatment on particle release by cleanroom personnel (data from L. U. Bieser, M.D., IBM Germany).

we show some data from IBM, Germany. The application of certain skin creams does reduce flake generation. Oil-based cosmetic creams last well, but the oil could contaminate. The water-soluble creams are better for the cleanroom, but wear off over the course of the day and have to be reapplied. The German reports indicate no problems with the cream, and it may well be that a system of this type might be of value in this country. Studies of skin-coating materials have been performed in the United States. Figure 11.6 shows results of such a study on a material called Hydrogel. We have found that if the skin coating is put on human hands and allowed to dry for a few minutes, it is possible to pick up a wafer and not leave any fingerprints. We are not suggesting that the skin coating replaces tweezers, but rather that it might well be used with products or equipment components that may be too heavy to pick up while wearing gloves. Another application would make use of the material on the face before putting on a mask. Masks are known to abrade the skin and to generate particles; with the skin coating, abrasion and particle generation are greatly reduced.

The skin coating has other advantages in that, while it allows water vapor and oxygen to pass, it apparently stops organic materials completely. This offers the possible opportunity for the protection of employees against chlorinated hydrocarbon vapors which can be very irritating to human skin. A full study has still to be performed to confirm the medical aspects of this. The skin-coating materials are soluble in warm water and can be removed quite easily when the employee is ready to leave the cleanroom. All the ingredients

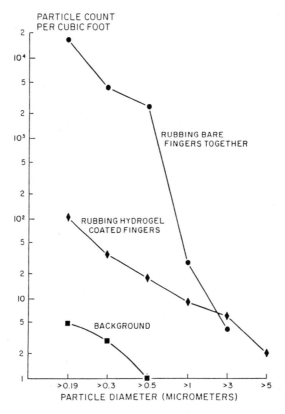

Figure 11-6 Experimental results effect of hydrogel on particle shedding by human hands (flow rate 1 cfm).

have approval for use as cosmetics, so there should be no problem in bringing them into the cleanroom.

Facemasks

To reduce the contamination from the nose and mouth, a facemask may be used. The principle is simple; a material barrier (which is permeable to air for obvious reasons) is placed across the nose and mouth to trap emissions such as moisture droplets. The mask style is generally similar to the surgeon's mask and the material should be non-linting.

This, unfortunately, is a gross simplification of the problem. It is a sad fact that many of the masks available at the time of writing of this book do not do a particularly good job of preventing contamination from the nose and mouth from reaching the product. For example, the surgeon's style mask, which is generally a rectangular piece of fairly dense soft filtering material

(a)

(b)

Figure 11-7 Micrographs of a typical spittle residue. (a) General view of spittle mark. (b) Close-up of KCl crystals. (c) Damage caused by spittle residue (courtesy of R. Thomas, RADC).

(c)

Figure 11-7   *(continued)*

with a metal clip to allow it to contour around the bridge of the nose, allows exhaled air to escape at the eyes, cheeks, and chin. Any particles or droplets which are present in this air may then be carried by the air currents which move around the body and dispersed to the environment. The stiffer preformed masks are not a particularly good solution either as, although they tend to make a better seal with the face, the exhaled air comes through the mask pores as high velocity jets. Contaminants may thus be ejected great distances from the face. Also, masks like these are dreadful at preventing contamination release during a cough or a sneeze.

A potential solution to the facemask problem is the use of advanced suit concepts as described later in this chapter. However, there is some question of employee comfort, both from a physical and a psychological point of view, with fully enclosed headgear. Therefore, the solution is not a simple one, but we will supply some suggestions later.

This is perhaps a good point to consider what is actually emitted from the nose and mouth and which factors affect emissions. Potassium chloride (KCl) is present in large quantities, dissolved in the water in the mouth. This chemical will be carried by the droplets of water which are continually splashed out of the mouth during talking and to a lesser extent during breathing (but especially during coughing or sneezing). As the water evaporates, solid KCl crystals will form. Figure 11.7 illustrates this problem perfectly in the context of IC manufacturing. Figure 11.7a shows a general view of the spittle contamination, which has a diameter of several hundred microns. Figure 11.7b shows a close-up view of the KCl crystals within the area. The crystals range from less than 1 micron to several microns in size. Figure 11.7c illustrates the damage that spittle deposits can do to an IC. There are also some complex proteins carried in the spittle. This is why it is a good idea to use spit shields on microscopes and to discourage personnel from talking while in close proximity to the product. Eating prior to going back to work seems to make little difference to the amount of material ejected, as any residual material seems to stick fairly well inside the mouth. However, one action which does make a large difference to the particle count is smoking.

Unfortunately, microcontamination control and smoking do not go together particularly well. The best illustration of this is in Figures 11.8 and 11.9, which show the particle counts from nonsmokers and smokers under various test conditions. Clearly, the particle emissions from smokers is worse but the emissions do decline with time after smoking stops. Fortunately, electrot fiber facemasks can help in this situation. The main question here is just what

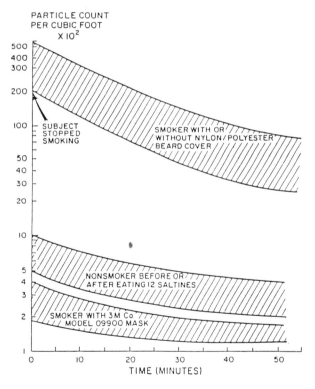

Figure 11-8 Experimental results particulate generation by smokers vs. nonsmokers. Size range 0.2 to 0.4 microns. Test distance 25 mm from mouth. Subjects were not allowed to drink coffee during test period.

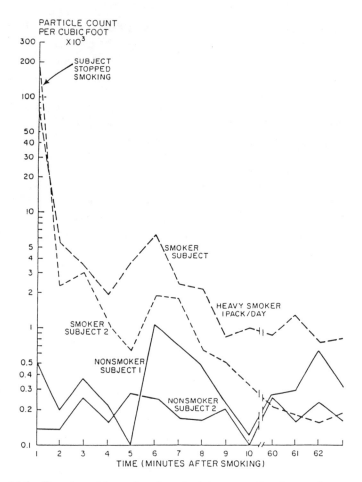

Figure 11-9 Experimental results of particulate generation by smokers vs. non-smokers. Particle size range 0.2 to 12 microns (the majority of the particles were in the range from 0.2 to 0.4 microns). Test probe 25 mm from subject's mouth. Subjects were allowed to drink coffee during the test period.

exactly is being emitted? Some smoke particles may be present for some time after smoking stops but there seems to be more than just these present. It is actually likely that the lining of the mouth is the culprit, as the cells lining the mouth tend to be killed and dried out with smoking and may detach to become ejected.

One should not forget the contributions from the eyes. They are coated with natural saline solution which is splashed out with each blink. Goggles are a good way of reducing this, besides being necessary for safety reasons.

Particle Dispersion and Apparel

Although we have discussed how cleanroom apparel protects the controlled environment from the wearer, we should bear in mind that in most cases it is not the ultimate barrier. Even with appropriate materials and suit styles, there is still the likelihood of particle escape. This may happen in one of two ways: through the material and at suit openings.

Figure 11.10 is of particular interest in that it presents information about the number of particulates passing through various types of fabric. The results will, in many cases, vary after the garments have been exposed to a series of wash and wear cycles. However, for the moment, this data has proved very useful in making decisions about the choice of fabrics for garments. In any case, it highlights the advantages of the primarily Japanese practice whereby employees put on clean, company-supplied underwear before donning the bunny suit. This undergarment is not made of the cleanroom suit materials discussed earlier, but the wearing of it eliminates street clothing as a source

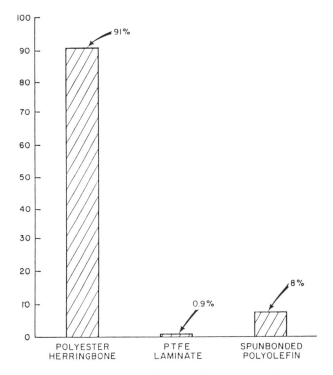

Figure 11-10 Percentage of particles from employee underclothing passing through various materials (from paper by Weaver and White, *Semiconductor International,* Oct. 1987, p. 30).

of contamination. We also show the numbers of particles escaping from different areas in two different suit styles in Figures 11.11 and 11.12. Quite obviously, the bunny suit design is considerably better than the smock at reducing human particle dispersion. However, we may see another significant point from these results in that the bare face in the bunny suit case is by far the biggest particle release area and hence should be the area of most concern.

One persistent problem with cleanroom suits is leakage from the cuff, neck, legs, front fastener, or anywhere that the fabric is not continuous, for example, seams. Leakage from open regions like the neck is particularly problematic due to what is called the *bellows effect*. Normal cloths are sufficiently porous so that they allow air trapped inside the suit to pass through when the wearer moves and collapses any air pockets. Our tightly woven cleanroom fabric will not allow this to happen and hence the air is shot out, at rather high velocity, from the open areas. Thus, contamination may be projected great distances from the body on this air jet. Unfortunately, the better the fabric is from the viewpoint of particulate transmission, the worse this problem is. Hence, we may reduce the problem by having a slightly porous cloth, by ensuring that

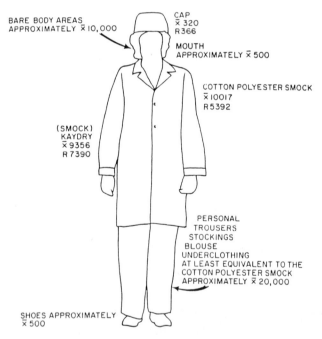

Figure 11-11 Particle generation data. $\bar{x}$ refers to the average particle count for the size range 0.3 to 7 micrometers. The factor "R" indicates the approximate range of the values above and below the average. An average value of 1900 with an "R" of 1000 would indicate that the actual numbers ranged from 900 to 2900 (from paper by M. S. Dahlstrom in *Semiconductor International*, April 1983, pp. 110–116).

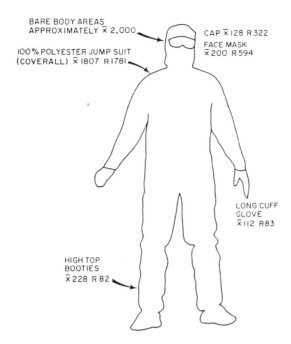

BARE BODY AREAS
APPROXIMATELY x̄ 2,000

100% POLYESTER JUMP SUIT
(COVERALL) x̄ 1807 R 1781

CAP x̄ 128 R 322

FACE MASK
x̄ 200 R 594

LONG CUFF
GLOVE
x̄ 112 R 83

HIGH TOP
BOOTIES
x̄ 228 R 82

Figure 11-12   Particle generation data. x̄ refers to the average particle count for the size range 0.3 to 7 micrometers. The factor "R" indicates the approximate range of the values above and below the average. An average value of 1900 with an "R" of 1000 would indicate that the actual numbers ranged from 900 to 2900 (from paper by M. S. Dahlstrom in *Semiconductor International*, April 1983, pp. 110–116).

the cuffs and legs are tucked into the gloves and booties respectively, or by taping the cuffs and legs. The latter is rather uncomfortable (especially if one suggests taping the neck too), and therefore it may be better to use filtering material at the open points to allow the air to be ejected but not the particles.

Laundering Practices

As a cleanroom garment as worn in a working situation, it will become soiled internally by the occupant, for example, by perspiration and associated salts, skin oils and flakes, and so forth. It will also become soiled externally by contact with contaminated water, process chemicals, pump oils, and so on. It is therefore obvious that if the garment is not disposable, it must be regularly cleaned by appropriate laundering. When one thinks of laundering, one normally thinks of water and detergent, washing machines, and hot air dryers. This is not too far from the truth in the cleanroom laundry except that the water is filtered and deionized, the detergent is sodium-free, and the washers and dryers are stainless steel and have appropriate water and air filters respectively. In addition, the loading and unloading areas are clean, that is, fed with HEPA-filtered air, and the staff are clothed in cleanroom apparel. The more advanced laundries do wet (water) washing first to remove water soluble salts and then dry clean, using solvent vapors such as carbon tetrachloride, to remove oils and greases. If you are in doubt of the quality of the equipment and procedures at your laundry, it will do no harm to check by regular visits.

Unfortunately, there seems to be very little information at this time on residual materials, detergent, skin scales, salts, and the like. Some laundries advertise that they rinse in 18 Mohm-cm deionized water. This is certainly a sign of low soluble salts, but one wonders about organics, silica particles, and lint which will not be detected by a resistivity measurement. The very question of how often a washable garment should be recycled also seems to have received little or no attention. Certainly garments wear as they are laundered and the quantity of lint increases with the number of laundering cycles. Is there a point of no return? The answer is yes. There is a whole world of problems in laundering/drying processes that have not really been investigated to any great extent. Monitoring the water from the last rinse cycle might well provide data on how much particulate material is present. If the particulate level is high, we might expect that the garments will be contaminated and vice versa. Observation of the lint/dust count in the air coming from a cleanroom dryer would provide information about the condition of the garments. If the air is contaminated, so are the garments. All laundries should use the ASTM vacuum filter test (see Chapter 3) on selected garments to determine how clean the material is in a gross sense. This will give an indication of how effective the laundering is and also how well the garment material is holding up. Unfortunately, this method may not be used safely on laminated materials like GORE-TEX as delamination may occur. However, there are alternative test methods. One such method is the Helmke Drum, which gently tumbles the garment in air. A particle count is taken of this air during tumbling to determine the level of particulate emissions. Other questions about the time/temperature cycle that should be used for garment drying may well be critical for the new fabrics, like GORE-TEX/PTFE laminate, that are quite expensive. It might be quite practical to change over to alternative (for example, ultrasonic) laundering systems in place of the usual tumbling process. Tumble drying is known to damage fabrics, but electrostatic drying systems can be used on garments that are hanging freely. These technologies are under investigation, primarily in the Far East.

There are a number of laundering-related practices which may readily be adopted which can have a profound effect on the quality of cleanroom garments. For instance, since garments will tend to degrade with laundering, only the newest (least laundered) garments should be used in the most demanding areas. For instance, once the garments used in a class 1 area have been laundered a predetermined number of times (as determined by inhouse experimentation), they may then be transferred to class 10 use and eventually to class 100, and so on. This may sound an expensive way to go, but the potential yield improvement should negate this argument. Also, garments worn by different people get contaminated in different ways. For instance, an inspection worker's garments should become considerably less contaminated than a maintenance technician's, the latter person spending most of his or her working day inside equipment, on the floor, near pump oils, and so forth. Under no circumstances should these two sets of garments be laundered together. In addition, if a photolithography worker gets photoresist spilled on his or her suit, it should be pre-sorted and given separate attention. These relatively simple procedures will not work unless someone is given direct responsibility for them. This could be one of the most important jobs in the industry.

Garment Static Control

We have discussed in detail why synthetic materials are required for controlled environment garment applications. However, the use of synthetics has one rather unfortunate drawback; normal wear tends to result in the fabric's gaining a net electrical charge by rubbing. Since synthetics are good insulators, the charge will not leak away to ground and will continue to build until it reaches the point where the potential difference with respect to ground is great enough, typically a few tens of thousands of volts, to allow a discharge (spark) at a high field (sharp) point across a gap of a centimeter or so to ground. Such discharges can be catastrophic to unprotected electronic components and equipment as these voltages will rupture gate oxides in MOS structures.

Therefore, since we are forced to use synthetics, we must attempt to reduce the problem of garment static by altering the material. This may be done in a number of ways. The simplest way is to spray the garments with a coating which will not allow electrons to be stripped from the polymeric material. Such sprays are readily available and relatively inexpensive. However, they have the disadvantage that the coating will wear off by abrasion and during laundering, and hence the garments have to be resprayed regularly. The need for retreatment can be determined by taking the garment, rubbing it on a surface which is known to charge an untreated garment, for example, polythene, and assessing the generated garment charge with an electrometer. There is also the (unanswered) question of contamination from these "antistatic" substances. Therefore this treatment should be closely monitored.

The other methods of garment static control involve making the cloth conductive to some degree, thereby allowing charge to flow to ground, through conductive wrist and ankle bands or conductive soles, before it can build up. A popular approach is to impregnate the cloth with carbon to reduce the resistance. Note that this does not make the garment conductivity too high: this is actually an advantage, as any charge will be dispersed slowly and a rapid spark-type discharge is avoided. Another way of making the cloth conduct is to weave conductors or "wires" into the cloth structure. These may be conductive plastic filaments or even fine stainless steel wires. They generally form a wide grid in the fabric and can thus collect and disperse charge. Both of these methods potentially suffer from contamination problems, as carbon particles could be expelled from the carbon-impregnated materials and brittle conductors could break and become dislodged. However, no data is currently available on this. A gleam of hope lies with the recent introduction of shaped conductors which are coated with nylon. The ridges which run along the length of these conductors create an increased electric field situation, and this allows charge leakage through the anticontaminative coating.

Advanced Suit Concepts

We have discussed some of the drawbacks of cleanroom apparel. However, in many situations the contamination associated with cleanroom garments is tolerable, that is, the contamination from a polyester bunny suit with hood and booties may not add significantly to the particulate count of a class 100 room. However, if we are to consider a so-called "world class" or class 1 room, we really cannot tolerate any suit contamination. This calls for a suit design which is beyond conventional styles.

First, the material is, of course, expanded PTFE laminate, due to its excellent particle retention properties. The suit style is a tight-fitting bunny suit to reduce the bellows effect and provide maximum body coverage. To circumvent the rigidity of the material, the suit has elasticated expandable panels at the waist, elbows, and knees, covered with the laminate material in ruffled form. The wrist and neck seals are also made of similar elasticated material, and the legs may have elastic seals or popper fasteners. The booties are primarily PTFE laminate with a reinforced sole, and the gloves, interestingly enough, are also the same material but with a thin plastic front to maintain tactility. The gloves and booties mate tightly with the bunny suit to prevent leakage.

The most important difference of these advanced suits is the face and head covering. This may be one of two styles: a PTFE laminate hood with a full face covering shield, or a head-enclosing helmet. In the former style, the hood covers the head and neck, sealing tightly to the neck section of the bunny suit by the elasticated section there. The transparent face shield is rather like an extended diver's mask, usually made of polycarbonate. The helmet type design is also a polycarbonate construction but is totally rigid (like a lightweight full-face motorcycle helmet). Both styles completely seal the face from the environment. One may note a small problem here, as the occupant still has to breathe. This function is taken care of by a battery-driven, belt-mounted pump which sucks air into the mask thereby supplying fresh air and preventing fogging. The exhaled air leaves at the back of the mask/helmet after being passed through a small HEPA filter. Some mask style variants allow detachment of a bubble visor to allow the wearer to use a standard optical microscope. Tests show that even coughs and sneezes do not increase the particle count beyond the masks.

As mentioned previously, some employees find the full-face shield or helmet very uncomfortable to wear for extended periods, and claustrophobia can be a problem with some individuals. For certain operations, for example, microscope work, the face covering can get in the way. The companies who make these systems also supply a hard or soft half-face shield, which leaves the eye region uncovered but still extracts the air from around the nose and mouth. We believe that a design of this type is an excellent compromise and could well be the facemask of the future.

USE OF APPAREL

Changing Room Layout

Since the employees cannot don their cleanroom apparel in the cleanroom without introducing significant contamination in the process, there has to be a clean changing room; a buffer region between the access corridor and the cleanroom proper. Unfortunately, in many cases little time is spent in the design of the changing room (many are an afterthought). However, if a company is going to spend over $2000 per square foot on the cleanroom, it is absolutely essential that a comparable investment is made in the design and construction of the changing room, as a poorly designed changing room could negate the whole contamination control concept. Some suggestions and topics to think about are listed below.

It is imperative that employees entering the gowning area in their street clothes do not come in contact with personnel leaving the clean room in bunny suits. The logic here is simple: street clothing will shed millions of particles per second. If these particles get onto a bunny suit that is going to be worn back in the cleanroom, contamination will be spread everywhere. In Figure 11.13 we show a proposed design for a gowning room where employee separation is enforced and space is provided for a gowning room supervisor (the "bunny mother" or BM). The BM is a significant factor in ensuring that proper gowning procedures are followed. He or she can take care of cleaning the facility, ordering, stocking, and evaluating garments, dealing with the laundry company: all the things which tend to be forgotten.

Garment Storage and Monitoring

The system shown in Figure 11.13 permits the storage and retrieval of bunny suits and booties from the exit to the entrance areas. Note that booties are stored below the bunny suits in separate compartments. That is the only way to ensure that dust from booties will not get onto the bunny suits.

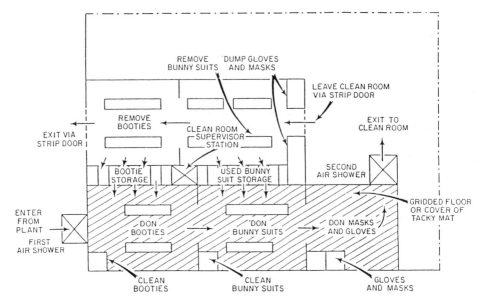

Figure 11-13   Schematic drawing gowning room layout to ensure separation between incoming and outgoing employees.

Suits may also be stored on hanging racks which are fed with HEPA-filtered air. This will ensure that the suits are continually bathed in clean air, but care should be taken to keep the suits separate to reduce cross contamination and allow airflow right to the bottom. This may be achieved by using panels or by spacing the suits several inches apart. The drawback with this approach is that it takes up a great deal of room.

As mentioned previously, the suits should be monitored regularly by a trained garment supervisor. This monitoring starts with an incoming quality control check of newly laundered suits by taking air samples from inside the sealed bags (a large hollow needle attached to a particle counter hose is useful here). Monitoring continues throughout use to determine wear and to separate badly contaminated garments from the rest. Vigilance is the order of the day here.

Changing Procedures

It should always be kept in mind that the best clean apparel in the world is totally useless if it is put on and worn improperly. Changing procedures are extremely important and should always be taken seriously by all cleanroom employees and management. The actual procedures will depend on the type of apparel worn, but there are a number of common operations.

Before the employee enters the changing room, it is a good idea to have some kind of precleaning operation to attempt to reduce the amount of outside contamination brought into the room. As far as the person inside the clothes is concerned, strict personal hygiene is mandatory for effective contamination control. This also extends to the application of unnecessary contaminating substances such as make-up. As mentioned previously, the best cleanrooms have their workers don company-supplied undergarments before they put on their outer clean apparel. This may be somewhat extreme for many cleanrooms, but it is a good idea in any case to discourage employees from taking unnecessary garments (for example, overcoats) and personal items (for example, purses) into the changing room. Some form of cleaning of the employees' clothing may also be used (shoe cleaning and airshowers are discussed in the next section).

The next problem is to don the cleanroom apparel without contaminating it. This is not easy. Some companies insist on their employees' wearing gloves while they handle the apparel. While this will prevent the transfer of oil and skin scales from the hands, it does nothing to reduce the contamination transferred from the employees' clothing onto the suit. The outside of the cleansuit will inevitably rub against the street clothes unless the suit is turned inside-out before handling (that is, after laundering). Unfortunately, an inside-out suit is not very easy to don, and it would have to be reinverted when the wearer takes it off. Therefore, the suits tend to be donned with as much care as possible and then cleaned as well as possible. This aspect is also discussed in the next section. In any case, it is good practice for the employee to don some form of head covering before putting on the suit to reduce contamination transfer from the hair. This is usually merely the bouffant cap or hood, but a hairnet may be used first. For bunny suits, some companies prefer the donning of the foot covering next to prevent contamination from the employee's shoes coating the inside of the suit. There is a counter argument which says that it does not matter what goes inside a bunny suit, and that it is more important to don the booties after the suit to prevent the outside of the booties becoming contaminated during their journey inside the legs. The answer to this dilemma depends on the efficiency of the shoe-cleaning process and the porosity of the garments. Alternatively, if overshoes or company-supplied shoes are used under the cleanroom booties, the booties should definitely be put on after the suit.

Changing rooms should always have a dirty side and a clean side, clearly delineated by some physical barrier such as a bench. No part of the suit, including the booties, should touch the floor on the dirty side, and no part of the employee's clothes should make contact with the clean side. The employee should be careful not to drag the suit along the floor on the dirty side during dressing. One item that might be useful is a gadget that allows employees to don the bunny suit without picking up contamination from the floor. The system is shown in Figure 11.14. The raised grid structure acts as a trap for particles, and the garments are thus less likely to become contam-

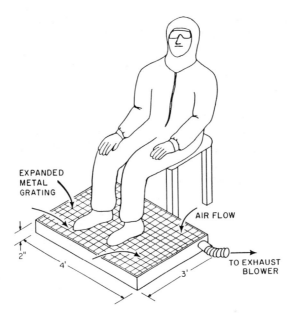

EXPANDED
METAL
GRATING

AIR FLOW

2"

4'

3'

TO EXHAUST
BLOWER

Figure 11-14   Expanded metal grating system to reduce contamination of cleanroom garments.

inated during dressing. If the ventilation blower is not available, it helps just to line the bottom of the box with a tacky mat (discussed in the next section) to collect any dust that falls through the grid. Trying a small unit of this type for a few days will show just how much dust falls through the grid to be caught by the tacky mat.

Shoe Cleaning and Air Showers

The generous use of sticky flooring material will help to reduce shoe contamination considerably. This material generally takes the form of thin plastic sheets with a strong adhesive compound on one side. Alternatively, a soft synthetic rubber-like material may be used. In the former scheme, the particles on the shoes are removed and held by the adhesive. When the sheet becomes saturated with particles, that is, when there is very little exposed adhesive left, the mat no longer works. These tacky mats are therefore sold as multi-layered mats so that when one layer becomes saturated it may be peeled off and discarded, leaving a fresh sheet exposed. The alternative mats are one layer and do not use an adhesive coating as such. The material used is inherently sticky and may be cleaned with water while retaining this stickiness. These tacky mats may be purchased in a variety of sizes and placed at strategic points so that employees have to walk on them. One question of some interest concerns the number of times that a person must step on "fresh" tacky mat in order to properly clean his or her shoes. Some data on this point are shown in Figure 11.15. Clearly more than one or two steps are required, and this was with a fresh mat. In many cases, the mats are not large enough or clean enough. Little wonder that we find so much contamination in the cleanroom.

A system which can help to extend the use life of tacky mats is the shoe cleaner. The shoe cleaner is a mechanical device, usually right at the entrance to the gowning room, with the purpose of cleaning the dirt off street shoes before people go any further. It consists of a number of rotating brushes and a vacuum system to carry away the brushed-off dirt. Unfortunately, very few employees have had training on how to use the unit, so the value in terms of contamination control can be very low in many cases. We have seen shoe cleaners that were not hooked up to a vacuum system or even a collection bag—they just ground up the dirt and put it into the air where it could drift into the gowning room. As mentioned previously, advanced changing procedures, such as those used in many facilities in Japan, involve leaving street shoes at the entrance to the plant and wearing "dedicated shoes" inside the plant. Upon going into the gowning room, employees shed these shoes and

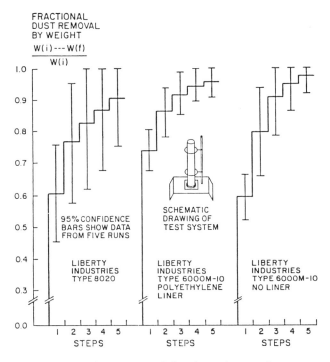

Figure 11-15 Removal of Arizona road dust from shoes with various types of adhesive matting.

put on cleanroom shoes. If you say, "I think it costs too much," we say "Look at the yield improvement possible and think again." If a shoe cleaner is used, it might be worthwhile to have a slide tape showing how to use the unit. Putting some fluorescent powder on the shoes before the cleaning process will provide graphic evidence of the results.

A contamination-reduction device which can be something of a source of controversy is the air shower. Air showers are, as the name implies, a cabinet in which the employee stands before entering the cleanroom (or even the changing room in some cases) and is showered from all sides with clean air. The air is extracted from the cabinet through vents near the floor. The idea is that the air will remove particles from the outside of the cleanroom (or street) clothing. Some studies have shown that the benefits of air showers can be more psychological than physical. Experiments have shown that the conventional air shower is at best 70 percent effective and then only on large (over 5 micron) particles. One would do well to remember that particles may be held on the surfaces of materials very firmly by static forces and will not be removed easily by air passing over them. If they are to be removed in an air shower, the employee must stimulate removal by agitating the fabric and ensure that air passes over most of the suit surface by moving body, arms, and legs appropriately. Air showers that permit blowing inside the garment are more effective (Figure 11.16), as they can eject the surface particles by the air jets created by the pores in the fabric. Unfortunately, these showers are not widely available and they require specially designed suits with internal air connections. In most cases the air showers function as visible barriers and reminders that the employee is entering a special area. This reminder is an important factor in employee motivation; cleanroom personnel should feel like special people who are permitted into an area forbidden to the unauthorized.

We may also be able to reduce the contamination from employee street clothing by using an air-supplied brush similar to that shown in Figure 11.17. If the employee uses the brush on his or her street clothing while standing in the air shower, the dust level will be greatly reduced, as shown in Figure 11.18. Problems with particles generated by human hair can be reduced by using the ventilated bonnet shown in Figure 11.19. If the employee wears this unit in the air shower, it will remove much of the contamination that might otherwise be released into the cleanroom (Figure 11.20).

WORK PRACTICES
Personnel Motivation and Training

Cleanrooms are unpleasant places to work, the required garments are uncomfortable, and in many companies cleanroom employee turnover is very high. If we agree that this is a genuine problem, common sense would suggest

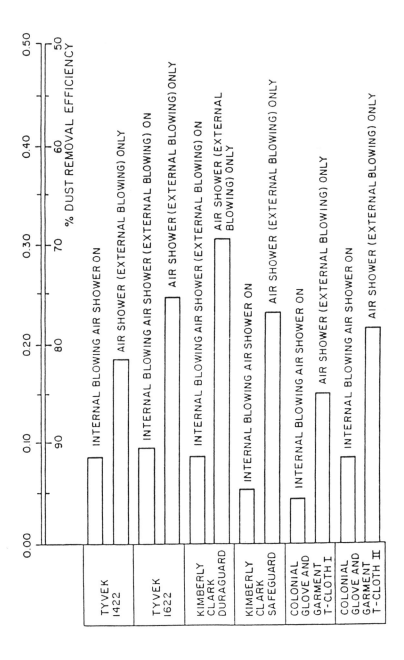

Figure 11-16 Experimental results of dust (AC Fine) removal from cleanroom garments with and without internal air blowing at 2.5 cu m/min. Air shower was "on" at all times. All testing times in air shower 60 sec.

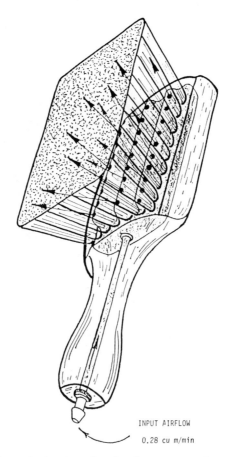

INPUT AIRFLOW
0.28 cu m/min

Figure 11-17 Schematic drawing of air brush to remove dust from street clothing.

the employees should receive extra pay and recognition. The costs here would not be very high; some companies have issued special T-shirts or have reserved tables in the cafeteria for cleanroom employees. The military have had long experience with elite forces that are expected to do more than the average GI. They get this extra performance by issuing special uniforms and generally making the personnel feel like an elite group. A system of this type might go a long way toward improving employee performance and reducing turnover in the cleanroom.

Cleanroom employee motivation and training is a critical area where there is little but lip service and an occasional slide show that often says, in effect, "Do it this way because we tell you to." In many cases, employees are not encouraged to report problems or try to improve work procedures. Engineering and management personnel are allowed to enter the area improperly gowned and to interfere with employee work schedules. In many cases clean-

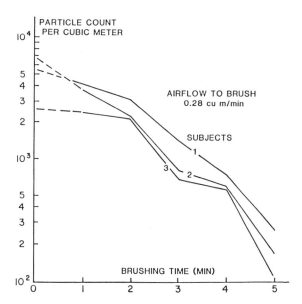

Figure 11-18 Experimental results of use of a hand-held air brush for removal of dust, lint, and so on from street clothing (airflow to brush 0.28 cu m/min).

rooms are just plain dirty. There are fingermarks on the walls, spilled photoresist on the floor, acid splashes on the equipment, and people wearing bunny suits that look like garage rags. If these problems are allowed to persist, there is no point in even having a training and motivation program; no matter what is done it won't work. Employee training and motivation procedures must be part of the design of the cleanroom. Questions about who will be responsible for proper operation of the gowning room, cleaning the cleanroom, and handling spilled chemicals are just as important as the choice of HEPA filters. What use is a chain with ten good links and one weak one? It is the weakest link that determines the strength of the chain.

A number of slide tapes and TV programs on the market are designed to train cleanroom employees. Some companies even send people to facilities that train the employees. Most of these systems or programs provide the proper information, but the attitude is one of "do this because we tell you to." Even the United States Marine Corps has learned that it is better to tell the troops why they are supposed to do something. An effective procedure involves first presenting some of the more obvious fundamentals via a taped program. If this tape had some graphic photos of fluorescent particles leaking out of cleanroom garments at the neck and sleeve, it would be particularly effective. The employees should have a chance to digest this material—it would not hurt to show it more than once. The next step would involve meeting with the employees on a regular basis, discussing problems, and

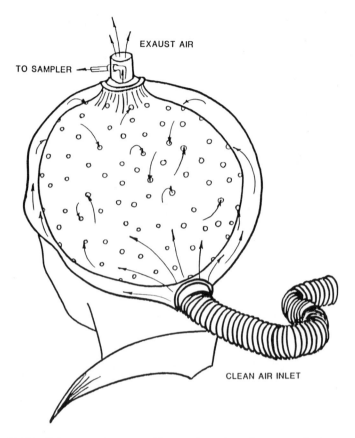

Figure 11-19   Schematic diagram of bonnet hair dryer for removal of particulates.

then reviewing possible solutions. Presenting the workers with hard data on the effects of cosmetics or smoking is far more effective than simply saying "no cosmetics or smoking." Some women employees are sensitive about going out on break or to lunch without make-up. All it takes is another employee saying, "You are so pale; have you been sick?" and the no-cosmetics program is in trouble. Cleanroom employees need to feel like special people, and this may require separate break areas and reserved tables in the cafeteria.

Another advantage gained from talking to employees is the good suggestions one gets. Employees know the processes better than any outside consultant and even many of the engineers. Cleanroom supervisors must be convinced by management that these discussions will not reduce their authority or responsibility. This sort of two-way communication can be called "quality circles" or employee participation—the point is that it works.

One personnel/training factor that seems to help is employee-activated slide

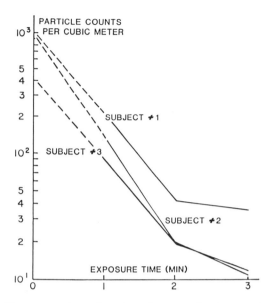

Figure 11-20 Experimental results of application of a bonnet hair dryer to the removal of particulates from hair before entrance into the cleanroom.

tapes in various locations. In many cases an employee will be unsure of a procedure and somewhat shy about asking, "How do I do this?" The presence of a nearby slide tape unit will often make the difference between something being done right or done wrong or not done at all. In many parts of the United States there will be employees whose basic language is not English. If slide tapes in other languages are available, the employees will use them. In one experiment the same procedure was supposed to be run on three shifts. On the first shift (where there was good supervision), the procedure was followed fairly well. On the second shift, there were a few minor deviations. On the third shift, the procedure was unrecognizable—small wonder that the yield was off. Investigation showed that on the third shift there was very little supervision and employees were unsure of what to do. A simple slide tape would have paid off in this instance.

The system used by one company to encourage employee interest in better cleanroom operation involved placing a plastic tube at every workstation. These plastic tubes led to a scanning valve and a dust counter. Each week every employee received a printout of the dust count hour by hour at his or her workstation. The company did not use the data to reward or punish the employees, but the effects were very apparent. Employees were proud of low dust counts and posted their dust count data for passersby to see. If there was a sudden increase in dust at a particular time, they would write in "wafer broke" or "maintenance man in area" on the chart. Clearly they did not want to get blamed for the high dust count. Another result of this program was interemployee pressure to keep clean. If an employee was thought to be a bit dirty, he or she was told to "stay out of my work area, you will raise my dust count." In other cases employees came to management with excellent suggestions for reducing dust. The whole process was a fantastic gain for the facility at very little cost. The tubing and scanning valve/counter system is sold commercially by at least one company in the United States. Another valuable idea involves having a sign at every workstation to tell employees the "value added" (how much worth the product gains after each step) at that point. If employees know that they are handling a $500 wafer, the rate of breakage and damage will certainly decrease.

The point of the above discussion is that training new employees is costly and the mistakes they make while learning are even more expensive. It would seem worthwhile to spend money on more effective training plus special T-shirts, reserved tables in the cafeteria, extra work breaks, and even a special cleanroom pay rate, if it results in motivated employees that produce a higher yield.

The best estimates are that the next generation of semiconductor fabs will cost about half a billion dollars. At that price, even the top companies in the world will have to think hard about training, particularly when we recognize that a few "monster" fabs can handle all the commodity chip production the world needs. If we are to meet the challenge of advanced semiconductor manufacture, we must look at contamination control as a system rather than a collection of miscellaneous, unrelated items. Equipment manufacturers, facilities personnel, cleanroom designers, contractors, suppliers, garment laundries, employees, and fab operators must recognize that they have a common goal. A person who says, "That is not my problem," is failing to recognize that the loss of an industry or a manufacturing company is everyone's problem. John Donne said "no man is an island entire unto himself . . .". In this industry we live in a true state of symbiosis where the failure of just one component (for example, the cleanroom laundry) can render all the efforts to reduce contamination worthless.

In this connection we might emphasize that all the training and motivation in the world will be worthless if management is not behind the program 100 percent. Management support means paying for good janitor service, keeping the cleanroom in order, changing the tacky mats regularly, providing all employees with time to clean up their workstations, and *most of all, not letting anyone go into the cleanroom unless properly gowned.* Just one instance of a manager or an engineer being allowed in the cleanroom improperly gowned will negate months of training.

One solution to the improper access problem is to have several wall-mounted TV cameras so that visitors can tour the cleanroom without going inside. At other times, cameras and associated telephone equipment can be used by engineers and technicians to talk over problems. Communication equipment is essential because if it is difficult to get into the cleanroom, engineering/ supervisory people will not go there. The line workers will begin to feel that no one cares what they do, and important procedures will simply not be followed. Cleanroom interactive television systems were pioneered by the Japanese. Some high-definition TV cameras are connected to microscopes to permit close-up viewing of the wafers. It is a technology well worth considering.

One question that frequently arises is, "Who implements these ideas; who does all of this work?" Simply asking a busy engineer or senior technician to "do this in your spare time" will result in its not being done or done very poorly. One effective procedure is to appoint a Contamination Control Coordinator (CCC) and have him or her take care of contamination control and other things. This position will require experience and diplomacy to allow the CCC to work effectively with a large number of people in and out of the fab area. If there is a question as to what garments, gloves, cleaning solutions, masks, and laundry procedures should be used, the CCC is the person to investigate the problem, read the literature, present management with the cost factors, and help them make a decision. If there is a problem in evaluation of incoming materials—contamination of the DI water, clean-up of chemical spills in the cleanroom—the CCC would coordinate the various personnel involved and make sure the problem is solved. It is critical to have one person whose primary concern is contamination control. Individuals with other areas of concern will lose interest when their own particular problem has been solved.

Reading the literature and keeping up with what is going on in contamination control is no small task. Rapid developments in the field require that someone in the company keep up with what is happening in terms of clean chemicals, garment laundering, particle detection, and robotics. The price of not keeping up is being left behind. No company can afford to let that happen.

The personnel training and motivation program might well be run and coordinated by the CCC because he or she will be concerned with other aspects of the problem. Here again we might note that contamination control is like a chain, no stronger than its weakest link. In a sense, the CCC is the guardian of the chain.

General Microcontamination Practices

Listed below are some points which may be introduced at an employee training program and implemented in contamination control. The list is by no means comprehensive, but does offer a basis for the readers' own lists.

1. Only properly gowned, masked, and gloved personnel will be allowed in the cleanroom. (This applies even to engineers who are only "running in for a moment.") This rule should be adhered to even when production

has been halted, as contamination introduced during a shutdown, for instance, can lie dormant and become reintroduced into the work environment at a later time.

2. Suits, masks, and so forth must be properly closed at all times. All snaps and ties should be fastened. (If they are on the suit, they are there for good reason and should not be ignored.) Visible hair (including beards) is evidence that the garments are not being worn properly. A mirror should be supplied in the changing room so that personnel can see if they have gowned properly or not.

3. Garments should never be allowed to drag on the floor during changing. It is especially important that the cuffs are kept clean. No outdoor shoes should ever be worn on the clean side of the changing room. Similarly, once the booties have been put on, walking on the dirty side is forbidden.

4. Aerosol spray cans, feather dusters, paper or cotton dust mops, non-cleanroom paper which has not been laminated (including notebooks, cardboard boxes, and the like), candy, gum, and so forth are not allowed at any time.

5. Wet vacuum cleaners may be used to pick up dust and nonhazardous liquid spills. Only vacuum cleaners with HEPA-filtered discharge may be used, otherwise a centralized house vacuum system is appropriate. Buffers and floor wax will not be used under any circumstances.

6. Open-toed shoes or shorts may not be worn; safety glasses must be used at all times. Chemical operations should have the appropriate safety equipment listed and supplied. Failure to wear all the necessary safety gear in the proper manner should be a firing offense.

7. Garments that have become contaminated by chemicals or dust should be changed immediately. In any case, no garment showing dust or stains or even suspected of being contaminated may be carried into the cleanroom.

8. Gloves with long cuffs will be worn at all times. The end of the glove must lap over the cuff of the bunny suit at least two inches to preclude particle leakage.

Cleaning the Cleanroom and Cleanroom Wipers

Housekeeping is a topic that seems to have received little attention in the West and yet some thought will indicate that is critical to successful cleanroom operations. In the most successful companies on the Pacific rim, cleanroom employees wipe down their own areas every day. Any chemical spills are mopped up before they can dry and form particles from residue or from reaction with the materials in the spill area. In some companies the floors are mopped with DI water every two hours to remove particles before they can get into the air. There does not seem to be anything like a standard cleaning procedure, but the system used in at least one large semiconductor company is worth repeating here.

Daily Procedures

1. Peel off and change sticky mats two to three times per shift. If a more permanent material (for example, Dycem℠) is used, it should be mopped with clean DI water at least three times per shift. All custodians must attend the cleanroom certification and safety program before entering the facility.
2. Empty solvent and acid trash. (Remember, this should be treated as hazardous waste.)
3. Clean floors with DI water or other approved sodium-free cleaner at least three times per shift. Note that only an approved cleanroom sponge may be used; string mops are not permitted. Water must be changed after 500 sq. ft. of area has been cleaned. Black marks or spilled photoresist may be removed with alcohol or other solvent.
4. Employees are responsible for their own areas: all equipment, pipes, sinks, lab benches, and so forth, are to be wiped down at least once per shift. Chemical splashes on equipment or the floor must be removed with DI water and an approved sponge. Any dried residues in an area are evidence that proper cleaning has not been done.

Periodic Procedures (Six Months)

1. Wet mop floors with DI water, machine scrub, and rinse.
2. Vacuum all overhead structures.
3. Wash and wipe walls and ceilings with DI water and solvent, for example, 10% IPA in water (if needed to remove contamination).
4. Wash all laminar flow curtains/partitions with DI water and solvent (if needed).
5. Mop floors with DI water again.

Worker Stress and Related Problems

There is great potential for worker stress in the cleanroom environment. An employee is asked to clothe himself from head to toe with only eyes visible to the world. Except for relative size, all people look alike. Individuals tend to lose their identity. This can be a traumatic experience, particularly for the new employee. Some people are claustrophobic and don't even know it. They suffer from vague feelings of uneasiness in the cleanroom because they have covered their nose and mouth with a mask. Symptoms usually manifest themselves as irritability, nausea, or shortness of breath. In severe cases, the employee may even hyperventilate and lose consciousness. The best course is to identify these people and counsel them or, in the extreme case, remove them from the area.

Other stressors come from the fears instilled in people by training sessions. A great deal of communication occurs with the new employee in order to assure that he or she understands that, for example, "This is a poisonous gas," or, "That material causes severe burns, cancer, birth defects, liver damage," or some other equally fearful effect. An employee entering the cleanroom after being told that the place is filled with potentially deadly things will feel stress. Similarly, an emergency involving hazardous material causes stress in the organization.

The key to controlling these stressors is communication. In the case of the vague fears caused by the knowledge that poisonous gases are used, fears are tempered by pointing out that safety systems control exposures. Fears over exposure potential to corrosive chemicals or organic solvents are allayed by discussions of local exhaust system controls, personal protective clothing, and procedural controls. When people understand that the risks are recognized and that controls are in place to help them deal with the threats, they are more willing to accept the remaining risk, provided they can be convinced that they have some control over their own bodies. Communication is also the key to stresses caused by upset situations. Chemicals spilled onto the workroom floor or gases or vapors released into the workroom atmosphere cause immediate reaction in people. They are rightly or wrongly fearful that they are going to be exposed to some element which will result in death or long-term physical problems. If an alarm goes off warning that a hydride gas is leaking, and I see no activity to correct the situation, I am concerned that I am exposed and could die. If, on the other hand, I am informed as soon as possible that the reason for the alarm was that the technicians were calibrating the system, or testing the system, or whatever the situation is, my fears are immediately calmed. A situation which requires evacuation of the area also requires communication of the facts to the people who are outside the area. They will assume that the situation is worse than it is if they are not told the truth. False stories have a way of moving through a crowd of people faster than the truth, particularly if the false story is sensational. The only way to fight this phenomenon is with the truth, and just as soon as it is known. It is frequently prudent to issue progress reports on the status of any emergency even if it is not resolved.

SUMMARY

In summary, since personnel are a major source of contamination within controlled environments, they deserve special attention both with respect to how we gown them and how they are trained to act within the environment. Various thread materials, cloths, and suit styles are available to control human contamination, but the best garments in terms of particle trapping and comfort tend to be the most expensive. Gloves and masks are extremely important because the hands and mouth are contamination sources which cause great problems by bringing contaminants right to the product. Proper laundering of nondisposable garments is critical, as is how the apparel is donned and stored. It is also important for employees to understand what the problems of contamination control are and how proper actions by the individual can make a real difference to the environment as a whole.

12

Safety Issues

When considering so-called high-technology manufacturing such as that carried out by the semiconductor industry, we would do well to remember that what goes on in our controlled environments is actually much more chemistry and physics than electrical engineering. Much of what we are trying to do lies in the area of contamination control: control of microcontamination in the environment, control of deliberate contamination (for example, dopants in semiconductors), and control of the contamination of the work force. Environmental contamination is the realm of the facilities engineers; dopant contaminants are handled by the process engineers; and human contamination is the business of the safety and health team. There is a very strong link between these areas, although many people in the industry do not realize it. For example, product health and work-force health are frequently affected by the same things. Silica dust in the air is a prime example of yield-reducing particulate contamination that is also very hazardous to humans, leading to severe respiratory ailments.

Many things we do for safety purposes may help in the area of yield enhancement and vice-versa. However, it goes without saying that safety is extremely important in its own right. Whenever life and limb are involved there can be no compromise—or can there? Unfortunately, most manufacturing companies are not in the business of safety, and although safety first has been an oft-quoted motto of many organizations for over fifty years, the priority may not be quite correct. For economic reasons, we frequently find ourselves using materials and processes that are inherently dangerous. If safety truly was "first," then we would not use them.

Materials in the workplace which have a degree of hazard associated with them are generally called *HPMs* (hazardous production materials). Since we have to use them, assuming no substitutions are currently available, we must do a good job of recognizing the hazards and protecting the work force (and product) against them. To do this, we must expand our knowledge of our materials and processes.

PROPERTIES OF HAZARDOUS PRODUCTION MATERIALS

Hazard Properties

When discussing the hazard properties of materials, we may use four categories to describe the hazards. These are

- Flammability
- Reactivity
- Corrosivity
- Toxicity

Flammable essentially means "will burn or support combustion." Legally defined, flammable materials are those which have a flash point below 100° F, whereas combustible materials have a flash point in excess of this. *Flash point* (expressed as a temperature) is the point at which the material in question will give off sufficient vapor at its surface to support combustion (for example, acetone—f.p. = 0° F). It is important to remember that burning is an exothermic chemical reaction; the reactants, fuel and oxidizer (for example, gasoline—f.p. −54° F and air), have to be in the correct form for the reaction to take place. Gasoline liquid will not burn whereas the vapor will. Also note that the third component which is generally required for burning is a source of ignition, unless the fuel material is *pyrophoric* (flames spontaneously in air).

The reactants also have to be in the correct proportions for burning to take place. If the fuel–oxidizer mixture is too *lean* (not enough fuel) or too *rich* (too much fuel), then it cannot burn. The lower explosive limit (LEL) or lower flammable limit (LFL), expressed as a percentage in air, represents the lowest concentration by volume of fuel which can support combustion (the term *explosive* is used because exothermic reactions of this type give off so much heat and hot gas that an explosion is actually taking place). The upper explosive limit (UEL) and upper flammable limit (UFL) represent the maximum volume in air of fuel gas/vapor which can support burning. For example, the LEL and UEL of hydrogen are 4 percent and 75 percent respectively, whereas gasoline has a much narrower range of 3 percent to 8 percent. Pyrophoric materials such as silane (SiH_4) are a little more unpredictable, as their auto-ignition characteristics can depend on discharge rates from orifices.

Reactivity describes how a material will react in the presence of other materials. A prime example of how important knowledge of reactivity can be is in the case of acids and solvents. Both types of materials are used in vast quantities in the high-technology industries. If, for example, nitric acid (an oxidizer) and iso-propyl alcohol (a fuel) are accidently mixed, the flash point of the mixture will drop below the auto-ignition temperature and there will be spontaneous combustion. These fires can be very difficult to extinguish as they have their own oxygen source. The National Fire Protection Association (NFPA) has collected a great deal of information on reactivity from accident scenes under a wide range of conditions.

Corrosivity refers to how quickly the material will etch another substance under particular conditions (temperature, concentration, pressure, and so forth). One other substance of great interest is human tissue, and corrosive material information will often include data on how aggressive the chemical is in this respect and how to protect against it. Also, corrosivity data will enable chemical users to design appropriate storage and use vessels. For example, whereas we can store nitric acid in a glass bottle, hydrofluoric acid has to be stored in plastic as it is extremely corrosive to glass.

The final category, toxicity, really relates to human (and sometimes animal and plant) responses to materials. This is the realm of toxicology and industrial hygiene.

Basic Toxicology

Toxicity is the ability of a substance to do harm once it reaches a susceptible site within the body. Toxicology is the science (or non-science depending upon your viewpoint) of the effects of toxic substances on living things. Industrial toxicology and industrial hygiene relate specifically to the human body. Toxicology is a very imprecise science, as no two creatures are exactly the same.

Discussion of toxicity inevitably leads to a somewhat awkward question: how much is too much? If we accept that there will always be work-force exposure, how do we assess what the upper limit should be so that our workers do not come to any short-term or long-term harm? This is also something of a philosophical question. In toxicology and the related field of industrial hygiene, the concept of *dose* is critical. Relatively innocuous substances (for example, water) can become deadly if the dose is high enough. So what we have to do is assess what dose is harmful and then scale down from there. This is not as easy as it sounds because we cannot deliberately subject humans in this day and age to possibly harmful quantities of chemicals in order to answer the question. Also, when discussing dose, we must include factors such as the duration of exposure, the dispersion of the substance (what form it arrives in), how it enters the body (or much more significantly how it enters the bloodstream), as well as the obvious factor of concentration of the substance. In the medical sense, absorption specifically means entry into the bloodstream, not merely into the body. This is an important distinction, as many substances can be taken into the body via the lungs or gastrointestinal tract but need not be taken into the bloodstream, that is, they may be exhaled or excreted before they have an opportunity to do harm.

The three routes of absorption are through the skin, the gastrointestinal

(GI) tract, and the lungs. The skin is quite permeable to gases and liquids; for instance, solvents find it rather easy to pass through the skin into the bloodstream. Gases can become dissolved in a surface layer of perspiration and become absorbed. Absorption by way of the GI tract is variable, as it depends very much on the solubility of the substance in question: if the material does not dissolve, it will pass through without absorption. The most significant route in an industrial sense is via the lungs as this is the most difficult area to protect. Once a material becomes dissolved, or held in the mucous that lines the lungs, it tends to be very difficult to remove.

When discussing toxicity, there are a number of other terms we should be aware of. The first of these are *acute* and *chronic*. These terms are frequently misused in everyday conversation to merely mean severe. Acute in the tox-icological sense means an exposure of short duration whereas chronic refers to long-term exposure. For example, a large inhaled quantity of extremely toxic phosphine (PH_3) gas (TLV-TWA = 0.3 ppm) will produce an acute effect (pulmonary edema) whereas arsenic ingested over a long period will create a chronic effect (cancer).

Two terms which may be used with those above are *local*, referring to an effect which occurs at the point of contact of the substance (that is, point of contact and site of action are the same), and *systemic*, which refers to an effect which occurs at a site other than the point of contact. For example, strong nitric acid will have a local (and acute) effect on the skin whereas solvents such as methanol may be absorbed through the skin to cause liver damage (actually, most solvents will also defat the skin and thereby also produce a local effect).

To these four terms we may also add a pseudo-quantitative component. These ratings are:

U—unknown. Either nothing is known about the substance or the infor-mation is questionable.

0—not toxic. This rating is given to materials which would cause harm only under unusual circumstances. For instance, water is toxic if you were to drink vast quantities in a short time (this qualifies as an unusual circum-stance).

1—slight toxicity. These materials would produce only a slight and highly reversible effect (for example, a temporary reddening of the skin or a scratchy throat).

2—moderate toxicity. Some of the effects may be irreversible but certainly not life-threatening or causing physical impairment.

3—severe toxicity. These materials are truly dangerous and can kill or maim.

This rating system is widely used and can be seen on chemical labels and in information sheets. The NFPA (National Fire Protection Association) also has a rating system for chemical storage and use-area labeling. The information is posted in the form of a color-coded, diamond-shaped notice outside the area. NFPA ratings go from 0 to 4 (3 essentially means severe harm and 4 means a high risk of death). In the uppermost 3 squares of the divided diamond, the numbers refer to flammability, reactivity, and toxicity. The fourth square contains other information (for example, radioactivity or "don't use water"—a "W" with a line through it). Note that the missing hazard property is corrosivity. In a disaster situation, firefighters are not particularly concerned with contact hazards as they are attired to prevent chemical contact.

TLVs and PELs are used to establish safe working limits. Other published parameters are of use in disaster planning and management. One such parameter is IDLH or the concentration of a material which is immediately dangerous to life and health. There are other (more curious) definitions such as the MLD or "minimal lethal dose," the amount per unit of body weight which has the ability to cause even one fatality in a group. Related measures are the LD50, which is the dose capable of killing 50 percent of a population, and the LC50, which is the airborne concentration capable of producing this degree of lethality. The latter two definitions lead to an interesting question: if we have a threshold concentration or dose which is fatal to one individual, why is it not fatal to all in a group? This is because of a factor known as individual susceptibility.

No two individuals are exactly alike. Even identical twins, although physiologically alike, will generally have different medical histories, habits, and so on which will set them apart from each other. A great many factors determine the response of an individual to a toxic substance. It is possible for individuals in a population to exhibit a wide range of symptoms after exposure, from no discernible response to death.

The factors which are believed to determine susceptibility are:

Anatomical structure—for example, size of lungs, structure of nose,

Physiological factors—for example, efficiency of lungs and other organs such as the kidneys,

Previous illness—certain diseases can weaken particular organs,

Obesity—being overweight puts a strain on the body,

Age—the very young and the very old tend to be more susceptible,

Sex—the anatomical differences between the sexes can make the response to substances radically different,

Previous exposures—previous encounters with substances usually result in a weakening of target organs, but in some cases a degree of immunity can be built up,

Work rate—people who work harder in the physical sense will breathe harder and faster, their pulse rate will be high, they may also perspire (allowing substances to become dissolved in the surface moisture), all of which aid in the absorption of airborne substances,

Diet—poor nutrition and regular ingestion of substances such as ethanol will alter susceptibility.

Effects of Common HPMs

A huge range of materials used within the high-technology manufacturing industries may be considered hazardous. We will briefly discuss the effects on humans of the most common of these materials in this section.

First we must define more terms which relate to the injury-causing nature of these substances:

- Carcinogen. This is a material which has the potential to cause cancer in a particular organ or area of the body; in hazardous material listings there are *suspected* and *confirmed* carcinogens.
- Mutagen. This can cause genetic mutation within cells, that is, it can interfere with cell reproduction by distorting the genetic code.
- Teratogen. This refers to a substance which can cause severe birth defects, producing offspring which are deformed in some manner.
- Fetotoxin. This material is specifically toxic to a growing fetus.
- Male reproductive system (MRS) toxin. This material tends to target the male reproductive organs.

The first classification we shall consider is the organic liquids, many of which are solvents used to clean and degrease materials and components. These liquids will be absorbed through the skin and usually lead to dermatitis by their defatting action. Absorption may also be by inhalation, where the mucous membranes can become irritated, especially in the case of the higher molecular weight substances. The lower molecular weight substances generally tend to be narcotic, inducing dizziness, blurred vision, and headaches.

Glycol ethers. This group includes *Cellosolve,* a material used in mixtures for photolithography. They have low acute toxicity but are irritants to the skin, eyes, and mucous membranes. The more severe effects include a depression of the central nervous system (CNS) in humans. (Lab animal tests have revealed other effects such as kidney and liver damage, lethargy, tremors, and anorexia.)

Esters. This group includes Cellosolve Acetate, also used in photolithography. Esters are thought to have low toxicity but may produce narcosis.

Ketones. This group includes acetone, methyl ethyl ketone (MEK), and methyl isobutyl ketone (MIBK), also used in lithography. These compounds produce narcotic effects, and exposure to large quantities of acetone can produce unconsciousness.

Aromatics. This group includes benzene, xylene, toluene, and so on, also commonly used for photolithography and cleaning. They are all skin irritants. Exposure to toluene will cause autonomic and peripheral nervous system disorder and prolonged exposure will lead to CNS damage. It is also thought to be a fetotoxin. Benzene is a confirmed carcinogen and a fetotoxin, and has been linked to leukemia.

Halogenated hydrocarbons. This group includes trichloroethylene (TCE), trichloroethane (TCA), fluorocarbons (Freon), and so on, which are (or were) used extensively as solvents. TCE has been linked to depression of the CNS and is a carcinogen. It is possibly also a mutagen. These factors tend to preclude uncontrolled use of TCE. 1,1,1-trichloroethane appears to be considerably safer than TCE, but the 1,1,2-TCA isomer is a suspected liver carcinogen. Also, since 1,1,1-TCA is corrosive to aluminum, inhibitors may be added and these may be toxic. Although fluorocarbons are generally considered to be a safe option within the industry, chlorinated fluorocarbons such as Freon 11 ($CFCl_3$) can produce constriction in the lungs. Chlorinated fluorocarbons (CFCs) such as Freon 11 and Freon 12 (CF_2Cl_2) are also thought to be responsible for the destruction of the earth's ozone layer and hence are environmentally undesirable. There is also growing evidence that fluorinated hydrocarbons may be mutagens. In addition, a little known fact about halogenated organic solvents is that they may cause the heart to become sensitized to epinephrine. This may not seem to be much of a problem until one considers that the adrenal glands produce large quantities of this compound in a fright situation—one could thus literally die of fright!

Hexamethyl disilizane (HMDS). This is used in photolithography as an adhesion promoter and is a respiratory tract irritant, but little else is known about this substance.

The next general category we will consider is acids, bases, and oxidizers. These are used for cleaning and etching operations and contact is generally through liquid splashes on the skin or the inhalation of vapors. Damage to the skin depends directly on the concentration of the liquid. Strong acids and oxidizers will literally corrode human tissue, but damage to the skin from bases tends to be more severe (these will gelatinize the skin to large depths). Boric acid (from dissolved boric oxide) and phosphoric acid (from dissolved phosphorus pentoxide) are the least corrosive of the inorganic acids, but like organic acids (acetic acid), they will irritate the mucous membranes. It is thought that boric and acetic acids, and the oxidizers ammonia, ammonium chloride, and hydrogen peroxide, are somewhat mutagenic.

Sulfuric and nitric acids. These are used for etching and cleaning. The principal problem here is a contact hazard as both of these will corrode human flesh very rapidly unless heavily diluted.

Hydrofluoric acid. This aqueous solution of hydrogen fluoride is used to etch silicon dioxide. It is a unique acid as the fluoride ions can penetrate through the skin and cause deep ulcers and degradation of the bone. As with all acids, the immediate treatment should be instant flushing of the affected area with water followed by a prolonged soak. Magnesium oxide cream is frequently use to stop the action of the acid, and subcutaneous injections are frequently necessary. There have been reports of the effects of prolonged exposure to the vapor (mainly HF). This apparently causes headaches, weakness, back pain, loss of memory, and incontinence.

Sodium hydroxide. Aqueous solutions of this material are used in photoresist developers. The only real hazard is by contact.

Hydrogen peroxide. This is used as a cleaning agent, usually mixed with sulfuric acid for this purpose. It is a strong oxidizer and will corrode the skin at high concentrations.

The next category to be discussed is metals and metallic compounds. The main route of absorption here is inhalation of dusts produced by sawing and polishing. Many exotic metals are now being used in compound semiconductor technology in substrate manufacture and as dopants.

Lead, cadmium, and tin. These metals are found in solders and in plating applications. Lead will produce abdominal cramps, anemia, and ultimately kidney damage. It is also a suspected carcinogen, an MRS toxin, a teratogen, and a fetotoxin. Cadmium is also thought to be a teratogen and a fetotoxin. It is certainly a respiratory irritant and can cause kidney damage. Tin is not as much of a problem but can cause lesions to the skin.

Gallium, indium, and antimony. These are used in the compound semiconductors gallium arsenide and indium antimonide. The dusts of these compounds can cause inflammation of the lungs. Gallium itself can cause kidney damage and indium could be a teratogen. Antimony trioxide, a dopant compound, will severely irritate the respiratory system.

Silver, gold, and tantalum. These materials are used in conductor systems and are generally regarded as harmless.

Nickel, cobalt, and platinum. These metals are also used in conductor systems (usually as silicides). Nickel is a carcinogen, a teratogen, a fetotoxin, and a MRS toxin. It is also a cutaneous sensitizer, creating an itching form of dermatitis. Cobalt tends to accumulate in the body and causes lesions in the heart muscles. Platinum causes bronchial asthma, and vomiting and diarrhea at high exposures.

Barium, yttrium, and beryllium. These metals are used heavily in the production of CRTs and barium and yttrium are components in superconductors.

Barium and its compounds tend to affect the CNS. Yttrium compounds can cause bronchial problems. Beryllium and its compounds are very toxic, destroying the lung tissue. It is also a suspected carcinogen.

Arsenic. This is a dopant and compound semiconductor component. It is a confirmed carcinogen, a teratogen, and a fetotoxin. Arsenic will tend to accumulate in the body and its effects are almost always chronic in nature. Arsenic oxide, on the other hand, is immediately poisonous.

Silicon dioxide (silica). Silicon is not a metal, but we will discuss its oxide here. Silica dust can be released from oxide deposition systems. This can cause severe lung problems after prolonged exposure.

The final category of materials is gases. Huge volumes of toxic gases are used in high-technology industries: some are so toxic that little is known about the subacute effects. They generally affect the upper or lower respiratory tract. Upper respiratory tract (mouth, throat, and so on) irritant gases include gases such as hydrogen chloride which dissolve in the mucous to form an acidic solution (hydrochloric acid in the case of HCl gas). This can actually be a life-saving factor as personnel subjected to these substances become warned by the irritation and can escape further exposure. The lower respiratory tract (mainly lungs) irritant gases can, in severe cases, cause pulmonary edema (a swelling of the lung tissues) and fibrosis (a change in the tissue which results in a choking of the lungs). These do not tend to produce a warning irritation and their more sinister effects may not be felt for several hours.

Boron gases. Boron trichloride, diborane (B_2H_6), and boron tribromide are dopant gases. They all tend to cause pulmonary edema but only boron trichloride creates hydrochloric acid on the mucous membranes, giving a warning irritation. Chronic exposure to diborane will cause coughing/wheezing and a tightness in the chest.

Metal hydrides. These include arsine (AsH_3), stibine (SbH_3), germane (GeH_4), and phosphine (PH_3). These are used as dopant gases. Arsine and its antimony counterpart, stibine, are extremely toxic, causing severe kidney damage (abdominal pain and bloody urine are symptomatic). Unfortunately, the symptoms can take as much as twenty-four hours to manifest themselves. Low-level exposures will allow arsine to break down in the blood, releasing elemental arsenic, which is a carcinogen. Germane, like arsine, can cause destruction of the red blood cells. Phosphine also creates pulmonary edema, damage to the heart muscles, and chronic hepatitis.

Nitrogen oxides and ozone. Nitrogen dioxide is used in chemical vapor deposition processes and is capable of causing acute and chronic pulmonary disorders, including fibrosis. Ozone is used as a biocide in ultrapure water systems and is also incidentally produced by the lamps used in photolithography systems. Ozone is extremely reactive and will readily ionize other chemicals. It is thus an extreme mutagen. Long-term exposure can also cause a narrowing of the breathing passages.

Table 12.1 Hazard Properties of Production Gases

Gas	Flam	Pyro	IDLH (ppm)	ODOR (ppm)	IRRI (ppm)	TLV (ppm)
NH_3	Y/N	N	30k	5	50	25
AsH_3	Very	Y/N	250	0.5	0	0.05
B_2H_6	Very	Very	160	3	?	0.1
HCl	N	N	100	1	5	5
PH_3	Very	Very	200	2	8	0.3
SiH_4	Very	Very	?	?	?	0.5

Silicon gases. These include silane (SiH_4), dichlorosilane (SiH_2Cl_2), trichlorosilane ($SiHCl_3$), and chlorosilane ($SiCl_4$). These gases are used in chemical vapor deposition systems. They are all severely irritating to the skin and mucous membranes and pulmonary edema is not unusual. Trichlorosilane is much less toxic than silane.

Asphyxiants. Inert gases such as argon and nitrogen are used in large quantities to provide nonreactive atmospheres in many processes. These will displace oxygen from the air in the event of a large leak, and if the concentration of O_2 in the air drops by as much as a few percent, unconsciousness can occur. Hydrogen, apart from being extremely flammable, is also an asphyxiant. Cyanide gas, produced when an acid is (accidently) mixed with cyanide-containing plating solutions, is an asphyxiant in the respect that it passes through the lungs or skin and stops the transfer of oxygen at the cellular level.

In addition to their toxic qualities, many of these gases have other unfortunate attributes. These are summarized with some toxicological information in Table 12.1. In this table, FLAM = flammability, PYRO = pyrophoricity, IDLH is in ppm, ODOR is the threshold at which the material can be smelled (ppm), IRRI is the threshold at which irritation occurs (ppm), and TLV is the TWA (ppm). As may be seen, many of these substances will be at concentrations considerably higher than the TLV before they can be smelled and many also present a considerable fire hazard.

CLEANROOM SAFETY PRACTICES

The Safety Team

The concept of the cleanroom requires a closer relationship between the users of the controlled environment and the safety organization than normally would exist in an industrial environment. Safety decisions, requirements, and procedures must be made with the mission of the organization in mind so that in solving a safety problem, contamination problems are not created. It is necessary for safety purposes, for instance, to require that employees who handle liquid chemicals wear personal protective clothing. That clothing must be selected with the cleanroom, and its sensitivity to contamination, in mind. For instance, rubber gloves or synthetic gloves for chemical protection are normally packaged in talc to assist the user to put them on easily. That talc can be a significant contributor to particles in the cleanroom. It is necessary for the safety equipment vendor who supplies the gloves to provide them without talc.

In order to integrate safety practices into the cleanroom smoothly, it is best to involve the cleanroom engineering staff in the safety decision-making processes at an early stage. The engineering staff needs to be educated to the safety requirements so that they can build around and upon them to preserve the cleanliness of the process as well as provide the necessary protection for the employees. Usually a committee is formed to integrate safety and processing requirements. The team sets about the formulation of safety/process control rules and procedures necessary for the cleanroom.

Formulating Regulations

Practices in the cleanroom which are related to safety require the same basic safety department activity that would prevail in any other industry. Someone, usually the safety engineer, studies each element of each job and identifies the individual hazards attendant to it. From that list of hazards, procedures will need to be devised which are detailed in the work document. The importance of the work document in the cleanroom area or any area handling hazardous materials cannot be overemphasized. This document is called by a number of different names (specification, procedure, job ticket, work order, and so forth) but its purpose is basically the same in all cases. It is the piece of paper that tells the employee, step by step, exactly what is to be done, in what order, where, and so on. It describes each operation one step at a time. The job safety analysis conducted by the safety engineer finds its way into this document. Normally, this document is used not only to control the process from an engineering standpoint, but is also the document used to teach a new employee on the job. It is used to cross-train employees in the cleanroom in various parts of the operation so that the supervisor has the luxury of moving people around to accommodate the needs of today's production schedule. It also allows him or her to use people on other jobs within the cleanroom while equipment is down for maintenance.

A job safety analysis, thoughtfully conducted, will disclose the majority of the hazards, or opportunities for accidents, in the operation. It also documents the sources of accidents in each job. If the job safety analysis is married to the work document, it generates a production-safety training tool. It will train employees how to do the job procedurally and contain appropriate cautions and warnings, protective requirements as well as instructions in event of problems. If the job safety analysis has been thoroughly carried out, the only real unknown potential is the upset situation, that is, an accident. The safety team must deal with emergency situations. There are rarely emergency procedures written for all possible situations, but a number of situations can be expected and must be preplanned. Under what circumstances will we evacuate the area? Who will make the decision to evacuate? How will we get to wherever we are supposed to go? Who is in charge? Who will count heads after we get there? Who gets notified that something is wrong in the cleanroom and who is responsible for notifying that person? A good planning session will provide the answers to these and other questions. The result is an emergency plan.

What cannot be preplanned? Not much, in reality. Perhaps explosion of a pressure vessel or hazards from outside the cleanroom such as total power failure, lightning strike, flood, earthquake, and so forth. Going through "thought exercises" or disaster scenarios is good for the emergency team as well as the cleanroom safety team. For instance, recognizing that a single piece of equipment is vulnerable to loss of power can be a major preventive factor. If we provide emergency power to this piece of equipment, the chances of major consequence are avoided. One of the things that preplanning does is to force us to examine all possibilities. We may brace shelving a little differently in a zone where earthquake potential is high enough to warrant it. We may isolate a piece of equipment which could cause major damage with structurally reinforced members in an effort to reduce the amount of damage potential in event of failure. Safety procedures in the cleanroom require the same engineering effort as they would in any other part of the business. Study, documentation, training, and replanning are the keys to any operation where safety is a concern.

Safety Training

In the safety professional's arsenal for accident prevention, training is one of the most powerful tools. Training, done properly, is the method of assuring that correct procedures are communicated to those who are in need of the information. However, it should be understood that training courses must move with the times and not stagnate. It is the safety professional's responsibility to see that procedural errors that caused accidents last week are not still causing accidents this week.

The modern cleanroom requires a great deal of training from a safety organization. This is not to say that the safety training requirements for a cleanroom are remarkably different from those of any other technology. Work areas have supervisory people and supervisory people need safety training whether they work in a cleanroom or not. They need to learn to recognize unsafe employee acts and how to effectively correct unsafe behavior. Work areas handle hazardous materials. The cleanroom situation makes little difference in terms of training needs. Those who handle hazardous material need to be trained in the hazard properties of the materials they handle, effects of exposure, how to prevent exposure, how to do things the right way, and what to do when things go wrong.

Supervisory training is probably the first priority, since attitudes of supervisors can have great impact upon the attitudes of the people who work for them. The safety policy of most industrial organizations places line responsibility for safety into the hands of supervisory personnel. That is, management expects that supervisory people will enforce safety requirements in the work area and correct unsafe conditions or acts that are causing or are likely to cause accidents. Management further expects that this safety responsibility will carry the same weight as the responsibility to perform to budget, quality, or quantity dictates. This is a very broad charter.

The mission of safety training for supervisors is to put some flesh on these bones: Just what is an unsafe condition? How do I recognize an unsafe act? Having found one, what do I do to get it corrected? What is the reason behind the regulations I am expected to enforce? Supervisory people generally respond better to safety rules and regulations if they have a clear understanding of the need. This understanding allows the supervisor to explain to an employee why this or that is required. The supervisor is not put in the position of making what may appear to be unreasonable demands on employees by having to say it is "policy." The supervisor is given an opportunity to participate in the safety program rather than simply to be a "cop" who writes people up who don't follow the rules. A supervisor is expected to enforce the use of protective clothing at a chemical handling operation. If the supervisor is to be expected to do this successfully, he or she must be given a rational explanation of the risk and be convinced that the requirement cancels or reduces the risk. It is a fact of human nature that people do not like to wear protective clothing. Given the slightest excuse, many will abandon it in favor of some rationalized excuse for the behavior, such as "The engineer doesn't wear protective clothing." The supervisor must be given the arguments that are needed to convince an employee that although it is understood that protective clothing may not be comfortable, it must be worn. The company policy statement or threats of disciplinary action rarely work in the long term. The employee knows that the supervisor is really not enforcing safety rules from a knowledge base, he is simply saying, "They make me do it." The effective supervisor points out to the engineering staff that their bad example causes problems and that the engineering staff must set good examples for the other employees to follow.

Employee safety training in the cleanroom must concentrate on hazard recognition for everyone who works there. The safety department may provide this training directly or may work through the training department. In some organizations the supervisors are expected to conduct training in addition to all of their other duties. In a small organization, this may be the only choice. In any case, the training needs to concentrate on teaching the employee how to recognize hazards, and having recognized them, what to do to protect against or avoid them. Hazards may come in the form of heat from furnace tubes, push rods, hot plates, and similar appliances. Extreme cold may come from a variety of cold traps using dry ice, liquid nitrogen, and a number of refrigerants. Acids are corrosive, toxic, and some are reactive. Other acids are flammable, or at least combustible. Most solvents are flammable, all have toxic properties to one degree or another, and are reactive with oxidizing materials. Many of the metals used are at least toxic, depending on their form. Each chemical brings a variety of hazards to the job.

In addition to hazardous materials, we have other hazard elements. Electricity poses a serious hazard wherever it is found, as does ultraviolet radiation from alignment tools, or soft x rays in some of the newer lithographic equipment. Radiation is also a potential in ion-implanters because of the high voltages involved. Lasers are used to scribe lot tracking numbers on wafers and bring their own hazards to the workplace. Finally there are the hazards of mechanical movement of machines or machine parts and the activities of the people themselves. Employees must be taught to recognize and deal with each of these potential injury sources if the safety program is to be successful in preventing injuries and illnesses. Given the long list of potential hazards and a reasonable turnover rate in the labor force, the task of training never ends; it simply resets and starts again.

There are some legally mandated training requirements as well. If the area is devoted to research, radioisotopes may be handled. A special program is required to certify each of the handlers of radioactive material by training. Specific training is also required if the people are required to use respiratory protection on the job, or if measurements indicate exposures are taking place to some toxic materials above an acceptable or safe action level. Some organizations train people regarding exposure monitoring, whether or not the results require the training, in the belief that if employees have a better understanding of the concepts involved, they will make a positive contribution to controls. U. S. Occupational Safety and Health Administration regulations require some training for all employees who are potentially exposed to hazardous materials in their Hazard Communication standard. This training program must explain employee rights and duties under the law and those of the employer as well. The program must also explain that the employer is required to generate a list of all hazardous material in use or stored on the site and to have available a compliance document on how the employer intends to comply with The Material Safety Data Sheet (see Chapter 8) collection and access provisions, labeling requirements, lists, non-routine work, and the training required by the standard. All of these documents must be available for employee inspection at any time that employees are working and are potentially exposed to hazardous material.

The employee training program itself must teach the employee several important things about the hazardous material he or she is required to handle. It must explain how the material can cause harm, how to tell if something is wrong and exposure is taking place, what to do to avoid or reduce exposure, and, finally, what to do in an emergency. A semiconductor cleanroom contains such a variety of chemicals that most of the industry has elected to handle this element of the training by chemical class rather than one compound at a time. It could take months to cover the required data, one compound at a time. Such an approach would also leave the employees totally bored, as it would be repetitive. Taking the training in segments is a much more logical approach. For instance, one segment might discuss metals, another solvents, and a third, acids and caustics. The metals module can treat the material whether it is in liquid, gas, or solid form. The hazards of exposure to the element can be covered with the differences between the various forms of the material pointed out. Once the toxicity of the metal is understood, for instance, that phosphorous is a normal metabolite of the body until toxic levels are reached, the differences are not too difficult to handle. If the re-mainder of the segments of the program are put together in this same logical fashion, the whole thing becomes digestible. It is a waste of time to impress one's audience by describing terratogenicity of a particular hazardous material. The audience may be impressed by the instructor's ability to use big words, but the real message is lost. If people are told that a material may have the potential to harm their unborn child, the message is understood. That state-ment is perhaps a little too blunt, but the point is that the training session is the student's place to learn the facts in a way he or she can best understand and apply upon return to the job.

Once the initial training is completed, it is the employer's responsibility to assure that a training gate is set up to catch all new employees who enter the system. In most larger semiconductor operations this will mean an addition to existing programs already given to new employees. Most larger companies commit the first forty or so hours on the job to training the new employee. This training is usually an introduction to the technology, the cleanroom and its procedures, and any initial safety training already in place. Fire codes also require some training. Fire-reporting procedures and evacuation procedures for the facility must be the subject of at least one part of the training schedule. A number of cities are also passing ordinances with training elements in them. Toxic gas ordinances recently passed in the State of California require formal training for employees who handle toxic gases in cylinders.

Training will also be required for both supervisors and employees in plant-wide disaster and emergency plans. Most members of the industry have developed formal disaster and emergency plans but are reluctant to practice them with any degree of regularity because when people leave the cleanroom under emergency circumstances, cleanroom clothing becomes contaminated and must be changed before re-entry. If one combines the costs associated with fresh cleanroom clothing with product losses because wafers were left in an acid bath, the cost factor becomes significant. This dilemma can be solved with some searching of the procedures in place in the cleanroom. Let us say, for instance, that the policy is to have all employees change all cleanroom clothing on Tuesday afternoon at lunchtime. That is the time to schedule an emergency evacuation drill. The clothing is to be changed anyway, and since it is planned for all personnel to leave the area at a given time, the production operations are timed accordingly. Care should be taken not to damage the garments, however. (For example, employees should never be allowed to sit outside on walls and so forth while still wearing apparel.) Another device is to stage an evacuation drill involving supervisory people only. This still allows the use of the evacuation signal and generates many questions for the supervisor, who is required to explain, "What's it all about?" A sign is placed at the plant entrance on the day of the drill announcing that the evacuation signal is to be sounded at 6 A.M., 10 A.M., and 6 P.M., however no one is to leave except supervisors. Supervisors are notified in advance by memo that the drill will be held. They are to report to a specific location where they are met by the Safety Engineer. He discusses evacuation routes and alternates and advises supervisors that they are responsible for getting their people out and accounting for them. Invariably, when this type of drill session is conducted, the employees are curious enough to inquire, and the supervisor is obliged to give an explanation of evacuation procedures, routes, muster points, and the overall purpose of planning for emergencies.

Several employers also hold at least quarterly drills involving only the management team assigned to coordinate activities during an emergency. The drill is a scripted scenario usually drafted by the safety department. The drill script will outline a potential disaster or emergency situation in sufficient detail so that the team members recognize the need for action. Each member of the management team is sent an envelope with instructions to open it at a precise time. When it is opened, the message is simple, *The security officer on duty has just called your office to inform you that . . . (scripted disaster) . . . Please follow procedure for the disaster plan.* It is up to the manager to know what to do and to do it. There are always a number of problems posed by the safety department during these exercises so that each member of the team is required to make decisions on the spot. The press in the story line arrives at the gate demanding information immediately. The Public Relations or Personnel representatives are dispatched to meet and advise them of the situation. The nurse calls in with a problem; an employee has had a heart attack at the assembly point and she has forgotten her oxygen. Each member of the management team is given several problems that are called in by radio or telephone from the problem area. Their ability to react under the stress of the moment is thus tested and they gain experience in making decisions under stressful conditions. Critiques, held after the exercises, examine performance. Problems are discussed and plans made to improve performance. It is not unusual to invite members of the local fire department and police department to participate in these drills. It gives them the opportunity to see the team in action and to gauge their competence in handling emergency situations.

Exposure Monitoring

The science of toxicology deals with the reaction of man to his environment. Of particular interest to us is the work environment of man; the cleanroom area. In this area, we bring together materials, chemicals (solid, liquid, and gas), and man. Man, if exposed, takes up material into the body by absorption, inhalation, and ingestion. The first two of these are the most significant routes of exposure industrially, although there are many cases of ingestion recorded as well. The response or reaction of the human body to any given material is a function of three things.

1. Dose—how much of the material gets into the body,
2. Time—over what duration of time does the dose occur,
3. Toxicity—the ability of the individual material to cause harmful effect upon the body.

A few grams of strychnine may be fatal or may be beneficial, depending upon the time frame within which the dose is taken. Strychnine is used medicinally in low dosages with iron and quinine as a tonic. Over a sufficient period of time, a few grams of strychnine can thus be consumed and have a beneficial effect. The LD Lo (lowest lethal dose) for this material is 30 milligrams per kilogram of body weight for an oral dose in humans.

A great deal of toxicological research has been conducted to establish toxic levels of response in humans. Much of this work uses laboratory animals for the research, since society frowns upon experimental use of people. War crimes commissions make careers out of finding and prosecuting those who use humans for this research. Animal studies assist us in predicting probable effect on people at a variety of doses and via all routes of exposure. These results are not directly applicable to human beings, but taken together with human experience, form the basis for Threshold Limit Values of the American Conference of Governmental Industrial Hygienists and Permissible Exposure Limits enforced by OSHA in the United States and its territories. These limits, based upon the forty-hour work week, allow us to control exposure to a level based upon good science to a level at which it is believed most working people are safe from illness or significant discomfort. This system has its limitations which must be recognized. It does not take into account that the worker may be a woman, or a pregnant woman. It does not provide limits to protect the community, only the worker. Based on these limits, or other limits the company may establish, the safety organization through an industrial hygiene program sets up regular monitoring programs. The purpose of these programs is to establish the actual exposure levels at the workplace while the work is being conducted.

Typically, every employee on every job is not monitored. Each job task is monitored and the results of a representative set of samples and observations are applied to all who do that job task. Actual numbers of samples taken to support representative sampling are determined by statistical confidences needed to support the program. This sampling allows for the construction of an exposure picture that is then applied to similar operations. Time-weighted exposure patterns are applied to people who may not be directly measured at all. This type of monitoring verifies the validity of engineering controls put into place to control exposure. It also discovers the adequacy of procedural controls and training programs. It discovers whether or not the personal protective clothing used is adequate to the task, is used correctly, and is job-suited. Finally, it demonstrates compliance with the regulatory requirement to control exposures to a specific level or below.

Area monitoring is also conducted in many semiconductor cleanrooms. Specific monitors for the hydride gases, chlorine, ammonia, hydrogen, solvents, and so forth are often used to monitor for upset situations. These monitors do not accurately reflect individual exposures, but they do serve two important purposes. First, they monitor for upset condition and take some action, such as alarms and so forth. Secondly, they record potential peak exposures within the cleanroom during an upset situation. All of this monitoring is put into place to protect people.

On the production side, we design sophisticated systems to provide engineering controls for contamination. We all understand that microelectronic circuits are vulnerable to particulate contamination. Areas are designed to control this contamination to class 10,000, class 1,000, class 100, or even class 10 or 1 levels of cleanliness. We verify this cleanliness level by initial certification of the area to accepted standards, and then we regularly monitor the area to verify that the controls work. Like the safety issues, controls for contamination control are *engineering* and *procedural*. Engineering controls are built into the air recirculation system as a series of increasingly efficient filters. Procedural controls are the requirements for cleanroom clothing, special paper, pens, air showers, cosmetic restrictions, facial hair restrictions, and so forth. A great deal of time, energy, and expense goes into the maintenance of the cleanroom atmosphere. Yet all of the energy is expended on particle number and size. Little or no attention is given to particle chemistry. Suppose there are 10 particles of contaminant per cubic foot in the cleanroom and those particles are all 0.3 microns or less in size. Also, suppose, for the sake of discussion, that all 10, or 8 of the 10, are phosphorus from a poorly designed scavenger exhaust on a diffusion furnace. The problem with this scenario is that there is no established limit for accidental exposure of silicon wafers to phosphorus. Nor, for that matter, to any of the other materials we handle routinely in the cleanroom.

There are two routes of exposure for the silicon wafers to whatever contaminant we release into the cleanroom. The respiratory route of exposure in humans is analogous to diffusion on the wafer. The skin contact route is similar to surface contamination in the case of the wafer. If phosphorus settles out on equipment where it is picked up by contact with the wafers, or is transferred to the wafers by contaminated clothing, tweezers, carriers, or vacuum pick-ups, the wafers are still affected. One of the industrial hygiene procedures routinely conducted during a monitoring cycle for the metals is a wipe test. A filter paper is moistened with DI water and a known area (usually 100 square centimeters) is wiped. This sample is then analyzed along with air-monitoring devices. This gives the industrial hygienist a two-dimensional picture of what is happening in the area. He or she knows how much material is airborne and how much has settled onto the surfaces. From safety's vantage point, the wipe test evaluates housekeeping procedures that impact human exposure potential. It allows the safety professional to evaluate and control human exposure to toxic materials. It looks not only at engineering controls built into the system, but is also a good evaluation of procedural control (cleanliness) that impacts our ability to control exposure.

Although there is no established limit of exposure for silicon wafers, the data is still potentially useful to the engineering manager in charge of the area. It allows the engineering staff to establish a baseline of cleanliness in the wafer area for those materials which are monitored by the safety program. These materials will normally include the dopant metals, solvents, acids, and caustics, for these are all toxic in some way and have human limits of exposure. If there is an increase in contamination levels in the wafer cleanroom area as evidenced by air monitoring or wipe samples, the engineering staff can investigate the cause. It may be that these sample data will correlate well with yield data. The area may be in the middle of a yield bust and this is the exact data needed to find and correct the problem. The industrial hygiene monitoring data provide one additional parameter of measurement that will assist in maintaining the cleanroom at a level necessary to produce quality product consistently. Product quality control is not its basic purpose, but there is no reason to restrict the use of the information.

In the evolution of the microelectronics industry, the cleanroom is at a relatively early stage of development. There is still a great deal to be learned about the dynamics of what is happening in the air filtration and conditioning system. One parameter that has had little or no exploration is the phenomenon of filter leakage. This concept is relatively well understood on a smaller scale. If one examines the cartridge filter provided for a respiratory protective device, it is fairly simple to find references in the industrial hygiene literature that discuss the breakthrough concept. The discussions are invariably tied to safety.

Given that a mechanical filter is used to protect employees from harmful exposure, and that a filter can only hold so much material, there comes a time when the efficiency of the filter is impaired and the worker is potentially exposed to the toxic material. What is the leakage or breakthrough point? What is the breakthrough mechanism? These questions are the subject of a number of papers presented in industrial hygiene literature and are of particular interest to the safety professional and the regulatory world. The answers to these questions are used to establish protection factors for wearers of respiratory protective devices. They are also used to prescribe maintenance procedures for the equipment. It becomes apparent that these data could have great impact in the cleanroom.

One of the leakage mechanisms is called *channeling*. This occurs when a fault in the filter system allows the contaminant to penetrate through the filter in a relatively narrow channel and break through on the downstream side. In the case of protective apparatus this results in exposure to an employee who thinks that because he is wearing a respirator, he is protected. Is the channeling effect possible in HEPA filters? We suspect it could be. Given routine particle counting done in the cleanroom area, are we likely to find it? Probably not. The methodology for conducting particle counts in cleanrooms is not all that well established. In some locations, a particle counter is pushed around from cleanroom to cleanroom. It goes through relatively dirty corridors, or worse, out of doors, where it picks up enough contaminants on its own to become the source of much of the contamination it is measuring at the next location. It thus skews the data that it was designed to collect.

When are particle counts taken? During the operating shift? When there are no people in the room? They have been done both ways. It depends on what we want to prove. If we want to prove that the area is clean (to a customer), we set the sampling in such a way that there is less chance of contamination. Conversely, if we see a need to prove the area is dirty, we stack the deck in that direction. Standardized approaches to measurement of particles in the cleanroom are necessary if we are to achieve uniform, repeatable results. The safety program's industrial hygiene effort is a standardized monitoring or measurement activity. We are not suggesting that industrial hygiene measurements should be substituted for particle counting or whatever is a valid measurement of cleanliness in the cleanroom. We are saying that we see order of methodology in industrial hygiene sampling that is not apparent in standard cleanroom monitoring as it is currently done. The industrial hygiene monitoring, if it is done correctly by standardized procedures, is probably a more valuable tool in terms of evaluating the overall cleanroom than most of the monitoring currently done. We are also indicating that the industrial hygienist has a great deal to offer in the area of contamination control and evaluation. Industrial hygiene monitoring is a resource that is all too frequently overlooked in the manufacturing or research operation.

General Safety Practices

One of the most difficult hurdles for the safety professional to get over in the cleanroom environment is the idea that the protective clothing issued for the cleanroom is not protective safety apparel. Employees who handle acids or solvents in open containers are not protected from accidental contact by cleanroom clothing. Specialized personal protective clothing is still required in the cleanroom over and above cleanroom apparel. The selection process for this chemical protective clothing is more difficult in the cleanroom situation than it is in more routine operation. Here, we not only have to consider the protective ability of the clothing, it also cannot add to the particulate burden in the room. For example, protective gloves, which are normally packed with talc to make use easier, must be specified as dry to prevent contamination.

Cleanroom clothing manufacturers regularly attest to the chemical resistance of their cleanroom garments. The industry often mistakenly sees this chemical resistance as imparting people-protective qualities to the clothing. All the supplier is trying to tell us is that a solution of, for instance, 10 percent hydrochloric acid in contact with the material for eight hours will not damage the fabric. In actual practice, the acid generally penetrates through the fabric and gets onto the skin of the employee working with the acid.

Cleanroom safety practices that are unique to the cleanroom are really few in number. For the most part they require establishing some procedures for reaction to emergency situations. If a container of chemicals is dropped onto the floor, how will we react? It is not likely that we would be willing to wait for an emergency response unit to come to the area from some other part of the building, don cleanroom clothing, and then respond to the spill cleanup. Time to cleanup is too long, and we would probably have to evacuate at least the immediate area to prevent exposure. It is more practical to develop a spill cleanup team within the wafer cleanroom. These people obviously need the training necessary in order to react properly to the spilled chemical, and they need the necessary supplies to handle the spill.

Protective clothing must be assigned on an individual basis, it must be maintained by the user, and must be tested for protection prior to each use. The protective clothing must protect the user against the hazard involved and must be available when and where it is needed. Many safety organizations have aggressive testing programs for personal protective clothing that include field testing for fitness and protective ability. Laboratory tests qualify the clothing in terms of its ability to resist penetration of the contaminant to the skin underneath. Field tests establish service remaining, comfort, dexterity, and ease of cleaning or decontamination. Once the tests are completed, the item of protective clothing is approved for use in the cleanroom and placed into stock for issue. This is not a procedure one would want to repeat with any amount of frequency, as it is at best time-consuming.

Once personal protective clothing is tested and approved, there is a certain reluctance to change it. One of the most powerful inducements to change an item of protective clothing is an accident that indicates the item to be inadequate to the needs of the job. Another is the buyer who locates an item of equivalent protection for less money. There is always pressure to reduce costs, and this pressure may keep the safety department in the protective clothing testing business for a considerable time. Since there are so many layers of clothing required to be worn by an employee who works with chemicals, there is some advantage to approaching this issue of personal protective clothing and equipment by trying to eliminate the need by building the protection into the equipment.

A single outlay to modify a piece of equipment results in a continuing savings in protective clothing. Eye protection devices are probably the easiest item to eliminate in the cleanroom by building eye-protective barriers into the equipment that uses wet chemicals. Acrylic shields are thus extensively used in the industry to provide eye protection. As an additional benefit, these shields frequently reduce the area which must be protected by local exhaust ventilation or reduce the volume of air which must be exhausted. This, too, reduces costs. Each cubic foot of air which must be removed from the cleanroom for exhaust, must be replaced by conditioned air that is filtered down to the specifications of cleanliness required for the area. The costs of conditioning a cubic foot of cleanroom air can be a significant contribution to the cost of building a product and can make a significant contribution to ultimate profitability.

Equipment Layout for Safety

There are obvious and perhaps not so obvious traps that can be laid for the unwary by placement of equipment. By traps we refer to accident- or potential accident-causing situations. A panel which protrudes into the walkspace and which must be open during operation is one of the most frequent encountered. We really need to discuss today's employees and some of the restrictions they bring or may bring to the work environment. People who live in the Western part of the United States are used to ranch-style homes. These are homes that are built on one level, few with basements. This regional person, who has spent most of his or her life in this part of the country has problems with changes of elevation. He or she has difficulty with stairs, ramps, curbs, and so forth. Therefore, the first trap we lay for our employees is to build the cafeteria on the second floor of the building, thus guaranteeing that virtually all employees will have to navigate stairs several times a day.

We know, even in the dark, that to turn lights *ON* requires an upward motion of a switch. *OFF* is downward. Increases of power or speed or rate are controlled by clockwise rotational movements and decreases of power or speed require counterclockwise movements. There is a long list of similar specifications for which we must plan if we are not to build in traps which will ultimately cause accidents in the workplace.

Some basic ground rules of what not to do from a layout standpoint will help with safety problems before they have a chance to cause accidents. It should be apparent that in a wafer processing area, whether or not it is a cleanroom, chemicals are handled. If an employee has the opportunity to handle chemicals, sooner or later someone will make a mistake and chemical contact will result. Given that an employee can get chemicals into an eye or both eyes, the emergency shower and eyebath station necessary to render first aid must be located so that it can be found by someone who cannot see. Indeed, someone with chemicals in both eyes is literally blind and must be able to find and operate this equipment without sight. This situation will probably be complicated by pain or discomfort. Obvious layout traps to guarantee that the emergency equipment is not available or difficult to find would be to place the equipment in a location where the injured employee needs to travel around several corners to find it.

The emergency shower and eyebath station needs to be located close to potential contact areas and in such a way that an employee does not need to negotiate corners, dangerous equipment, or other hazards along the way. By dangerous equipment, we mean things that move by power, are heated, or generally are a hazard to touch. Touch is probably the way that the temporarily impaired person will find the way to the relief of the water in the eyebath. The general rule is that the emergency shower and eyebath station should be located within 25 feet of the potential source of contact and in a direction away from furnace tubes, glasswear storage, or similar hazards, and finally in as straight a line as the space will permit.

Emergency equipment such as showers and eyebath stations do not prevent injuries. Once contact with a chemical is made, the sooner one can get to the equipment and begin irrigation, the less severe will be the ultimate injury. The layout does not need to further complicate an injury by forcing an injured employee to run an obstacle course to get help. The placing of chemical stations such as hoods or wet benches directly opposite each other, with chemical-handling employees working back to back, is another example of a trap. Let anything go wrong at one station, and the operator will do the most natural thing, jump back. Who will get bumped into the chemical equipment on the opposite side of the aisle? His coworker. One injury has probably resulted in two in this case.

Chemical storage cabinets need to be convenient to the point where chemicals are used. If an employee is expected to carry more than one container of chemicals from storage to use point, it is another trap to place obstacles such as doors between the cabinet and the use point. Over and above the obvious physical hazard posed by a door for an employee with two chemical containers in hand, there are hazards to others. We normally require an employee handling chemical containers to wear protective gloves. If the employee has not thoroughly washed the gloves before the trip to the chemical cabinet, we now have potentially contaminated several surfaces along the route, including door knobs, push plates, handles, and so forth.

Layout needs to take into consideration that chemical protective clothing, unlike cleanroom clothing, rarely leaves the immediate area of use. It needs a storage location. That storage location needs to be vented so that chemical clothing can be washed down as it is removed and put away in a location where it can dry and be ready for use again.

It must be understood in the layout exercise that equipment will require periodic maintenance or adjustment. Sufficient room must be provided so that the technicians responsible for these operations can remove panels, open doors, and remove parts of the equipment to gain access to the maintenance or adjustment points. Further, these maintenance people frequently have the need to bring portable equipment such as meters to the equipment to do the job. If we do not provide convenient or sufficient electrical connections for them, they will be forced to string extension cords from outside the immediate area. Sufficient space to pull diffusion tubes is required. Where doors are needed, care must be exercised for direction of door swing, or that doors do not swing into each other or into cross-flow traffic. All of these things must be taken into consideration in the layout process, so that accident situations are not built into the work area.

A key to successful layout is communications. The safety department must set up a line of communication with the department responsible for layout of the work area. In most larger organizations this is performed by a layout section of the Industrial Engineering Department. The safety engineer must remember that the basic charter of an industrial engineer is the efficient utilization of the space available. That simple concept is not necessarily compatible with the aims of the safety program. The industrial engineer frequently brings several concepts with him to the layout job. These concepts are what the safety engineer must be prepared to argue against with accident facts.

In some smaller locations, layout of the area is done by the production staff. The set of rules that these people bring to the task is the most efficient process flow for the particular product being built. Again, as with the industrial engineer, they must be convinced that the requirements of safety must be taken into account during the layout phase of the project. The message is that the industrial engineer and the production engineer are not wrong. They each bring their own priorities to the layout task. The safety engineer also brings priorities, must recognize that the other considerations are valid, and be prepared to convince the parties that safety needs to be treated on an equal footing. Probably the most successful layout is one that gives equal value to efficient utilization of space, process flow requirements, and the efficient use of equipment to do the job, while providing for the safety of the users of the space and equipment.

One simple method of assuring that all parties have their concerns addressed is to have a layout sign-off system which requires plan review by all three parties. This way the production engineer can assure the most advantageous placement of equipment to assure a logical product flow through the area. The industrial engineer can assure that the space is utilized in the most efficient manner consistent with the flow requirements. Finally, the safety department can assure that those traps which we have allowed to cause past accident situations are not repeated. In conflict situations, a simple conference where each of the parties can present his reasons for some requirement can be called. The necessary arguments can be heard, justified, and adjustments in the layout can be made to provide the needed changes in the area.

There are times when the safety engineer cannot give in to the logic of an argument. Legal duties imposed by safety regulations, building codes, fire codes, electrical codes, and so on are relatively inflexible. For instance, the National Electrical Code requires that there be three feet clear in front of all electrical panels. These regulations are usually enforced by the city or county electrical inspectors. If the rule is violated, the result will be a citation against the employer. There is simply no chance to compromise. This is also frequently the case regarding the width of aisles, particularly those leading to exits. The safety organization cannot compromise code requirements without running the risk of penalty to the employer. The standards are based on minimums, and the employer who violates minimum standards runs the risk of having to rebuild the wafer area to correct a problem after it is discovered. Most of the disputes can be settled by a little open discussion and negotiation. Each party must respect the priorities of the others and recognize that each is dedicated to the ultimate success of the effort. If this attitude is established and honored throughout the negotiations, the art of compromise will solve most problems.

Many organizations are now making their maintenance supervisor or engineer a reviewer of layout drawings. He usually has a history of the frequency and type of maintenance required for each piece of equipment. He knows what part of the equipment is usually responsible for breakdown and how long it will take to repair and get back into operation. This is extremely valuable input to the layout if maintenance is to be done with the least impact to the production operation in terms of downtime, safety, and efficient utilization of the asset. Frequently, the simple expedient of placing this piece of equipment at the end of the line can solve problems later on. The maintenance supervisor or engineer wants to be able to do his job efficiently, thus he brings another dimension to the layout activity which should be considered seriously.

Maintenance Safety

Including the equipment maintenance department in the initial layout and even the conceptual design of the cleanroom can resolve many of the safety and cleanliness problems brought about by opening equipment. Basic safety rules for the cleanroom require the maintenance people to be able to lock out power sources before starting to work on the equipment. This basic requirement stems from the need to protect the worker from accidental movement of machinery during preventive or breakdown maintenance procedures. Electrical power needs to have a single source for each piece of equipment so that the maintenance technician can turn off a single power source without voltages being present in the machinery. Modern cleanroom equipment frequently has several voltages used within the equipment, such as 220V to operate heaters, 120V to operate lights, and DC voltages to operate logic and control systems. If each of these is powered separately, the maintenance technician is forced to search for multiple disconnects or is forced to work systems "hot." This creates the opportunity for electrical shock injuries. This same philosophy is true with regard to pneumatic or hydraulic power sources. The source of the power must be capable of being shut down without residual power to protect the maintenance technician during his work.

Typically, safety interlocks are defeated by the technician during maintenance procedures. Extraordinary precautions are necessary to prevent injury while he is doing maintenance or repair. Pneumatically or hydraulically operated systems must be blocked open so that parts of the equipment do not activate while the technician is vulnerable. The maintenance technician has a greater opportunity for exposure to hazardous chemicals during maintenance operations than the operator has during operation of the equipment. A wet bench, for instance is designed to draw the proper amount of local exhaust ventilation to protect the operator during regular operation of the wet bench. A maintenance technician who is required to place his body into the equipment loses the protective ability of the exhaust. He must rely either on personal protective clothing to prevent exposure to hazardous chemicals or he must ensure that the chemicals are removed from the wet bench before the maintenance operation is begun.

Vacuum systems are generally operated as closed systems and prevent exposure to toxic elements during operation by being closed. The maintenance technician who is required to change oil on a vacuum pump, for instance, is required to break the system open to drain the oil. Any toxic material in the oil presents an exposure potential to the maintenance technician unless he can devise a closed system to change the oil or is willing to wear protective clothing during the operation. Additionally, this opening of systems presents an opportunity for contamination of the cleanroom unless the changing operation is conducted in a thoughtful manner with this problem in mind.

Facilities maintenance presents many of the same problems as does equipment maintenance. If an electrical system is overloaded and blows a fuse, a facility electrician is required to troubleshoot and discover, first the location of the blown fuse or circuit breaker, then the cause. Operations which require the technician to work with potential energy sources such as electricity mandate that the procedural control of lock-out be used and that personal protective clothing be employed to prevent injury in event of contact.

Unless cleanroom personnel assume the duties of building and equipment maintenance, it must be done by others. The equipment necessary to do the work must be brought into the cleanroom. This includes lamps for replacement in the lighting system, ladders to gain access to the lighting fixtures, and meters to assure that power to these systems is available. It also includes the necessary protective clothing to do the job in a safe manner without injury. The lamp changer would routinely require protective eyewear and gloves. It is the maintenance technician's responsibility to return all systems to their normal operating mode. This means that all safety interlocks are functioning, ground wires are attached, and protective panels are back in place prior to leaving the area.

Some operations use the precaution of a safety inspection of the equipment after maintenance and before allowing an operator to resume operation of the equipment. This inspection can be conducted by the maintenance supervisor alone or by the maintenance technician with the area supervisor. In either case, the idea is to assure that the equipment which has been the subject of a maintenance procedure is back in full operation including all safety systems, and the operator can concentrate on the production operation without having to worry about equipment failure.

Emergency Response

There are three schools of thought about responding to emergencies in the cleanroom:

1. Leave it to the professionals. Any time there is an emergency in the cleanroom, call the fire department or paramedics.
2. Develop an internal response organization at the plant level to respond to an emergency situation on the site. If the emergency is beyond that organization's ability to handle, then bring in outside help.
3. Require the people who occupy the cleanrooms to be responsible for response to emergencies. Each operation has to decide, based on its own circumstances, company policies, and so forth which response is correct for them.

A site with four or five cleanrooms may decide that the internal response organization is the best choice, based on the necessity for training the team to a high skill level. Those employees who respond to chemical spill emergencies are legally required to have a minimum of twenty-four hours of training annually. The distance to the local fire department, its response time and skill levels may also be determining factors in making the decision.

The emergency may be a chemical spill, gas leak, power failure, fire, explosion, or medical emergency. An employee who goes into diabetic shock in the cleanroom is just as upsetting to the operation as a chemical spill and must be handled as efficiently. The first step in planning emergency response is to evaluate the potential emergencies that could happen in the cleanroom. The search must look both within and outside of the cleanroom. A fire in an office area in the next building is as much of an emergency as a fire in the cleanroom to the people who are in the cleanroom operation. The search must look for man-caused possible emergencies as well as naturally occurring emergency situations such as earthquakes, tornadoes, floods, hurricanes or other natural disasters. Once the potential emergencies are identified, it is necessary to make the basic decision about the response organization.

The first priority must be the safety of people. Will we evacuate the immediate area, the cleanroom, the building? Who is authorized to give the order to evacuate? How will the alarm be given to evacuate? Where are we to go and how are we to get there? Once we have evacuated the area, room, or building, where are we to meet? Who is responsible to confirm that everyone is accounted for in the evacuation process? Who can give the all clear to allow people back into the area? As one can see, there are many questions which must be answered. The answers to these few basic questions form the first part of an overall plan for responding to emergencies.

Assume you have chosen the most frequent option regarding emergency response, that of a site-wide response team. The response team, most often called ERT (Emergency Response Team), must be given information regarding the location and nature of the emergency. How is this to happen? Who is in charge of the ERT? Where are they to meet for assignments? What is their priority—rescue or disaster control? The ERT will need some supplies to handle nearly any kind of an emergency in the cleanroom. They will need chemical protection, respiratory protection, chemical neutralizer or absorbent material, special tools with which to shut down gases, water, electricity, and so forth. Where will these supplies be kept? If they are responsible for monitoring the area to give the all clear signal, how will they do this? This forms another basic element of the emergency response plan. Part of this process must tell the ERT when to say, "NO, this is beyond our skill level." This response forms the basis for calling in additional help from professional sources such as paramedics, fire department, hazardous materials units, and so on.

If we make the decision to call in outside help, what is the role of the ERT? Emergency situations require management decisions just as do day-to-day operations. Someone must be in charge. The first thing the fire department does at the scene of a major fire is to establish a command post to coordinate the activities of the various fire companies and other responders. If there is no command post, chaos will result. If your organization is to successfully respond to an emergency situation, someone must be in charge, make responsible decisions, and be accountable for those decisions. Since the timing of emergencies cannot be predicted, your plan must assume that the Plant Manager, C.E.O., or other senior manager will not be in the building when something goes wrong. Who will be in overall charge of managing the emergency situation? We subscribe to naming the senior person on site at the time of the emergency as being in overall charge until replaced by senior management. This requires that call lists be produced to locate senior managers, safety professionals, or whoever is needed to respond to an emergency at whatever odd hour. This plan also requires that someone be responsible for locating these people in an emergency. The call list must be kept current and available so that whoever is responsible for making the calls is not delayed.

Where will the Emergency Manager manage from? What resources will he or she need to make decisions? How will communications be established so that information can be transmitted? These are the basic elements of an emergency response plan. The plan must be worked out in as much detail as possible and communicated to those who are expected to take action. Once the plan is worked out, it must be practiced. An emergency drill or two will quickly point out holes in the plan which must be filled with other pre-planning steps. Once the plan is acceptable, integrate it with the community resources expected to respond to your location in an emergency. The fire department, paramedics, police department, hazardous materials unit, and local hospitals all have a need to know what your plans are and how they can expect you to react to an emergency situation. They can then formulate their own plans around yours or make suggestions to your plan to better integrate the two plans.

SAFETY EQUIPMENT

Protective Clothing

Most cleanrooms handle hazardous materials in one form or another. The handling of hazardous materials in other than closed systems will require that personal protective clothing and equipment is available and used during chemical operations. One of the most common problems with personal protective clothing in the cleanroom environment is its own suitability for the environment. The protective clothing cannot contribute to particulate contamination or it becomes a major problem. The user of protective clothing can solve this problem either by testing all protective clothing for particle contamination potential or specifying such tests by the supplier. If the supplier performs the tests, he must know what you are trying to accomplish in terms of cleanliness.

Another problem is that vendors of cleanroom clothing frequently extoll the virtues of their clothing in terms of chemical resistance. The unwary buyer can assume that this means the cleanroom clothing offers some degree of protection to people who are handling chemicals. This is not the case. Suppliers test for fabric resistance to chemical attack. They testify that material X can stand up to eight hours of exposure to 10 percent hydrochloric acid (for instance) without degrading the fabric. They do not mean that even 10 percent hydrochloric acid will not penetrate the material and burn whatever is underneath. The use of cleanroom clothing does not remove the need for personal protective clothing when chemical handling is required by the task at hand. Personal protective clothing must be suited to the job.

We must also note that all chemical gloves are not equal. The suppliers of protective clothing generally are quite willing to attest to the ability of a particular glove to resist exposure to a particular chemical. It is important to select protective clothing that is suited to the exposure expected. Protective clothing should be assigned to an individual, not to a job. By this assignment, the individual can be held responsible for decontamination and any maintenance required to assure the protective ability of the item. Many chemical contact accidents have occurred because employees thoughtlessly shared protective clothing, with no one being responsible for cleaning and maintenance.

Protective clothing, like cleanroom clothing, requires storage when not in use. The users must also recognize that protective clothing has a service life and, after a time, will no longer protect the wearer. Protective clothing also has its limitations. A bulletproof vest will not protect the wearer against armor-piercing bullets. Neither can a protective glove be expected to protect an employee who reaches into a container of heated, concentrated acid. Protective clothing is designed to protect against accidental and incidental contact. It was never intended to protect the operator against deliberate dunking into a chemical.

Personal Protection Systems

The cleanroom, particularly the semiconductor cleanroom, is a series of systems designed to accomplish a particular task. The air system is designed to deliver air of an absolute quality and quantity to a defined space and at a specified rate. A diffusion system is designed to deliver an exact amount of a specified contaminant to the surface of a silicon wafer under very specific conditions to obtain an exact electrical effect. Each one of these systems presents an opportunity for the safety organization to prevent injury or illness by using a systems approach to accident prevention.

Fault-tree and *failure mode-effect* are two forms of system safety analysis frequently used with systems which require interaction of components to accomplish some task. These two methods of accident potential analysis were born in the space program, but are applicable on a smaller scale to nearly any system. A fault-tree could be run on a gas hot water heater system to discover which components in the system could cause an undesired effect, such as a gas explosion. Failure mode-effect analysis is basically a what-if exercise. In either case, they identify systems or parts of systems which are capable of producing some undesired effect (illness or injury), and the user has only to devise a back-up or alternative to prevent injury.

This type of system safety engineering is what requires two-hand trip controls on much of the equipment in use. If both operator hands are occupied elsewhere, they cannot be at a danger point. The system safety analysis also permits the use of the system itself to control and prevent exposure to hazards. For instance, using a normally closed valve on a flammable gas and a normally open valve on an inert gas in a furnace system will prevent explosion in event of electrical failure. The normally closed valve closes upon loss of power and shuts off the flammable gas supply, while the normally open valve assures a purge of inert gas under the same conditions. The system is thus said to be fail-safe. Having built hazardous systems to be fail-safe, it is then possible to build a great deal of safety into the softwear programs which run much of the equipment in the semiconductor cleanroom. The if-then capability of the computer softwear program is the most useful tool for system safety approaches to accident prevention. "If interlock A does not energize, then fail all power to the system. . . ." The protective power of this statement should be apparent.

The safety organization and the equipment and facilities people must work closely together in order to integrate system safety approaches into the equipment and basic building components of the cleanroom. Specification preparation and review and system analysis are the tools most frequently employed during this exercise. It affords the engineering staff the best opportunity to build safety into the building and equipment without reliance on procedural controls. It takes advantage of the ability of systems to analyze situations and react to them with very short reaction times and has the capability to prevent injuries.

HANDLING HAZARDOUS PRODUCTION MATERIALS
Safety Controls

There are essentially two areas of interest with respect to the control of workforce exposure to HPMs. These are *medical control* and *engineering control*. Medical control deals with the personnel whereas engineering control is concerned with the environment.

There are four main elements to medical control:

Preplacement physicals

Regular check-ups

Protective garments/devices

Education

The execution of medical control is the responsibility of the industrial health team, which generally should consist of a full- or part-time industrial physician, a full-time industrial nurse (or nurses), and an industrial hygienist.

Preplacement physical examinations, handled by the physician and/or nurses, are critical, as they not only set a baseline by which any physiological changes can be measured, but they also facilitate screening of particularly susceptible individuals. If a worker is found to be (or is likely to be) susceptible to particular chemicals, he or she should not be given a job dealing with these. Regular physical examinations performed at intervals of one year, or six months in particularly hazardous environments, are also necessary to detect changes. The information from these examinations should be combined with casual visits to the nurse or physician for treatment of minor ailments (headaches, rashes, and so forth), and data on previous accidents and known exposures. Workers who may be exposed to materials such as arsenic are generally required to have two physicals per year, which include blood testing and urinalysis.

Protective garments, such as acid-proof aprons and suits, and personal protective devices such as respirators, really should be a last resort when other controls are not totally efficient. The industrial hygienist will advise on the use of these in many cases. Education by the industrial hygienist as to what the hazards are is extremely critical, as an awareness of the severity of potential exposure problems will often reduce carelessness.

Engineering control is the responsibility of the safety engineering team. The team should consist of safety engineers and industrial hygienists who work closely with production and R&D personnel, and management. Engineering control has many aspects, but there are eight readily definable categories:

Ventilation

Enclosure

Monitoring

Segregation

Substitution

Scrubbing

Neutralization

Maintenance

Ventilation is the most important engineering control measure. There are two types of ventilation; *general,* in which dilution by huge quantities of fresh air is used, and *local,* in which contaminated air is removed near the point of generation and discarded. General ventilation will allow the dispersion of lower levels of chemicals and thus cannot be used for highly toxic materials. There is also a problem with this approach in controlled environments as the new air has to be heavily preconditioned, and this is expensive (300 cfm of make-up air generally requires about 1 ton of refrigeration capacity). Local ventilation only takes air from the volume around the generator and is thus more economic. However, this form of ventilation presents more engineering problems as the local air velocities have to be high in order to capture the contaminant, we cannot mix all exhausts together because of chemical reactivity factors (for example, acid and solvents), and in many cases the effluent cannot merely be dumped in the environment without treatment. We will return to this latter factor later. The recommended minimum face velocities (in ft per min across the open face of the area/bench/fume hood to be ventilated), depending on the toxicity of the substance (TLV-TWA in ppm), are given in Table 12.2.

Good engineering practice requires measurements of the face velocity at 9 points across the open face of a fume hood or cleanroom wet bench to ensure even capture. Sources of turbulence which could affect the airflow and allow outflowing of contaminants should be avoided. In gas cabinets, used for holding gas cylinders while in use, codes (discussed later) dictate that the face velocity at the window should be a minimum of 200 fpm. Exhaust systems should also have features such as back-up fans, emergency power connection, indication at the chemical use/storage area of failure of low flow, and so forth. Centrifugal blowers are typically used on exhaust systems due to their ability to move large quantities of air with a relatively low fan speed. Only the fan blades, not the motor, are immersed in the effluent, an important fact if the exhausted substance is flammable. The fan should be located at the head of the system so that the ducts are at a slight negative pressure to negate leaks. A zero pressure stackhead (no coverings to deflect or retard the airflow) is generally recommended. The exhausts should be at least 7 feet above the tallest part of the building structure and situated downwind of the

Table 12.2   Minimum Face Velocities for Fume Hoods

Extract Class	Face Vel. (ft/min)	TLV (ppm)
A	125–150	<10
B	100	10–500
C	75–80	>500

air intakes. (The local meteorologists will tell you which prevailing wind direction to expect.) Rain which falls into the open stackhead should be drained and treated as hazardous effluent.

Another form of ventilation is the venting of hazardous gas delivery systems prior to breaking the seals for cylinder changeout. This is achieved by the use of an appropriate gas panel which generally contains the purge apparatus as well as the pressure regulators (cylinder and line pressure), excess flow shutoff, and cylinder connection via an appropriate CGA (Compressed Gas Association) fitting. The advanced panels will flush the lines right up to the cylinder valve with an inert gas (N_2) and will also switch to apply a vacuum to remove the hazardous gas. It should be kept in mind that it can take many tens of flush/vacuum cycles before the most toxic materials are brought to below the TLV within the lines. Also, the source of inert gas should always be a cylinder rather than a house delivery system as check valves can fail, allowing backflow of the HPM (the flushing gas cylinders should thus be treated with great care just in case).

Another form of control by dilution is the purging and balasting of equipment vacuum pumps. Rotary pumps generally contain oil which forms a mist at the outlet, and the combination of this mist, air, and flammable gases is rather dangerous. Nitrogen is frequently injected into the pump and exhaust line to dilute the mixture to below the LEL. The mist is also removed by means of a demistor. Occasionally water- or nitrogen-cooled cold traps are used to condense hazardous substances, but it should be kept in mind that these traps and oil demistors can collect hazardous condensates and should be treated with the utmost care.

Enclosure of storage and use volumes is also extremely important. It is much easier to control, monitor, and ventilate an enclosed unit than an open workbench. Equipment which uses HPMs is generally enclosed by means of a vented cabinet (with interlocks to prevent opening during processing), but the most frequently encountered enclosure systems are the gas cabinets which hold the gas cylinders while in use. There are some simple rules of separation which should be followed. There are essentially seven categories of gases:

Flammable

Pyrophoric

Toxic

Acid

Base

Oxidizer

Inert

None of these should ever be put in the same cabinet as a gas from another category other than inert.

Environmental monitoring is mandatory by law (hence the existence of PELs) and is also good microcontamination practice. We have already discussed aspects of detection and monitoring and therefore we will not discuss them further in this section.

Segregation is always a good idea for HPMs, and many plants today are being designed with storage buildings away from the main plant to isolate disasters.

Substitution of a less hazardous material is not always possible but the chemical companies are seeking alternatives to many HPMs. Examples of successful substitutes are TCA for TCE and solid source phosphorus (phosphorus compounds contained within a refractory matrix) for phosphine.

Wetting and scrubbing is a very large topic in itself. These techniques tend to apply to acid and base vapors and gases, dusts, mists, and fumes but not solvents, as having dangerous solvents in large amounts of water merely makes the disposal problem larger. Industrial scrubbers for large volumes of air are generally of the packed tower type in which the gas to be scrubbed is made to flow up a column in which water is percolating down through some type of inert packing (for example, PTFE) such that the hazardous substance becomes dissolved or at the very least trapped in the liquid. The contaminated water is then neutralized (for example, by adding a base solution in the case of acid contamination) or filtered and recycled. This method is acceptable for materials which are not very toxic as some will escape from the tower. Typical removal efficiencies are in the region of 99 percent. It is likely that dry scrubbers which rely on the capture of the chemical by a dry absorber will become more popular in the future as these are simpler to operate and maintain (no moving parts).

The next engineering control topic is neutralization. Neutralization essentially involves rendering the toxic material harmless by means of a chemical reaction. As mentioned previously, the water from acid and base gas scrubbing can be neutralized by the addition of base and acid solutions respectively in the appropriate amounts. Highly toxic gases are another matter as water scrubbing alone is generally insufficient. A much more efficient method is to use "combustion, decomposition, and oxidation" in a CDO unit. This involves the combustion of the gas in an oxygen (usually air) ambient at high temperature. This method tends to be used only for equipment and gas panel exhausts as the maximum flow rates tend to be in the order of a few liters per minute. This is especially good for metal hydrides as they may be transformed from highly toxic gases to somewhat less toxic oxides. For instance, phosphine may be oxidized to phosphorus pentoxide. This then may be water-scrubbed to form phosphoric acid, which can subsequently be neutralized. Arsine poses a problem as arsenic oxide is an extremely toxic solid, but we can scrub this with an appropriate chemical solution to render it safe.

Care must be taken when oxidizing silane as silica is insoluble in water and therefore can clog the system if not filtered out.

The final category in this discussion is general maintenance. There is no substitute for care in installation and maintenance of equipment and systems. If care is not taken here, all other safety controls can be negated.

Gas Safety Revisited

For the user, gas safety begins during the ordering process. Volumes of gases stored on the site are of concern from a safety standpoint and a quality assurance standpoint as well. The material manager simply cannot rotate large volumes of gas cylinders through his system to assure that the users get first in—first out service. Restrictions of the numbers of cylinders of hazardous gases in cylinders are part of the Fire Codes. Purchase orders for gases should routinely be routed through the Safety and Environmental organizations for their approvals before orders are placed. This assures that Material Safety Data Sheets are on hand, that the planned user has the necessary engineering controls in place to deal with the hazards of the gas anticipated, and that treatment systems can handle any expected discharges. The department responsible for delivery and storage of the cylinders must be on the approval cycle as well. This is simply to assure that the storage area is equipped to store the volumes of gases ordered in a safe and compatible manner.

Semiconductor cleanrooms use gases in gas cabinets as a general rule. The location of the gas cabinet in relation to the cleanroom has become a major concern in recent years. The industry experience, particularly with the pyrophoric gases, has not been particularly good if the gas cabinets are located within the cleanroom air stream. Any leak or failure to contain the gases will result in recirculation of the gas or products of combustion through the cleanroom air filtration system. Relatively small fires, as a result of leaks at pipe joints, have resulted in major losses in terms of damage to the cleanroom environment. Placing the gas cabinets in the return air space or plenum sets up a situation where any failure, either human or mechanical, will guarantee that the gases or smoke from a fire are sent directly into the cleanroom. One approach to this problem is the blister building which is separated structurally from the cleanroom by a wall. The air-conditioning and exhaust systems for this separated structure are also completely separated from the cleanroom ventilation and exhaust systems. In this manner the gases are isolated to the extent that a leak or fire will be restricted to the gas building and not extend to the cleanroom.

Unfortunately, gas systems become more and more complicated in our efforts to control every potential problem area. The gas piping systems within the gas cabinets are referred to as the piping "jungle," with good reason. High pressure shut-off valves, pressure-reducing regulators, purge block assemblies, flow switches, vacuum generators, and so on, all occupy the space at the upper rear portion of the gas cabinet. It is necessary for the technician or engineer who changes out cylinders to understand the piping system for each gas cabinet. Understanding may not be as easy as it sounds. There may be several generations of gas cabinets in a single cleanroom, depending on the dynamics of the process changes which have taken place over time.

This situation raises the question of who should be allowed to change out gas cylinders in gas cabinets. The same people who store and deliver cylinders to the users are allowed to do the change-out in some locations. Others require that it be done by process engineering people from the user area only. Some operations have no ground rules at all; anyone who needs to change out a cylinder may do so. The current school of thought is that there probably should be a dedicated team of experts on the site who are responsible for this activity. The potential variety of gas cabinets and their internal piping layout certainly seem to point to the need for someone who has specialized knowledge to do the cylinder changing. Whereas local exhaust ventilation in gas cabinets is certainly capable of handling a minor leak which might occur during the changing process, these systems are not capable of guaranteeing that there will be no significant exposure to employees in the event of a more serious problem. This thought process has led to the general practice of having the cylinder change personnel wear supplied air respiratory protection during the changing process. The use of a dedicated team of experts to do the changing significantly eases the burden on the safety system if respiratory protection is used. There are fewer people to monitor through the mandatory respiratory protection standard of OSHA. The training activities are simplified as well.

Bitter experience has taught the industry that cylinders must be examined for leaks before they are brought into potential contact with people. Some locations require that the gas supplier leak-check each gas cylinder prior to delivery to the site and that he provide a certificate that the cylinder does not have any leaks. Other locations conduct their own leak-checks during the receiving process, and still others leak-check prior to delivery. The point is that after a cylinder has sat in storage for a time, been manhandled through a building or several buildings, and installed in a gas cabinet, it is not the time to discover that it has been leaking for some unknown length of time. The numbers of potentially exposed employees in a situation such as this become astronomical. Leak-checking of toxic gas cylinders in particular must be accomplished before there is opportunity for exposure to people.

Cylinders which are removed from gas delivery systems are traditionally marked MT so they are not reused. This has lead to some confusion in the manifesting of cylinders back to suppliers out of state by common carrier. This MT cylinder in fact probably has on the order of several hundred pounds of residual gas pressure if it is a high-pressure cylinder. The cylinder cannot be manifest as empty under Department of Transportation regulations if it contains more than 25 p.s.i. residual pressure. Fires involving common carrier vehicles have resulted in citations from transportation authorities for errors in the manifesting process.

Wet Chemical Safety Revisited

There is too much wet chemical safety information available for this subject to be treated here in any detail. The handling, storage, use, and disposal of liquid chemicals is the subject of entire volumes. However, there are some special problems with the handling of wet chemicals which are unique to the cleanroom and which do need to be addressed. As mentioned previously, the air recirculation system in a cleanroom literally guarantees that any accident involving the discharge of liquid chemicals onto the cleanroom floor will extend to at least part of the cleanroom not directly involved in the accident. The laminar flow air patterns pick up the vapors of the liquid which has been spilled onto the floor and recirculate them via the HEPA filters to other parts of the cleanroom. The extent of the area involved in the broadcast of chemical vapors will be dependent upon the design of the air return and circulation system installed in the cleanroom. If the entire cleanroom is served by a single charged plenum, the entire cleanroom will be involved in the chemical spill.

The design of the air circulation system must be examined to predict the extent of potential exposure zones in event of a spill. Once this is determined, it is necessary to put procedures into place to address the handling of the emergency. How much of the area needs to be evacuated for the safety of the occupants? Do we evacuate the entire cleanroom? These decisions must be addressed *before* the first accident. It is practical to assign and train an internal team of experts to deal with such an emergency—first, because they are usually there. They are close enough to the problem to respond in a timely manner to start the stabilizing and clean-up process before damage to equipment can occur. Secondly, if the people in the area know that they will be ultimately responsible for cleaning up their own messes, there is less likelihood they will be messy to start with.

Some locations use the plant emergency response team to handle chemical spills. It can be made to work, but there are several logistical problems to be overcome. Someone in the area must still be trained to recognize the problem and order evacuation. The call for help must be issued via some system so that the response team can be mustered.

Even in an emergency situation, the cleanroom management can be very zealous about cleanroom clothing, and delays before stabilization and clean-up have been caused while waiting for the emergency response team to get into cleanroom clothing and then protective clothing so they can enter the cleanroom. This attitude of being clean even in an emergency can appear to be rather silly to most people. If chemicals have been spilled onto the floor of the cleanroom and are being broadcast throughout the cleanroom causing damage, it seems absolutely ridiculous to insist that the responder be required to put on cleanroom garb. However, in the case of a small spill, the presence of an uncovered person can do untold damage to product which is in process. The decision must be made in advance of any potential spill in the cleanroom as to what the ground rules are. Rules in place, cleanroom clothing and protective clothing can be strategically placed to accommodate quick entry, if it is determined that the emergency response people will not be allowed in the cleanroom without it.

Multilevel cleanrooms pose some specialized problems with regard to the potential for chemical spills. Pumps and power supplies for equipment in the cleanroom are frequently located in the subfloor area under the active cleanroom level. The location of this equipment and the potential for equipment maintenance or facilities maintenance causes some special problems. If an employee is working on a vacuum pump under an etch bay at the time an accidental chemical spill takes place, the results could be disastrous. One location works maintenance on such systems by way of hazardous work permits. They require that the maintenance people obtain a work permit from the production operation to authorize the work to be done, in writing. This allows the work to be scheduled, and will permit the individual bay over the maintenance operation to be shut down while the maintenance is carried out.

The semiconductor industry is moving more and more to delivery of wet chemicals via piping systems directly to the use point. This solves many of the materials-handling problems and storage problems attendant with the handling of chemicals in individual containers. It does, however, bring a whole new set of potential problems to the front. The potential for piping system leaks, valves which will not open or close, container overflow, and so on, all must be addressed. No matter how much automation is introduced, there will always be some chemicals in individual containers. It simply is not economical to pipe in all chemicals used in the cleanroom. So we will continue to be faced with the problems or potential problems of both manual handling and systems handling of chemicals in the cleanroom.

Disposal

Disposal is an area which is very heavily regulated by local municipalities as well as the federal government under the EPA. Local officials generally issue pretreatment standards detailing exactly how much of each class of material may be discharged into the sanitary system. Treatment systems must be designed to meet these discharge limits either using a continuous or batch process. If the standard cannot be met, we are faced with collecting the waste material and arranging for disposal via other avenues.

Disposal may be by seeking another potential user for the material. Solvents from the semiconductor industry have been successfully sold to reprocessors who clean them up and sell them to others. Each waste stream must be examined as a potential for recycling or reuse. Failing to find a market for a waste material, it becomes necessary to dispose of the material via an approved hazardous materials disposal site. Since Love Canal the trend seems to be that the generator of hazardous waste can be held liable for problems even after disposal. Because of this philosophy, it is necessary that the contractor who hauls waste, and the location receiving the waste, be examined very carefully. If there are leaks or problems in future years, you may have to fund a project to dig up and remove your waste from one location or another.

On site, a disposal pad is usually provided so that hazardous waste coming from the cleanroom operation can be properly segregated and held for disposal. It is important that all parties in the cleanroom understand that the waste must be characterized for disposal. It is no longer acceptable to simply get rid of this or that. Waste must be properly identified so that it can be classed properly at the disposal site. For instance, chlorinated solvents should always be kept separate from nonchlorinated varieties. If these two types are mixed, the resulting mixture is considered to be chlorinated and hence represents a bigger (more expensive) disposal problem. Unidentified waste must frequently be analyzed to characterize it for disposal. Analysis is an expensive proposition and should be avoidable if the waste generators at the site manage their chemical inventories properly.

Environmental experts usually arrange to visit the proposed disposal sites to qualify them as responsible managers of hazardous material. Qualification involves examining the financial status of the operation as well as the physical layout of the site, handling procedures, paperwork procedures, and emergency capability. Most semiconductor manufacturers try to have three or four disposal sites qualified to receive hazardous waste so that if some untoward thing happens there will be options.

SUMMARY

To summarize, potentially dangerous substances are commonly used in high-technology manufacturing or research. It is imperative that the hazard properties of these substances are known before the materials even enter the facility. Hazardous materials can be flammable, reactive, corrosive, or toxic (or all of the above). Toxicity is of particular importance to the personnel working in the environment and also to the people who live and work near the facility. The hydride gases are particularly dangerous to humans as they are extremely toxic and are readily absorbed through the lungs. Also, skin protection, particularly in the form of appropriate gloves, is important for solvents, as this group of chemicals may be absorbed through the skin. Safety within the facility is the responsibility of a team consisting of engineering and industrial hygiene staff who are responsible for engineering and medical controls. Environmental monitoring, proper equipment layout, coordinated emergency response, and appropriate safety equipment are all critical in the safety scheme. Employee safety training is particularly important, as understanding is the key to success in maintaining a safe workplace.

13

The Future of Controlled Environments

We have seen how controlled environments can provide us with an appropriate place for the production of or research into sensitive materials, processes, or artifacts. As we discussed in Chapter 1, however, the very nature of high-technology industry creates a drive toward increasing complexity, and this more often than not increases the sensitivity factor. To use the semiconductor industry as an example, at the time of writing this book, more than 50 percent of the wafer starts in the United States involved technologies with linewidths below 2 microns. This suggests particle control at 0.2 microns and below in leading edge high-volume facilities. In addition, we are also witnessing the emergence on the market of super-high density components such as 4 megabit dynamic RAMs (memories) with claims that 64 megabit components are near to fruition.

Unfortunately, merely being able to make these components might be fine in the research arena, but it is never going to be enough in market terms. If yield, or reliability, is poor and volume is low, costs will increase, profits will be reduced, and a company's market share could be completely lost forever. In this realm, manufacturing begins to take over from physics.

A knee jerk reaction from the facilities camp to this drive in technology, coupled with the increased lean toward manufacturing priorities, might be to pour more money into the facilities themselves. This has been the trend up to this point in time and has led to the evolutionary style of controlled environment development. However, the cost of facilities is currently averaging $40M, with $100M not being unusual for larger plants. Of course, the expense of having a state-of-the-art facility does not stop with the capital outlay for the building and plant systems. The estimated air-handling cost for a modern facility is close to $100 per square foot, which means a $2M annual cost for the air alone (before HEPA filter costs) in a 20,000 ft^2 facility.

If we are to continue with this trend, only a few companies worldwide will be able to stay in the game, and we may lose much of the small company innovation which fuels the technology drive. Does this mean that the development of high-technology industry is self-limiting? We believe not, but it will take a revolutionary rather than evolutionary approach to controlled environment design and implementation. We will briefly discuss some forward-looking ideas in this chapter.

ADVANCED CLEANROOM CONCEPTS

In attempting to define what an advanced cleanroom actually is, we could take two routes. The most obvious definition perhaps takes us in the direction in which we are already travelling. Better cleanroom materials, garment materials and styles, equipment, filters, and so on, are constantly being developed. These bring down the contamination generation factor in our room, allowing a much cleaner environment and hence a better class of cleanroom, which is exactly what we need. Unfortunately, there is still a great deal of work required to develop appropriate cleanroom practices for advanced facilities of this type (we hope that this book helps in this respect). The West has generally been slow to react to the upgrading of practices, whereas the East, and Japan in particular, has moved ahead in areas like facilities construction, operator behavior, maintenance techniques, and so on. Success in the reform of practices requires a change in thinking by both workers and management, which can be attainable to a great extent through increased understanding by training of both groups. Remember that no matter how much money is spent on the facility, the beneficial effects of advanced cleanroom technology can be negated by a small percentage of employees who do not understand how critical their part is in the entire manufacturing scheme. Therefore our advanced cleanroom has to become more than just a facility in which we control particles, gases, vapors, vibration, static, temperature, humidity, and other physical factors; more than ever, we also have to control attitude.

An alternative way of defining what makes an advanced facility is to look at the revolutionary approach we mentioned before. Here we consider facilities which differ radically from those we have discussed in previous chapters. For example, since people are such a large source of contamination, either by being naturally dirty or by acting in a dirty fashion, we could remove or reduce the personnel factor by implementing a full or partially automated production line. This is not nearly as simple as it sounds, as robots are still not as versatile or smart as people. We will discuss this further later in this chapter. An alternative example of an advanced facility concept is one which is designed to maintain the highest possible class at the lowest possible cost. This may sound like mutually exclusive concepts being thrust together but is possible if we maintain the class only where it is required, that is, at the product. This class is notionally achievable by putting together a reduced volume room, or more correctly an extended chamber, in which the material travels between clean process stations, or by making a mini-cleanroom follow the product around the facility. Our greatly reduced cleanroom volume means much smaller air handling costs, and we also get the personnel out of the clean area. We will carry on this discussion of enclosed transfer systems in the next section.

Perhaps the ultimate advanced controlled environment is one which embodies a high degree of automation and enclosed transfer. However, in order to implement an advanced facility of this nature, we have to re-think our ideas of cleanroom layout. For instance, a cleanroom layout designed for people and manual product or lot movements is not necessarily the optimum configuration for an automated production line. For example, we currently tend to build our production lines as linear entities to accommodate what used to be a linear flow of product through the line from one piece of equipment to the next. However, in a highly automated plant, we may wish a higher degree of flexibility to allow a genuine product mix on the line. A

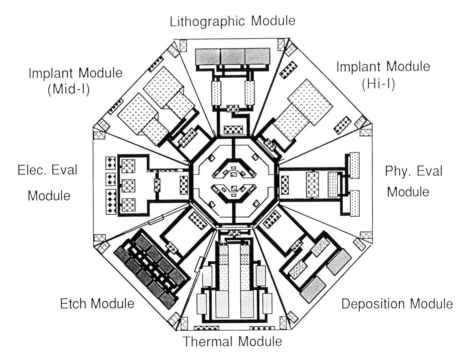

Figure 13-1 Advanced production facility layout. Heavy black lines are automated material handling paths (Courtesy of J. Coulthard).

linear configuration for one product type is not necessarily optimal for another, and hence we may have to reserve the linear concept for fixed process lines turning out large quantities of a commodity product. A good alternative is an annular layout in which all the production modules are arranged around a central hub, through which all product passes on its way to the next module. An example of this kind of facility is shown in Figure 13.1. This lends itself well to flexibility in automation as fixed position robots or circular track transfer systems can be used for intermodule transfer. This approach is very time-efficient and overcomes many of the transfer scheduling problems of a linear line. Note that expansion in such ring-like facilities is achieved by duplicating the entire line (for capacity) or by expanding to an outer ring of modules (for increased capability in some process steps).

It is also likely in advanced cleanroom facilities that we will see the emergence of micro-factories which will be small in size and in terms of the number of components produced but will be highly flexible to allow a large variety of products to be produced. This flexibility will be gained through automation, a subject which is discussed briefly in Chapter 12.

ENCLOSED TRANSFER SYSTEMS

As we mentioned briefly in the previous section, it is a rather attractive idea to reduce the requirement of a super-high-class cleanroom by the use of some type of enclosed transfer system which keeps the immediate environment around the materials and product clean during the journey from one piece of equipment to another. For this concept to be realistic, we have to assume that the equipment has an internal clean environment, too, and that the product may move from the transfer system to the equipment without risk of contamination.

Many pieces of equipment in the semiconductor industry are already at least partially compatible with this concept in the respect that they are internally relatively clean. For example, water steppers put the wafers in a low-contamination ambient during processing by ensuring that all system materials and mechanical assemblies are made to be as clean as possible. By supplying HEPA-filtered air into the system and developing a slight positive pressure we can have a mini-cleanroom within the stepper itself. Note that we still have to be careful not to have severe dead spots where contamination can gather. This kind of problem may be reduced by appropriate routing of the air or by shielding the product from relatively dirty areas. The air supplied to the HEPA filters and to the rest of the room still has to be preconditioned to maintain optimal temperature and humidity conditions. Open systems such as track coaters for photolithography are more of a problem in that a clean box or canopy has to be built around them and supplied with HEPA-filtered air. This can significantly add to the cost of the equipment, but once again reduces the facilities costs by providing a mini-cleanroom around the equipment.

One of the big problems of this clean equipment approach is interfacing the transfer system with the equipment in such a manner that the product does not get contaminated during the *handoff,* or transfer between pieces of equipment. There are several enclosed transfer systems which solve this problem. The system which is currently utilized in the semiconductor industry to some extent is the Standard Mechanical Interface (SMIF) system. In effect, the SMIF system controls only the immediate wafer environment with the idea that it is the wafer that counts, but it also includes equipment interfaces. Transfer between equipment is by way of sealed containers as shown in Figure 13.2. The wafers are held in a carrier in the bottom-loading SMIF Pod.

The system starts with clean wafers being loaded into a SMIF Pod by the

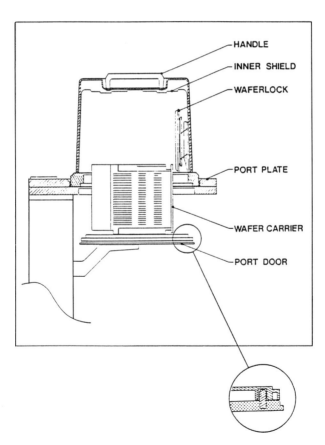

Figure 13-2 Cross-sectional detail of cassette being loaded into SMIF pod.

wafer manufacturer. The pod serves as the shipping container to get the wafers to the user. In this process, the outside of the SMIF box will inevitably get dirty. However, as we will see, this dirt cannot get onto the wafer or into the processing system due to the sealing techniques used. When the wafers are ready for processing, the SMIF box is placed on a SMIF elevator that is programmed to open the door, remove the cassette of wafers, and place them in the processing cycle (Figure 13.3). Note that the wafers are never exposed to the outside environment (including the outside of the box), nor is the equipment contaminated by the environment, as the pod seals off the loading port so that it may be opened. Just as important is that any dirt on the base of the pod is sandwiched between the pod and the port door, and therefore we may achieve transfer without contamination. When the wafers come out of the equipment, they go back into a SMIF box and are stored or transported to the next location where the removal-process-replace cycle will be repeated. In Figures 13.4 and 13.5 we show some results with operational SMIF systems. Even in a class 20,000 environment, the interior of a SMIF box is apparently cleaner than a class 10 room. In fact, the quality in terms of particulate contamination of the environment seems to make little difference to the internal environment.

The SMIF system is custom-designed for semiconductor substrates but potentially could be modified for other flat substrates or small components, for example, compact discs, computer discs, or flat panel display elements. Other items, in particular large components or irregularly shaped equipment parts, may not readily be accommodated by SMIF, but similar concepts do exist. For instance, the European La Calhene system will allow transfer of

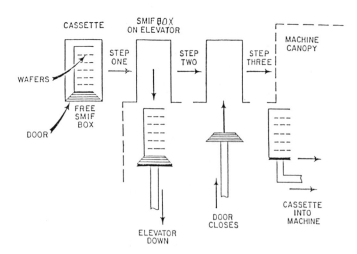

Figure 13-3 Schematic drawing of SMIF system in operation (scale none) (from Asyst Technologies, Inc., Fremont, CA).

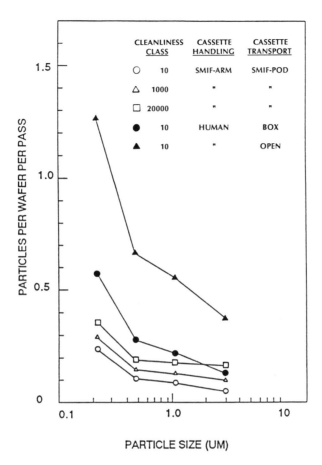

Figure 13-4 Experimental results with SMIF vs. gowned and masked human operators under a variety of cleanroom conditions. 100 cycles involving wafer loading, unloading, and travel for 30 meters (data provided by Mihir Parikh, Asyst Technologies, Inc.).

many aerospace, precision mechanical, biomedical, and so on items. This system may also, of course, be used in the semiconductor industry. In fact, the experimental Oasis Project utilizes a combination of transfer techniques to allow a range of semiconductor processing steps (but not the whole process as yet) to be performed in an office environment. The results so far are proving to be extremely promising.

The SMIF and La Calhene systems are excellent for batch transfer, which is how a great deal of high-technology production is performed. Note that the containers take the place of standard lot boxes and are, of course, a much cleaner form of storage. As an alternative, we could utilize systems which use a continuous transfer technique. We can still keep to the concept of sur-

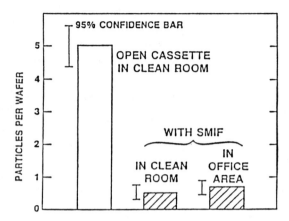

Figure 13-5   Comparison of SMIF performance in different external environments.

rounding the product with a mini-cleanroom, but in this instance the clean-room is actually an elongated tunnel structure with a cross section just large enough to take whatever is being transferred plus the carrying mechanism. Generally, these systems have small plenum-fed HEPA filters as their top to supply clean air into the tunnel and have trolleys driven by a variety of methods to carry the product. The trolleys may be wheeled using low-friction/low-contamination bearings. Magnetic levitation is also being used in some cases to reduce any friction-produced contamination. The drive for the trolleys can be by way of a flexible worm drive at the bottom of the tunnel or supplied from the exterior via a slot (positive tunnel pressurization helps to prevent the ingress of contamination). Magnetic coupling through a sealed tunnel envelope is potentially the best trolley drive as friction is reduced in addition to maintaining seal integrity. The tracks for these systems are generally run above head-height so that equipment and personnel movement is unim-paired. Elevator units allow materials to move from the equipment to these overhead tracks and vice-versa.

There are systems available, developed for the semiconductor industry in Japan, which combine the above concepts to some degree. These systems use a wafer carrier which is not sealed like the SMIF pod but is actually open to the environment. Control of contamination around the wafers is achieved by supplying HEPA-filtered clean air driven by a battery-powered blower on the carrier itself. The batteries are recharged when the carriers are placed in special storage racks. This concept is good in theory but is still at the practical evaluation stage in one large company.

So far, the transfer systems we have examined have utilized clean air which has been sealed in or continuously put in our product-surrounding clean zone. One extension of the enclosed transfer concept is to remove not only the particulate contamination but also the bulk of the gaseous contamination by using an inert gas instead of air or even removing all (or nearly all) gas and using a vacuum environment instead. There are many materials and structures to which even the oxygen in the air, as well as other gaseous contaminants, can be harmful. Storage and transfer (and processing when possible) in nitrogen or argon should be used to reduce this problem. When sensitive materials and structures are created in a vacuum or ultra-high vacuum (UHV) environment, ideally transfer should also be done in a vacuum environment; that is, the transfer environment should be as compatible with the equipment environment as possible. A good example of this technique may be seen in molecular beam epitaxy (MBE) systems which are linked to other such systems or other processing chambers (evaporators, etchers, and so on). In this case, a vacuum tunnel can be used for transfer, with materials being carried on trolleys as before.

Since we have brought up the subject of vacuum systems, we should be quite clear about one point—although most of the air and associated gaseous contamination have been removed, the environment need not be particularly clean. For one thing, the surfaces within the vacuum will outgas water vapor and hydrocarbons for some time, unless they are baked-out using elevated temperatures beforehand (for materials which can withstand high temperatures only). As we may see from Figure 13.6, ultraviolet light will also help

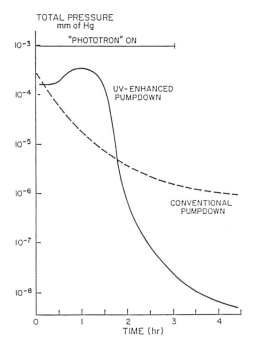

Figure 13-6   Effect of ultraviolet exposure on pumpdown of a vacuum system (from *Research and Development*, Vol. 28, No. 5, p. 101, 1986).

to rapidly desorb the outgassing species. Low levels of gas in vacuum systems can be measured using a residual gas analyzer (RGA), which is really a compact quadrupole mass spectrometer. However, our problems are far from over if we only reduce the amounts of residual gases in the environment as there will still be particles present. It is a common misconception to assume that if we remove the air, particles cannot be transferred from surfaces and mechanisms to sensitive areas. Unfortunately, particles can move in vacuum systems with great ease and rapidity, partly because there is little air to impede their progress. We cannot use a standard light-scattering particle counter to detect particles in a vacuum as these rely on air carrying the particles through the measurement cell. However, we can bring the light into the vacuum by using a beam net set-up as shown in Figure 13.7. Any particle passing through this net will disturb the light reaching the detector and hence we can detect and count them.

Naturally, the times of greatest particle generation and movement are during pump-down and when the system is brought back to atmospheric pressure, as there will tend to be great turbulence during these times. Figure 13.8 shows the effects of the speed of pump-down and ventilation on the amounts of particles present. As the time constant of the process comes down (that is, the speed goes up), the numbers of particles rise dramatically. However, if we reduce the turbulent effects by keeping the time constant above 200 seconds, particle generation is considerably reduced. Unfortunately, this might not be very practical as it would take a very long time to evacuate a system at this rate, that is, a 200-second value means that it would take this time for the pressure to drop to 1/e of its initial level. In addition, once a high vacuum level has been reached, particles can travel by forces created by static charges. Also, many vacuum processes will generate particles. For instance, ion implantation is known to generate particles although we are unsure at this time exactly why.

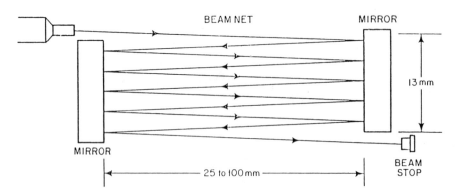

Figure 13-7 Schematic drawing particle fallout detector (Source: *High Yield Technology*).

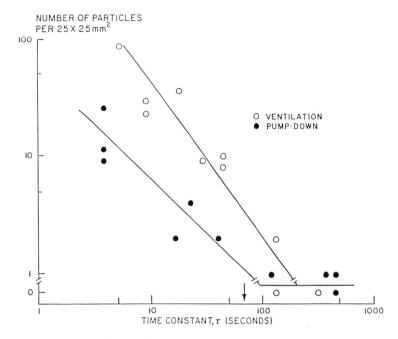

Figure 13-8 Effect of pumpdown or ventilation time on particle generation and deposition in a vacuum system, minimum particle size 0.5 microns (H. Yamakawa, ULVAC Corporation Japan).

One of the biggest potential problems with enclosed transfer systems is maintenance. Since the idea is that we surround the product with clean space and the rest of the room can be relatively uncontrolled, we will have contamination problems if we have to break the system for maintenance purposes. In the event of a sudden breakdown, any materials within the system will become contaminated by the external environment and will have to be cleaned, if possible, or discarded. Even scheduled maintenance has its problems as even though there may be no product to become contaminated, the internals of the transfer system or processing equipment will become contaminated and will have to be cleaned, otherwise the contamination will reduce the internal cleanliness for some time after the system is resealed. We can reduce these problems by having a cleanroom as the external environment. This may sound like we are defeating the purpose of our enclosed systems, but the external environment need not be super-clean. For instance, we may attain a sub-class 1 environment where we need it but have a class 1000 room for maintenance purposes. A more economic alternative is to supply a clean maintenance environment only where it is required. For instance, if we have to break a fixed transfer system at a particular point, we could erect a clean maintenance tent (CMT) over the region. The CMT is fed with HEPA-filtered air to flush out contaminants and pressurize the region around the break. The tent could be of a transparent vinyl or even an electret, and personnel working in the region should be gowned appropriately.

ROBOTICS AND AUTOMATION

It is exciting to many to consider the complete removal of humans and their associated contamination from the cleanroom and to have all processing and material handling performed by robotic systems. This would certainly reduce the contamination generation rate within the cleanroom and would also have other benefits; for example, the human stress factors would virtually be eliminated, shift changes and their contaminating influence would be removed, and so forth. We would do well to remember that in the automation of wafer movements, we should consider a layout which is optimal for automation (for example, the annular facility described earlier in this chapter). Also, automation goes far beyond the mere moving of materials and work-in-process from one work station to another. Information gathering, storage, and retrieval must also be automated with a high degree of integration between databases and high levels of communication between every area of the production environment.

Where do we currently stand as far as automation is concerned? Automated material movement between work cells is a well-developed technology with conveyors and guided vehicles being used in a number of advanced plants. This is a particularly good application of robotics, as much contamination within the cleanroom is generated or stirred up by personnel movements. Another use of robotics is in wafer transfer in the semiconductor industry. When wafers are transferred manually from carriers to processing equipment, contamination from operators or wafer damage (chipping, scratching, or dropping) is highly possible. Robotic systems can reduce the effects dramatically, not to mention saving time as they may operate precisely even at high operating speeds. Information automation is a dynamic field, with advances being made and new products appearing on a regular basis. Lot tracking programs, in conjunction with factory schedulers and factory and process simulators, are now becoming commonplace. We do not at this stage have a fully automated and integrated factory, but this dream should be realizable in the not too distant future.

As a parting negative comment, the use of robotics does not totally remove contamination, as our new operators are not 100 percent contamination-free themselves. As we mentioned near the beginning of this text, any moving components in equipment can generate contamination by frictional effects. This contamination is not nearly as bad as that of human origin but nevertheless is still there. Figure 13.9 illustrates the levels of contamination asso-

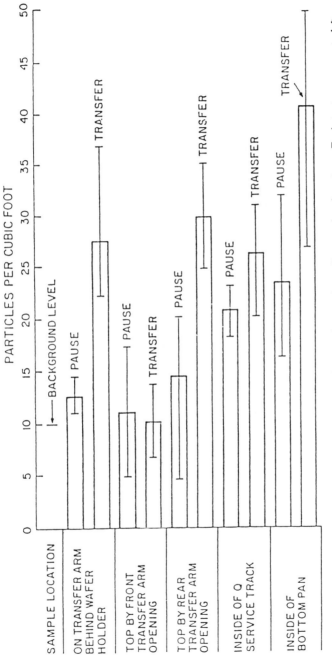

Figure 13-9 Contamination measurements on a wafer transfer machine. Data at five locations. Each test was repeated five times.

305

ciated with a wafer transfer robot in a class 10 room. As we can readily see, transfer operations generate the most contamination as it is during these that gears will be meshing, motors turning, pistons sliding, and so on. Many cleanroom robots wear their own cleanroom suits to cover their generating areas, but it should be kept in mind that these will become dirty and will wear and therefore must be cleaned or replaced periodically.

IN-SITU PROCESSING

Since it is possible to make the inside of many pieces of equipment relatively clean, can we dispose of the need to transfer product between systems and instead do all processing in one chamber? This would eliminate the requirement for a cleanroom, except perhaps for maintenance purposes, as the product would never leave the equipment. The dream of a "sand in, chips out" machine has been around for a long time but is far from realization. However, research has taken us a little way down this road.

Current in-situ processing systems are based on cells or a limited sequence of events within a process. For instance, in one scheme vacuum-compatible robots are placed at the hub of a multiple chamber deposition system and pick-and-place product from chamber to chamber. This is not exactly the same as all-vacuum transfer as this is really one piece of equipment, capable of multiple process steps in-situ. The great benefit of this concept is that not only is particulate contamination reduced, gaseous contamination, including the oxidizing effects of the air, is also greatly reduced. This is critical in certain multilayer metallization processes as oxygen can severely alter the characteristics of the structure being produced. There is no better way of tackling this problem than to use an in-situ/in-vacuum system such as this.

A further example, still at the development stage of the time of the writing of this book, is in-situ thermal processing. Here we have multiple high-temperature steps such as diffusion, oxidation, and chemical vapor deposition in one chamber or a series of independently heated interconnected chambers. Once again, the advantage of this is that the contamination of critical interfaces is greatly reduced, and this can have a profound effect on product stability and yield/reliability.

One of the main problems with a total in-situ processing system is that lithography using organic resist materials cannot readily be performed in-situ. This is because photoresists are almost impossible to apply without contaminating the immediate environment to some degree with excess lithographic material. An alternative to this may be inorganic resist materials which may be applied, exposed, developed, and stripped in vacuum-compatible conditions. This is very possible but is still a long way off from being seen in production systems.

EPILOG—THE CLEANROOM AS A SYSTEM

We have studied many aspects of the broad subject of controlled environments in this elementary text. Unfortunately, time and space constraints do not allow us to go into any great depth in any one particular area; specialized texts exist which do this for subjects like contamination, HVAC, and so forth. However, it is important for the reader to realize that the cleanroom is not merely one "clean room" but is actually a highly complex system consisting of multiple interacting parts. It is very true in our case that the whole is very much greater than the sum of the parts.

Remember also that the parts which make up our system are not just the physical elements like the filters, gas delivery systems, reverse osmosis units, and so forth; they also include safety practices, employee and management attitudes, training, and so on. Every part is effectively a link in a long chain. The chain can be no stronger than its weakest link which is why every "component" we have discussed in this book is so important.

It may be that some companies are headed toward so-called advanced facilities with clean transfer systems and robotics. However, many cleanroom applications will not accommodate these techniques. Therefore the basic problems of controlled environments will probably always be with us.

Appendix 1

Federal Standard 209D

FEDERAL STANDARD

CLEAN ROOM AND WORK STATION

REQUIREMENTS, CONTROLLED ENVIRONMENT

This standard is approved by the Commissioner, Federal
Supply Service, General Services Administration, for the
use of all Federal Agencies.

CONTENTS

I

APPENDIX A

PARTICLE MONITORING - MANUAL COUNTING AND SIZING METHODS

APPENDIX B

OPERATION OF OPTICAL PARTICLE COUNTERS

1. Scope and limitations.

1.1 Scope. This document establishes standard classes of air
cleanliness for airborne particulate levels in cleanrooms and
clean zones. It prescribes methods for class verification and
monitoring of air cleanliness. It also addresses certain other
factors, but only as they affect control of airborne particulate
contamination.

1.2 Limitations. The requirements of this document do not apply
to equipment or supplies for use within cleanrooms or clean
zones. Except for size classification and population, this
document is not intended to characterize the physical, chemical,
radiological, or viable nature of airborne particulate
contamination. No definitive relationship between airborne
particulate cleanliness classifications and the level of viable
airborne particles has been established. In addition to the need
for a clean air supply monitored for total particulate
contamination and meeting established limits, special
requirements are necessary for monitoring and controlling
microbial contamination.

2. Referenced document.

For further information on Student's t, see: Johnson, Norman L.
and Leone, Fred C., Statistics and Experimental Design in
Engineering and the Physical Sciences, Volume I (New York,
London, Sydney: John Wiley & Sons, Inc., 1964).

3. Definitions.

3.1 Airborne particulate cleanliness class. The statistically
allowable number of particles equal to or larger than 0.5
micrometer in size per cubic foot of air.

3.2 Calibration. Comparison of a measurement standard or
instrument of unknown accuracy with another standard or
instrument of known accuracy to detect, correlate, report, or
eliminate by adjustment any variation in the accuracy of the
unknown standard or instrument.

1

3.3 <u>Clean zone.</u> A defined space in which the concentration of airborne particles is controlled to specified limits.

3.4 <u>Cleanroom.</u> A room in which the concentration of airborne particles is controlled to specified limits.

3.4.1 <u>As-built cleanroom (facility).</u> A cleanroom (facility) that is complete and ready for operation, with all services connected and functional, but without production equipment or personnel within the facility.

3.4.2 <u>At-rest cleanroom (facility).</u> A cleanroom (facility) that is complete and has the production equipment installed and operating, but without personnel within the facility.

3.4.3 <u>Operational cleanroom (facility).</u> A cleanroom (facility) in normal operation with all services functioning and with production equipment and personnel present and performing their normal work functions in the facility.

3.5 <u>Unidirectional airflow.</u> (commonly known as laminar flow) Air flowing in a single pass in a single direction through a cleanroom or clean zone with generally parallel streamlines.

3.6 <u>Nonunidirectional airflow.</u> (commonly known as turbulent flow) Airflow which does not meet the definition of unidirectional airflow by having either multiple pass circulating characteristics or a nonparallel flow direction.

3.7 <u>Condensation nucleus counter.</u> An instrument for counting small airborne particles, approximately 0.01 micrometer and larger, by optically detecting droplets formed by condensation of a vapor upon the small particles.

3.8 <u>Optical particle counter.</u> A light-scattering instrument with display and/or recording means to count and size discrete particles in air.

3.9 <u>Particle.</u> A solid or liquid object generally between 0.001 and 1000 micrometers in size.

2

3.10 <u>Particle size.</u> The apparent maximum linear dimension of the particle in the plane of observation as observed with an optical microscope, or the equivalent diameter of a particle detected by automatic instrumentation. The equivalent diameter is the diameter of a reference sphere having known properties and producing the same response in the sensing instrument as the particle being measured.

3.11 <u>Particle concentration.</u> The number of individual particles per unit volume of air.

3.12 <u>Student's t distribution.</u> The distribution:

t = [(population mean) - (sample mean)]/[standard error]

obtained from sampling a Gaussian ("normal") distribution. (Available in tables in statistics texts.)

3.13 <u>Upper confidence limit (UCL).</u> An upper limit of the estimate of the mean, calculated in such a way that in a given percentage of cases (here, 95%) the upper limit of the estimate would be more than the true mean, if the means were sampled from a Gaussian ("normal") distribution.

4. <u>Airborne particulate cleanliness classes.</u>

4.1 <u>Determination of class limits.</u> Airborne particulate cleanliness classes listed in Table I shall be determined as follows:

3

TABLE I

Class limits in particles per cubic foot of size equal
to or greater than particle sizes shown (micrometers)[a]

Class	Measured Particle Size (Micrometers)				
	0.1	0.2	0.3	0.5	5.0
1	35	7.5	3	1	NA.
10	350	75	30	10	NA.
100	NA.	750	300	100	NA.
1,000	NA.	NA.	NA.	1,000	7
10,000	NA.	NA.	NA.	10,000	70
100,000	NA.	NA.	NA.	100,000	700

(NA. - not applicable)

[a]The class limit particle concentrations shown in Table I
and Figure 1 are defined for class purposes only and do not
necessarily represent the size distribution to be found
in any particular situation.

4.2   Particle sizes measured to determine Classes 100 and
greater.   Airborne particulate cleanliness classes shall be
determined by measurement at any one of the particle sizes listed
for the class in Table I.   The class is considered met if the
measured particle concentration is within the limits specified,
at any one of the particle sizes shown in Table I, as determined
by the statistical analysis of Paragraph 5.4.

4.3   Particle sizes measured to determine Classes less than 100.
Airborne particulate cleanliness classes shall be determined by
measurement at one or more of the particle sizes in Table I, as
specified[1], and determined by the statistical analysis of
Paragraph 5.4.

---

[1]When the terms "as specified" or "shall be specified" are
used without further reference, the degree of control
needed to meet requirements will be specified by the user
or contracting agency.

4

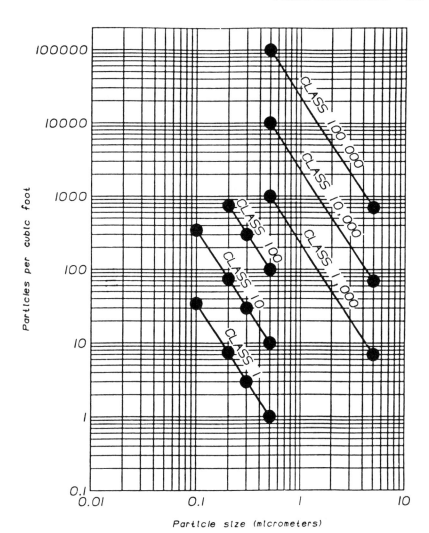

Figure 1. Class limits in particles per cubic foot of size equal
to or greater than particle sizes shown*

*The class limit particle concentrations shown on Table I and
Figure 1 are defined for class purposes only and do not
necessarily represent the size distribution to be found in any
particular situation.

5

4.4 Classification by measurement at other particle sizes. A classification of air cleanliness by measurement at particle sizes other than those specified herein may be performed with the following limitation: Particle counts may be interpolated between points but under no conditions may counts be extrapolated beyond the end points of Table I or Figure 1. The particle count limit for the next larger particle size in Table I must not be exceeded.

4.5 Provision for alternative airborne particulate cleanliness classes. Classes other than those stated in Table I (for example, 50, 300, 50,000, etc.) may be defined where special conditions dictate their use. Such classes will be defined by the intercept point on the 0.5-micrometer line in Figure 1, with a curve parallel to the established curves. Any curves that are used for these other classifications shall not be extrapolated to indicate concentrations for particles outside the following limits:

   (a)   Classes greater than 1,000 shall be determined by measurement at either 0.5 or 5 micrometers, as specified[1].

   (b)   Classes greater than 10 and less than 1,000 shall be determined by measurement at 0.2, 0.3, or 0.5 micrometer, as specified[1].

   (c)   Classes less than 10 shall be determined by measurement at one or more of the following particle sizes: 0.1, 0.2, 0.3, or 0.5 micrometer, as specified[1].

4.6 Particle counts for the determination of cleanliness classes. To determine an airborne particulate cleanliness class, particle counts shall be made in accordance with Section 5.

---

[1]When the terms "as specified" or "shall be specified" are used without further reference, the degree of control needed to meet requirements will be specified by the user or contracting agency.

6

5. Verification and monitoring of airborne particulate
cleanliness classes.

5.1 Verification of airborne particulate cleanliness classes.
The airborne particulate cleanliness class as defined in Section
4 shall be verified for a cleanroom or clean zone by measurement
of airborne particle concentration under the following
conditions.

5.1.1 Frequency. Verification tests shall be performed
initially and at periodic intervals, or as specified[1].

5.1.2 Environmental test conditions. Verification of air
cleanliness class shall be determined by particle concentration
measurement under specified[1] operating conditions.

5.1.2.1 Conditions of test. The conditions of test of the
cleanroom or clean zone shall be recorded as "as-built," "at-
rest," "operational," or as otherwise specified[1].

5.1.2.2 Environmental and use parameters. The applicable
environmental and use parameters of the cleanroom or clean zone
shall be recorded. These conditions of measurement may include
(but are not limited to) air velocity, air volume change rate,
room air pressure, makeup air volume, unidirectional airflow
parallelism, temperature, humidity, vibration, equipment, and
personnel activity.

5.1.3 Particle counting. Particle counting shall be performed
using a method specified in Paragraph 5.3 for verification of all
classifications of cleanrooms and clean zones.

5.1.3.1 Sample locations and number - unidirectional airflow.
For unidirectional airflow, the clean zone is identified by an
entrance and an exit plane perpendicular to the airflow. The
entrance plane shall be immediately upstream of the work activity

---

[1]When the terms "as specified" or "shall be specified" are
used without further reference, the degree of control
needed to meet requirements will be specified by the user
or contracting agency.

area within the clean zone. The minimum number of sample
locations required for classification of a clean zone shall be
the lesser of (a) the area of the entrance plane (in square feet)
divided by 25, or (b) the area of the entrance plane (in square
feet) divided by the square root of the airborne particulate
cleanliness class designation.

5.1.3.2 Sample locations and number - nonunidirectional airflow.
For nonunidirectional airflow, the number of sample locations[1]
shall be uniformly spaced horizontally, and as specified
vertically, throughout the clean zone, except as limited by
equipment within the clean zone. The minimum number of sample
locations shall be equal to the square feet of floor area of the
clean zone divided by the square root of the airborne particulate
cleanliness class designation.

5.1.3.3 Sample location restrictions. No fewer than two
locations shall be sampled for any clean zone. The number of
sample locations shall be uniformly spaced throughout the clean
zone except as limited by equipment within the clean zone. At
least one sample shall be taken at each of the sampling locations
specified in Paragraph 5.1.3.1 or 5.1.3.2. A total of at least
five samples shall be taken. More than one sample may be taken
at each location and different numbers of samples may be taken at
different locations.

5.1.3.4 Sample volume and sampling time. Table II lists the
minimum volume per sample for various airborne particulate
cleanliness classes and measured particle sizes. The sample time
is calculated by dividing the sample volume by the sample flow
rate. A larger sample volume will improve the precision of the
concentration measurements, decreasing the amount of variation
between samples; however, the volume should not be so large as to
render the sample time impractical. The particle concentration
shall be reported in terms of particles per cubic foot of air
regardless of the sample volume size. The sample volume size
shall also be reported.

---

[1]When the terms "as specified" or "shall be specified" are
used without further reference, the degree of control
needed to meet requirements will be specified by the user
or contracting agency.

TABLE II

Minimum volume per sample in cubic feet for the air
cleanliness class and measured particle size shown

Class	Measured Particle Size (Micrometers)				
	0.1	0.2	0.3	0.5	5.0
1	0.6	3.0	7.0	20.0	NA.
10	0.1	0.3	0.7	2.0	NA.
100	NA.	0.1	0.1	0.2	NA.
1,000	NA.	NA.	NA.	0.1	3.0
10,000	NA.	NA.	NA.	0.1	0.3
100,000	NA.	NA.	NA.	0.1	0.3

(NA. - not applicable)

5.1.3.5 Sample volume at other classes or particle sizes.
Sample volume for other classes or particle sizes not specified
herein shall be the same as that specified for the next lower
class or particle size.

5.1.4 Interpretation of the data. A statistical evaluation of
particle concentration measurement data shall be performed
according to Paragraph 5.4 to verify the airborne particulate
cleanliness class level.

5.2 Monitoring of airborne particulate cleanliness. After
verification, if specified[1], the airborne particulate cleanliness
shall be monitored during operations. Monitoring shall consist
of particle concentration measurements. Other environmental
parameters as suggested in Paragraph 5.1.2.2 may also be
monitored as specified[1] to indicate trends in airborne
particulate cleanliness.

[1]When the terms "as specified" or "shall be specified" are
used without further reference, the degree of control
needed to meet requirements will be specified by the user
or contracting agency.

9

5.2.1 Monitoring plan. A monitoring plan shall be established based on the airborne particulate cleanliness class and the degree of cleanliness control necessary for work activity or product protection. The monitoring plan shall specify frequency, operating conditions, the method of counting particles, the locations, number, and volume of samples, and some method for interpretation of the sample data.

5.2.2 Particle counting. Particle counting shall be performed using one of the test methods in Paragraph 5.3, as specified[1]. Particle concentration measurements shall be taken at locations throughout the clean zone or where the cleanliness level is particularly critical or where the higher particle concentration levels are found during verification testing. The air shall be sampled as it reaches the clean zone.

5.3 Methods and equipment for measuring airborne particle concentration. The method and equipment to be used for measuring the airborne particle concentration shall be selected on the basis of the particle size of interest. The following methods are suitable for class verification and monitoring of air cleanliness unless otherwise specified[1]. Other particle counting methods and equipment may be used if demonstrated to have accuracy and repeatability equal to or better than the methods listed below[2,3]

-----

[1] When the terms "as specified" or "shall be specified" are used without further reference, the degree of control needed to meet requirements will be specified by the user or contracting agency.

[2] For example, for particle size approximately 0.01 micrometer in diameter and larger, a condensation nucleus counter, which optically detects particles which have been grown by condensation of a supersaturated vapor, may be used. The counter must detect single particles.

[3] For monitoring purposes only, evaluation of particles by sedimentation methods may be carried out by allowing the particles to deposit on the surface of an appropriate medium and then counting them using optical microscopy.

5.3.1 <u>Counting particles 5 micrometers and larger.</u> For particle sizes 5 micrometers and larger, a manual sizing and counting method or an optical particle counting instrument shall be used. The manual sizing and counting method shall be in accordance with Appendix A, and the optical particle counting instrument shall be in accordance with Appendix B.

5.3.2 <u>Counting particles 0.1 micrometer and larger.</u> For particle sizes 0.1 micrometer and larger, an optical particle counting instrument shall be used in accordance with Appendix B. The instrument must count single particles. Only information obtained with a periodically calibrated and properly maintained particle counter shall be used in conducting airborne particle concentration measurements. Particle size data shall be reported in terms of equivalent diameter as calibrated against reference standard particles.

5.3.3 <u>Limitations of particle counting methods.</u>

5.3.3.1 <u>Optical particle counters.</u> Optical particle counters with unlike geometry or different operating principles may give different results when counting the same particles. Even recently calibrated instruments of like design may show differences in measurement results when sampling the same air. Caution should be used when comparing measurements from different instruments.

5.3.3.2 <u>Microscopic evaluation.</u> Since the microscopically measured size of a particle is the apparent longest linear dimension, and the size of particles measured by optical particle counters is based upon the diameter of a reference particle, microscopic counts will generally differ from counts obtained by optical particle counters.

5.3.3.3 <u>Upper limits.</u> Particle counters shall not be used to count particle concentrations or particle sizes greater than the upper limits specified by the manufacturer.

5.3.4 <u>Calibration of particle counting instrumentation.</u> All instruments shall be calibrated against known reference standards at regular intervals as specified[1]. Parameters which may need calibration include, but are not limited to, air flow rate and particle size.

5.4 <u>Statistical analysis.</u> The collection and analysis of airborne particle concentration data for verification of an airborne particulate cleanliness class shall be performed in accordance with the following requirements. This statistical analysis deals only with random errors (lack of precision), not errors of a nonrandom nature ("bias"), such as erroneous calibration.

5.4.1 <u>Acceptance criteria.</u> The cleanroom or clean zone shall meet the acceptance criteria for an airborne particulate cleanliness class if 1) the average of the particle concentrations (see Table I) measured at each location falls at or below the class limit, and 2) the mean of these averages falls at or below the class limit with a 95% confidence limit. The confidence limit shall be based on a one-tailed Student's t distribution, as follows.

5.4.1.1 <u>Average particle concentration.</u> The average particle concentration (A) at a location is the sum of the individual sample particle counts (C_i) divided by the number of samples taken at the location (N), as shown in Equation (5-1). If only one sample is taken, the average particle concentration is the same as the particle count measured.

$$A = (C_1 + C_2 + \ldots + C_N) \, / \, N \qquad\qquad (5\text{-}1)$$

---

[1]When the terms "as specified" or "shall be specified" are used without further reference, the degree of control needed to meet the requirements will be specified by the user or contracting agency.

5.4.1.2 <u>Mean of the averages.</u> The mean of the averages (M) is the sum of the individual averages (A_i) divided by the number of locations (L), as shown in Equation (5-2). All locations are weighted equally, regardless of the number of samples taken.

$$M = (A_1 + A_2 + \ldots + A_L) / L \qquad (5\text{-}2)$$

5.4.1.3 <u>Standard deviation.</u> The standard deviation (SD) of the averages is the square root of the sum of the squares of differences between each of the individual averages and the mean of the averages $(A_i\text{-}M)^2$ divided by the number of locations (L) minus one, as shown in Equation (5-3).

$$SD = \sqrt{\frac{(A_1 - M)^2 + (A_2 - M)^2 + \ldots + (A_L - M)^2}{L - 1}} \qquad (5\text{-}3)$$

5.4.1.4 <u>Standard error.</u> The standard error (SE) of the mean of the averages (M) is determined by dividing the standard deviation (SD) by the square root of the number of locations, as shown in Equation (5-4).

$$SE = SD / \sqrt{L} \qquad (5\text{-}4)$$

5.4.1.5 <u>Upper confidence limit (UCL).</u> The 95% UCL of the mean of averages (M) is determined by adding to the mean the appropriate UCL factor (see Table III for UCL factor) times the standard error (SE), as shown in Equation (5-5).

$$UCL = M + (UCL \text{ Factor} \times SE) \qquad (5\text{-}5)$$

TABLE III

UCL factor for 95% upper control limit

No. of locations(L)	2	3	4	5-6	7-9	10-16	17-29	>29
95% UCL factor	6.3	2.9	2.4	2.1	1.9	1.8	1.7	1.65

13

5.4.1.6 <u>Sample calculation.</u>   A sample calculation is shown in Appendix C.

6.   <u>Changes.</u>   When a Federal agency considers that this standard does not provide for its essential needs, written request for changing or adding to the standard, supported by adequate justification, shall be sent to the Administration.   This justification shall explain wherein the standard does not provide for esssential needs.   The request shall be sent to the General Services Administration, Federal Supply Service, Engineering Division, 819 Taylor Street, Fort Worth, TX 76102.   The Administration will determine the appropriate action to be taken and will notify the agency.

7.   <u>Conflict with referenced specifications.</u>   Where the requirements stated in this standard conflict with any requirement in a referenced specification, the requirements of this standard shall apply.   The nature of conflict between the standard and the referenced specification shall be submitted in duplicate to the General Services Administration, Federal Supply Service, Engineering Division, 819 Taylor Street, Fort Worth, TX 76102.

8.   <u>Federal agency interests.</u>

    Department of Commerce
    Department of Defense, Office of the Assistant Secretary
      of Defense (Installations and Logistics)
        Army
        Navy
        Air Force
    Department of Energy
    Department of Health and Human Services
    Department of Transportation
    General Services Administration
    National Aeronautics and Space Administration
    Nuclear Regulatory Commission

APPENDIX A

PARTICLE MONITORING - MANUAL COUNTING AND SIZING METHODS

A10. Scope. This appendix describes procedures for determining airborne particulate contamination levels of particles 5 micrometers and greater in size in cleanrooms and clean zones by a membrane filtration and particle count method.

A20. Summary of method.

A20.1 Description of the basic method. At the sampling point, air is passed through a membrane filter using a vacuum to effect the filtration. The air flow rate is controlled by means of a limiting orifice or an air flowmeter, and the total volume of air sampled is controlled by the sampling time. The membrane filter is examined microscopically, using a high-intensity oblique incident light source, to determine the number of particles 5 micrometers and greater collected from the air sample.

A20.2 Alternatives to optical microscopy. Image analysis or projection microscopy can replace direct optical microscopy for sizing and counting, provided that the accuracy and reproducibility are equal to or better than those of the direct optical microscopic method.

A20.3 Acceptable sampling procedures. There are two acceptable procedures for this method as described herein: (a) Aerosol Monitor Method, and (b) Open Filter Holder Method. They differ primarily in the apparatus used and in the time required for performance.

A30. Equipment.

A30.1 Equipment common to both methods.

A30.1.1 Microscope. Binocular microscope with ocular-objective combinations for 100X to 250X magnifications. These combinations are chosen such that the ultimate smallest division of the ocular reticle, at the highest magnification, is less than or equal to 5 micrometers. The latter objective should have a numerical aperture of at least 0.25.

A30.1.2 Ocular micrometer scale: 5- or 10-millimeter linear scale with 100 divisions, dependent upon ocular-objective combinations, or micrometer eyepiece with movable scale.

A30.1.3   Stage micrometer: standard 0.01- to 0.1-millimeter-per-division scale.

A30.1.4   External microscope illuminator.

A30.1.5   Vacuum pump capable of maintaining a vacuum of 500 torr while pumping at a rate of at least 1 cubic foot per minute.

A30.1.6   Electrical timer or timing device, 60-minute range.

A30.1.7   Flowmeter or limiting orifice calibrated with the vacuum pump, filter holder, and filter to collect a sample of sufficient volume.   See Paragraph A50.1.

A30.1.8   Manual tally counter.

A30.1.9   Filter storage holders for membrane filters after sampling; Petri plates or Petri slides with covers.

A30.1.10 Rinse fluid: purified water prefiltered to 0.45 to 1.2 micrometers.

A30.1.11 Forceps: flat, with unserrated tips.

A30.2   <u>Equipment for aerosol monitor method.</u>

A30.2.1   Aerosol monitors: dark, 0.8-micrometer mean pore size, with imprinted grid.

A30.2.2   Aerosol adapter.

A30.3   <u>Equipment for open filter holder method.</u>

A30.3.1   Filter holder: aerosol, open type.

A30.3.2   Membrane filter: dark, 0.8-micrometer or smaller pore size, with imprinted grid.

A30.3.3   Membrane filters: white (for evaluating dark particles), 0.8-micrometer or smaller pore size, with imprinted grid.

A30.4   <u>Optional equipment.</u>

A30.4.1   Image analyzer.

16

A30.4.2 Projection microscope and screen.

A40. Preparation of equipment.

A40.1 Preparation for both methods.

A40.1.1 All equipment preparation should be performed within a clean zone having an airborne particulate cleanliness class equal to or less than that of the clean zone to be monitored.

A40.1.2 All equipment should be maintained at maximum cleanliness and should be stored with protective covers, cases, or other suitable enclosures when not in use in a location having an airborne particulate cleanliness class equal to or less than that of the clean zone of lowest class number where sampling is performed.

A40.1.3 Personnel performing sampling, sizing, and counting operations should be equipped with garments consistent with the airborne particulate cleanliness class of the clean zone being monitored.

A40.1.4 Thoroughly rinse with purified water all internal surfaces of the Petri slide holders or Petri plates used to hold the exposed membranes for counting. Rinse in cascading action, as with the membrane holders. After rinsing, leave the lid open in a unidirectional airflow clean zone until the interior surfaces are dry.

A40.2 Preparation for aerosol monitor method.

A40.2.1 Establish a filter background count in the following manner. Where the manufacturer of aerosol monitors has indicated an average background count for a package of monitors (in the particle size ranges of concern), examine and establish the average background count for 5% of the filters in the package. If the average background count determined is equal to or less than the manufacturer's indication, use the indication as the background count for all filters in the package. If the determined count is higher than the manufacturer's indication, or if there is no such indication, establish a background count for each filter used.

17

A40.2.2 Background counts for individual filters are determined by following the microscopic procedures of Paragraph A70.

A40.2.3 After the background count has been established, package the aerosol monitors in a particle-free container or place into their appropriate sampling devices and transport them to the sampling location.

A40.2.4 Except for purposes of making background counts, aerosol monitors should be opened only when in the sampling location or the counting area.

A40.3 Preparation for open filter holder method.

A40.3.1 Disassemble the filter holder and wash in liquid soap and water. After washing, rinse and store in the unidirectional airflow clean zone until dry. (DO NOT WIPE DRY.) Deionized water or distilled water are the rinse media of choice.

A40.3.2 After the filter holder is completely dry, mount a membrane filter in the filter holder, with the grid exposed. After mounting the membrane filter, invert the filter holder assembly and thoroughly flush the filter surface area and exposed filter holder parts with purified water using a cascade rinsing action, starting at the top and progressing to the bottom of the filter face. Place in the unidirectional airflow clean zone and allow to dry.

A40.3.3 Establish a filter background count for each membrane filter to be used by following the procedures of Paragraph A70.

A40.3.4 After the interior surfaces of the filter storage holders are dry, apply a small piece of double-sided cellophane tape or stopcock grease to the bottom surface.

A40.3.5 After the filter holder and membrane are clean and dry, package them in a particle-free container.

A40.3.6 Transport the prepared filter holder, with membrane filter and vacuum source, to the sampling location. DO NOT EXPOSE THE FILTER SURFACE UNTIL THE APPARATUS IS ASSEMBLED AND READY FOR SAMPLING.

18

A50.   Sampling.

A50.1   Sampling orientation and flow.   For unidirectional airflow
cleanrooms and clean zones, the aerosol monitor or filter holder
should be oriented to face into the airflow.   For non-
unidirectional airflow cleanrooms and clean zones, orient the
aerosol monitor or filter holder so that the opening faces
upward, unless otherwise specified[1].   Airflow into the filter
should be adjusted to be isokinetic for unidirectional airflow.
For nonunidirectional airflow, the airflow into the filter should
be adjusted to be 0.25 cubic foot per minute for a 25-millimeter
filter or 1 cubic foot per minute for a 47-millimeter filter.
The minimum sample volume should be 10 cubic feet for Class 1000
and 1 cubic foot for Class 10,000 and greater.

A50.2   Sampling by aerosol monitor method.

A50.2.1   At the sampling location, attach the aerosol monitor to
the aerosol adapter and the adapter to the vacuum source.   Have
in line either a limiting orifice or a flowmeter.   Isolate the
vacuum pump exhaust from the area being sampled, as it may be a
source of extraneous airborne contamination.

A50.2.2   Adjust the flowmeter, if used, for the flow rate at the
operating vacuum pressure where it is used.

A50.2.3   Connect a timer to the vacuum pump power source.

A50.2.4   Remove the bottom plug from the aerosol monitor and
attach it to the free end of the aerosol adapter.   Position the
aerosol monitor as required, pry off the top portion of the
aerosol monitor, and store it in a clean location.

A50.2.5   Turn on the pump, adjust the flowmeter, and operate for
a time which will provide the required sample at the chosen flow
rate.

---

[1]When the terms "as specified" or "shall be specified" are
used without further reference, the degree of control
needed to meet requirements will be specified by the user
or contracting agency.

19

A50.2.6 When the sampling time has elapsed, release the vacuum, replace the top portion of the aerosol monitor, and remove the aerosol monitor from the aerosol adapter. The bottom plug need not be replaced. Identify the aerosol monitor with a sample identification tag. Transport the aerosol monitor to a counting area which should be a clean zone of airborne particulate cleanliness class at least equal to that of the clean zone sampled.

A50.3 Sampling by open filter holder method.

A50.3.1 When in the sampling area, place the filter holder in position. With the aid of vacuum tubing, connect the filter holder to the vacuum train which includes the filter holder, either a limiting orifice or a flowmeter, and a source of vacuum (vented outside the sampling area or filtered to prevent contamination of the area sampled).

A50.3.2 Adjust the flowmeter, if used, for the flow rate at the operating vacuum pressure where it is used.

A50.3.3 Remove the protective cover from the membrane filter holder and turn on the vacuum source. Turn on the pump, adjust the flowmeter, and operate for a time which provides the required sample at the chosen flow rate.

A50.3.4 At the end of the sampling period, turn off the vacuum source and carefully re-cover the filter holder with a precleaned cover. Return the covered sample filter holder to the counting area, which should be a clean zone of airborne particulate cleanliness class at least equal to that of the clean zone sampled.

A60.   Microscope calibration.

A60.1 IF CALIBRATION OF THE MICROSCOPE HAS BEEN PERFORMED PREVIOUSLY BY THE OPERATOR, OMIT THIS SECTION.

A60.2 Place the stage micrometer on the mechanical stage; focus and adjust the light to give an even and full illumination in the field of view.

A60.3 Verify that the proper eyepiece and objective combination is in place to provide total magnification equal to 100X to 250X, as required.

A60.4 Assure that the microscope is properly focused by focusing each eyepiece to achieve a sharp stage micrometer image.

A60.5 If an image analyzer or projection microscope is used, perform a similar calibration.

A60.6 Using the entire length of the ocular reticle scale, record the number of stage micrometer divisions the eyepiece reticle covers.

(a) Compute the ocular micrometer scale calibration for a particular magnification by the formula:

Micrometers per ocular scale division =

$$\frac{\text{(No. of Stage Micrometer Div.) x (Size of One Stage Micrometer Div.)}}{\text{(No. of Eyepiece Divisions)}}$$

Example:

At 100X: 100 eyepiece divisions equals 100 stage divisions, each 5.0 micrometers in length.

Thus: Micrometers/Eyepiece Division =

$$\frac{\text{(100 Divisions) x (5.0 Micrometers)}}{\text{(100 Divisions)}} = 5.0 \text{ Micrometers}$$

(b) Calculate the number of linear divisions required to measure each range.

Example:

At 100X: each eyepiece division equals 5 micrometers, so for a 16- to 20-micrometer range, 3 to 4 divisions would be examined.

Note: If the microscope is equipped with a zoom adjustment, this may be employed to adjust the calibration to the nearest integer (X micrometers/division, instead of X.Y micrometers/division), provided the adjustment is noted in the calculations.

NOTE: A CHANGE IN INTERPUPILLARY DISTANCE BETWEEN OPERATORS CHANGE FOCAL DISTANCE, HENCE CALIBRATION.

21

A70.   Microscopic counting and sizing of particles.

A70.1 In the clean zone where the particles upon the membrane filters are counted and sized, remove the membrane filter from the aerosol monitor or the open filter holder with unserrated flat forceps.

A70.2 Place the membrane filter, grid side up, in a precleaned Petri slide holder or Petri plate, allowing the filter to adhere to the sticky surface of the storage holder. Tightly seal the carrier to prevent contamination of the sample filter.

A70.3 The microscope should be clean so as not to add particulate contamination to the sample. Carefully place the covered Petri slide or Petri plate on the microscope stage and adjust the angle and focus of the illuminator to provide optimum particle definition at the magnification used for counting. Use an oblique lighting angle of 10 to 20 degrees to cast a shadow of the particle, thereby effectively separating the particle image from the filter background.

A70.4 Select a field size so that there are no more than about 50 particles larger than 5 micrometers in the field. Optional fields are: a grid square; a rectangle defined by the width of a grid square and the calibrated length of the ocular micrometer scale; a rectangle defined by the width of the grid square and a portion of the length of the ocular micrometer scale.

A70.5 Estimate the number of particles in the greater-than-5-micrometer range over the effective filtering area by scanning one unit area of the field size selected. If the total number of particles in this range is estimated to be less than 500, count the number of particles in this range over the entire effective filtering area. If the number is greater, the counting procedure in Paragraph A70.8.1 applies.

A70.6 In scanning for particles, manipulate the stage so that particles to be counted pass under the ocular scale. Only the maximum dimension of the particle is regarded as significant. The eyepiece containing the ocular micrometer may be rotated to accommodate specific particles, if necessary.

A70.7 Using a manual tally counter, record all particles in the selected field that are equal to or exceed the dimension as indicated by the ocular micrometer scale. Record the number of particles in each field counted, in order to establish uniformity of distribution and to have a record of the number of fields counted.

A70.8 Statistical particle counting.

A70.8.1 When the estimated number of particles over the effective filtering area exceeds 500, the method entails the selection of a unit area for statistical counting, counting all particles in the unit area, and then similarly counting additional unit areas until the following statistical requirement is met:

$$F \times N > 500$$

where:

F = number of grid squares or unit areas counted, and

N = total number of particles counted in F areas.

A70.8.2 Calculate the total number of particles on the filter as follows:

$$P = N \times \frac{A}{n \times a}$$

where:

P = total number of particles of a size range on the filter.

(When a background count is obtained, subtract this from the P value after calculation, but prior to dividing by sample volume.)

N = total number of particles counted in n unit areas.

n = number of unit areas counted.

a = unit area in square millimeters.

A = effective total filter area in square millimeters.

23

A80.   <u>Reporting.</u>

A80.1   Subtract the background count for a filter from the total count obtained for the filter in accordance with Paragraph A70.

A80.2   Results should be expressed for each size range of specific interest, including 5-micrometer particles, in particles per cubic foot of sample by dividing the number of particles, P, by the sample volume (V).

$$\text{Particles per cubic foot} = P/V$$

A80.3   Final results are expressed in particles per cubic foot of sampled air, 5 micrometers and greater.

A90.   <u>Factors affecting precision and accuracy.</u>

A90.1   The precision and accuracy of this method can be no higher than the sum total of the variables. In order to minimize the variables attributable to an operator, a trained microscopist technician is required. Variables of equipment are recognized by the experienced operator, thus further reducing possible error. The operator should have adequate basic training in microscopy and the techniques of particle sizing and counting.

A90.2   For training personnel, low- to medium-concentration specimens may be prepared on a grid filter and preserved between microslides as standards for a given laboratory. Standard counting specimens are available for this purpose.

A90.3   Accuracy for a sampling location can be increased by increasing the number of samples taken and processed at that sampling location.

A90.4   Accuracy for a sampling location can be increased by increasing the volume of air per sample and by increasing the time of sampling.

A90.5   Accuracy can be increased by establishing and using background counts for filters.

APPENDIX B

OPERATION OF OPTICAL PARTICLE COUNTERS

B10. Scope.

B10.1 Application. Optical particle counters provide data on airborne particle concentration and size distribution on a near-real-time basis. This appendix describes methods for the operation, use, and testing of optical particle counters used to satisfy requirements of this Federal Standard. Guidelines are given which should aid in standardization of optical airborne particle monitoring procedures for defining air cleanliness.

B10.2 Limitations. Particle size data are referenced to the particle system used to calibrate the optical particle counter; however, differences in optical, electronic, and sample handling systems among the various optical particle counters may contribute to variations in counting results. Care must be exercised in attempting to compare data from samples which vary significantly in particle composition or shape from the calibration base. Variations may also occur between instruments using particle sensing systems with different operating parameters. These effects should be recognized and minimized by using standardized methods for counter calibration and operation.

B10.3 Qualifications of personnel. Individuals performing the procedures described herein should be trained in the use of the optical particle counter and understand its operation, capabilities, and limitations.

B20. Applicable references.

B20. 1 ASTM F 328 Determining Counting and Sizing Accuracy of an Airborne Particle Counter Using Near-Monodisperse Spherical Particulate Materials.

B20.2 ASTM F 649 Secondary Calibration of Airborne Particle Counter Using Comparison Procedures, American Society for Testing and Materials, 1916 Race Street, Philadelphia, PA 19103.

25

B20.3   IES-RP-CC-013 Recommended Practice for Equipment
Calibration or Validation Procedures, Institute of Environmental
Sciences, 940 East Northwest Highway, Mt. Prospect, IL 60056.

B30.   Summary of method.

B30.1   Calibration.   Primary calibration of optical particle
counters is performed with spherical isotropic particles of
refractive index 1.6.   Secondary calibration may be performed
with atmospheric particles for correlation with a reference
particle counter.   In addition, stable operation should be
assured by standardizing against internal references built into
the counter or by other approved methods.

B30.2   Operation.   The air to be classified is sampled at a known
flow rate from the sample point or points of concern.   Particles
contained in the sampled air pass through the sensing zone of the
optical particle counter and produce a signal which is related to
particle size.   An electronic discriminator circuit sorts and
counts the pulses in relation to particle size and displays or
prints out the particle count in the sample volume.

B40.   Apparatus and related documentation.

B40.1   Optical particle counting system.   The optical particle
counting system may include a recorder or printer; alternatively,
data may be transmitted to a remote location for additional
processing and computing.

B40.2   Sample air flow system.   The sample air flow system
consists of an intake tube, a sensing chamber, an air flow
metering or control system, and an exhaust system.   The exhaust
system may consist of either a built-in vacuum source or an
external vacuum supply with a separate flow control element for
the optical particle counter in use.   If a built-in vacuum source
is used, and the optical particle counter is to be used where the
exhaust air could affect either the particle counts being
measured or operations in the cleanroom or clean zone, then the
exhaust should be suitably filtered.

26

B40.3 <u>Sensing system.</u> The sensing system of the optical particle counter is formed by intersecting the sample air flow with a fixed sensing volume of such dimension so that the probability of more than one particle being present at any time (the coincidence error) is less than 5%. The signal produced from each particle passage through the sensing volume is received and processed by the electronic system in real time. The instrument should be designed to maintain its stated accuracy despite variations in the specified operating line voltage and ambient temperature. The operating line voltage and temperature ranges should be specified.

B40.4 <u>Electronic system.</u> The electronic system includes a pulse analyzer and counter, along with a system for registering particle counts in relation to particle size.

B40.4.1 The pulse analyzer may operate in either one or both of two modes: (1) in response to all particles within discrete size ranges, or (2) in response to all particles larger than the predetermined lower threshold size limit(s). Air cleanliness classification data, however, should be reported in terms of mode (2). Particle size ranges or limits may be either selectable or fixed.

B40.4.2 The counting circuits during a known time interval may accumulate information generated by the pulse analyzer in response to particle passages. Pulse count accumulation in one or more size ranges may be provided.

B40.4.3 For determination of the airborne particulate cleanliness class, the counting circuit is allowed to accumulate data for a preset time interval before reporting. The time interval is selected to yield a known sample volume, so that particle concentration can be readily calculated.

B40.4.4 The registering system indicates the number of particles or the particle concentration with respect to the selected particle size range(s) or size limit(s). Counts may be recorded or displayed on the optical particle counter, or may be transmitted to a remote location for recording, display, or computer processing.

B40.5 Calibration. An internal secondary calibration system or a means of ensuring stability should be provided in the instrument. The internal secondary calibration system should be capable of validation with respect to primary calibration in accordance with the methods of ASTM F 328 and ASTM F 649. The secondary calibration system is used for checking the sizing and counting stability of the optical particle counter and to provide a stable reference for any necessary sensitivity adjustment of the instrument.

B40.6 Documentation. Instructions which should be supplied with the instrument by the manufacturer include:

(a) Brief description of the operating principles of the instrument.

(b) Description of major components.

(c) Environmental conditions (ambient temperature, relative humidity, and pressure) and line voltage range required for stable operation.

(d) Particle size and concentration ranges for accurate measurement.

(e) Suggested maintenance procedure and recommended intervals for routine maintenance.

(f) Operating procedure for particle counting and sizing.

(g) Secondary calibration procedure (where applicable).

(h) Primary calibration procedure (a factory primary calibration facility should be available for calibration of the counter upon customer request), and field primary calibration capability and procedures.

(i) Suggested intervals for primary calibration.

28

B50. Preparations for sampling and counting. The procedures
described in the following paragraphs should be performed or
verified before the optical particle counter is used for
determination of airborne particulate cleanliness classes. Each
of the procedures has its own requirements regarding frequency
interval.

B50.1 Primary calibration. Particle sizing and air sample
volume require primary calibration. The comments in the
following paragraphs are intended as a general guideline for
primary calibration to be considered in interpreting ASTM F 328,
ASTM F 649, or IES-RP-CC-013. Deviations may be necessary to
achieve a specific objective. It is, however, the duty of the
manufacturer to include in the operating instructions a
description of the appropriate primary calibration method for the
optical particle counter.

B50.1.1 Particle sizing. Primary calibration of the particle
sizing function of the optical particle counter is carried out by
registering the response of the counter to a monodisperse
homogeneous and isotropic controlled aerosol containing
predominantly spherical particles of known size and refractive
index, and by adjusting the calibration control until the correct
sizing response is obtained. Thereafter, the internal secondary
calibration system is adjusted, if necessary, for correct
response to the reference aerosol. Nonspherical particles may be
used for primary calibration for specific applications. In these
cases, the particle size is defined in terms of an appropriate
dimension for the reference particles. Means of generating the
reference particles has been extensively described in the
literature[1].

B50.1.2 Air sample volume. The air sample volume is calibrated
by measuring the flow rate and the duration of the sampling

---

[1]For instance, Liu, B.Y.H., "Methods for Generating Monodisperse Aerosols." 1967. Publication #104, Particle Technology Laboratory, Department of Mechanical Engineering,
University of Minnesota.

29

interval[2]. To avoid erroneous readings, equipment used for this measurement should not introduce an additional static pressure drop to the optical particle counter flow system. All flow measurements should be referenced to ambient conditions of temperature and pressure or as otherwise specified.

B50.2 Sampling setup.

B50.2.1 Sample location. In-place sample locations and orientation of the sample inlet tube should be established in accordance with Section 5 of this standard.

B50.2.2 Extension of sample inlet tube. Any extension of the sample inlet tube may affect the sampling results. The effects may be of little significance for particles in the size range from approximately 0.1 to 1 micrometer for sample tube extensions up to approximately 30 meters. Outside of this range, extensions of the sample inlet tube are used only if no other method is possible for sample acquisition. The sample tube extensions should be configured to maintain the sample flow Reynolds number in the range from 5,000 to 10,000 and the sample residence time in the extension below 5 seconds; no radius of curvature below 10 centimeters should be used. Where air sampling requires data on particles larger than 3 micrometers in diameter, no extension tube longer than 3 meters should be used.

B50.2.3 Particle counter exhaust air. The particle counter should be located and used so that air vented does not contaminate the sample or clean zone. The exhaust air should be filtered to a level consistent with the ambient airborne particulate cleanliness class or else vented outside of the cleanroom.

B50.3 Field calibration procedure. Perform secondary calibration or standardization (if applicable) in accordance with the manufacturer's instructions.

---

[2]See Baker, W.C. and Pouchot, J.F., "The Measurement of Gas Flow Part I," 1983, Journal of Air Pollution Control Association, January, Vol. 33, No. 1 and Baker, W.C. and Pouchot, J.F., "The Measurement of Gas Flow Part II," 1983, Journal of Air Pollution Control Association, February, Vol. 33, No. 2.

B50.4 <u>Zero count check.</u>   The absence of spurious counts is verified by a zero count check, as described in the following paragraphs.

B50.4.1 Place an appropriate filter on the counter sample inlet tube to prevent the passage of particles larger than the smallest size particle the counter can count.

B50.4.2 Turn on the sample air flow system; adjust for the specified sample air flow rate, if necessary.

B50.4.3 Turn on the counting circuits.

B50.4.4 Verify that the instrument reads zero counts for particles 0.5 micrometer and larger. If counts are registered, permit the counter to purge itself with the filter in place until a zero count level is reached.

B50.4.5 For counters capable of detecting particles smaller than 0.5 micrometer, zero counts may not be achievable for the smallest particles detectable. For such instruments, a tare concentration less than 10% of the airborne particulate cleanliness class concentration of such smaller particles (e.g., 0.1, 0.2, 0.3 micrometer) should be achieved.

B60. <u>Counting procedure.</u>

B60.1 Perform the field (secondary) calibration and zero count check, in accordance with Paragraphs B50.3 and B50.4.

B60.2 Check and adjust to the specified air flow rate (if applicable).

B60.3 Turn on the counting circuits, if necessary; read and record the particle count displayed for the particle size(s) of interest.

B70. <u>Reporting.</u>

B70.1 Record the particle size range(s), the volume of air sampled, the particle count, the time, and the sample point location.

B70.2 Report particle count data in terms of the number of particles per cubic foot of air sampled.

APPENDIX C

STATISTICAL ANALYSIS

C10.   Sample calculation.   The data and calculations presented in
the following paragraphs are intended to serve as a working
example, illustrating the statistical procedures involved in
determination of acceptance criteria for cleanrooms and clean
zones.   The data and calculations are based upon a 1-cubic-foot
sample volume and testing at 0.3-micrometer measured particle
size for Class 10.   (Note: Table 1 indicates that the UCL is to
be less than or equal to 30 particles per cubic foot 0.3
micrometer and larger to meet Class 10.)

C10.1   Tabulation of particle count data.

Location	Particle Counts (C_i)					Total No. of Samples (N)	$(\sum C_i)$ Total Count	(A_i) Average Counts
	1	2	3	4	5			
A	15	NR	NR	NR	NR	1	15	15.00
B	33	24	9	15	NR	4	81	20.25
C	18	3	12	24	NR	4	57	14.25
D	39	18	9	33	6	5	105	21.00
E	0	27	6	0	NR	4	33	8.25

(NR - no reading taken)

C10.2   Mean of averages (M).

$$M = (A_1 + A_2 + \ldots + A_L) / L \qquad \text{(Equation 5-2)}$$

$$M = (15.00 + 20.25 + 14.25 + 21.00 + 8.25)/5 = 15.75$$

L = (Number of sample locations)

C10.3　Standard deviation of averages (SD).

(Equation 5-3)

$$SD = \sqrt{\frac{(A_1 - M)^2 + (A_2 - M)^2 + \ldots + (A_L - M)^2}{L - 1}}$$

$$SD = \sqrt{[(15.00-15.75)^2 + (20.25-15.75)^2 + (14.25-15.75)^2 + (21.00-15.75)^2 + (8.25-15.75)^2] / [5-1]}$$

SD = 5.17

C10.4　Standard error of mean of averages (SE).

$$SE = SD/\sqrt{L} \qquad \text{(Equation 5-4)}$$
$$SE = 5.17/\sqrt{5} \quad = 2.31$$

C10.5　Upper 95% confidence limit (UCL).

For 5 locations, UCL factor = 2.1

UCL = M + (UCL Factor x SE)　　　(Equation 5-5)

UCL = 15.75 + (2.1 x 2.31) = 20.6

C20.　Conclusion.　Since the upper 95% confidence limit (UCL) is less than 30 and all location average particle concentrations (A_i) were less than 30, the above data meet the acceptance criteria for Class 10, although some of the individual particle counts were above 30.

33

APPENDIX D

SOURCES OF SUPPLEMENTAL INFORMATION

D10. Scope. The purpose of this appendix is to list references
for supplemental information which may provide instruction or
guidance in the preparation of documents related to the design,
acquisition, testing, operation, and maintenance of cleanrooms
and clean zones. This listing of sources and documents
emphasizes that information contained in such sources is not part
of this standard and is not mandatory for compliance with this
standard.

D20.   Source references.

D20.1   AFWP - Headquarters, AFLC/DAPD, Wright-Patterson AFB,
         OH 45433

D20.2   AFWR - Warner Robins ALD/MMEDT, Robins AFB, GA 31098

D20.3   ANSI - American National Standards Institute, 1430
         Broadway, New York, NY 10018

D20.4   ASHRAE - American Society of Heating, Refrigerating,
         and Air-Conditioning Engineers, 1791 Tullie Circle
         Northeast, Atlanta, GA 30329

D20.5   ASME - American Society of Mechanical Engineers, 345
         East 47th Street, New York, NY 10017

D20.6   ASTM - American Society for Testing and Materials,
         1916 Race Street, Philadelphia, PA 19103

D20.7   DOE - Nuclear Standards Management Center, Oak Ridge
         National Laboratory, Building 9204.1, Room 321, MS/10,
         P. O. Box Y, Oak Ridge, TN 37830

D20.8   IES - Institute of Environmental Sciences, 940 East
         Northwest Highway, Mount Prospect, IL 60056

D20.9   MSFC - Marshall Space Flight Center, NASA, Marshall
         Space Flight Center, AL 35812

D20.10   NPFC - The Naval Publications and Forms Center, 5801
         Tabor Avenue, Philadelphia, PA 19120

34

D20.11   NRC - U.S. Nuclear Regulatory Commission, Attn:
         Director, Division of Document Control, P-130A,
         Washington, DC 20555

D20.12   NSF - National Sanitation Foundation, 3465 Plymouth
         Road, P. O. Box 1468, Ann Arbor, MI 48106

D20.13   NTIS - National Technical Information Service,
         U.S. Department of Commerce, 5285 Port Royal Road,
         Springfield, VA 22161

D30.   Document references.

D30.1   Document No.   AFM 88-4 Chapter 5 Source AFWP & AFWR
         Title   Criteria for Air Force Clean Facility
                        Design and Construction
         Abstract   Prescribes criteria for the design and
                        construction of Air Force clean
                        facilities. It specifies the real
                        property standards for meeting the
                        requirements of Air Force T.O. 00-25-
                        203.

D30.2   Document No.   T.O. 00-25-203   Source AFWP & AFWR
         Title   Contamination Control of Aerospace
                        Facilities, U.S. Air Force
         Abstract   This document specifies cleanroom
                        design, operating, and test procedures.
                        It also includes recommended cleanliness
                        levels for typical operations.

D30.3   Document No.   ASHRAE Std. 52-76   Source ASHRAE
         Title   Method of Testing Air-Cleaning Devices
                        Used in General Ventilation for
                        Removing Particulate Matter
         Abstract   This standard defines unified test
                        procedures and apparatus for evaluating
                        filters with efficiencies below that of
                        HEPA filters.

D30.4   Document No.   F 25   Source ASTM
         Title   Standard Method for Sizing and Counting
                        Airborne Particulate Contamination in
                        Clean Rooms and Other Dust-Controlled
                        Areas Designed for Electronic and
                        Similar Applications

35

	Abstract	Procedures are given for membrane filter sampling and microscope counting in clean areas.
D30.5	Document No.	F 50 Source ASTM
	Title	Standard Practice for Continuous Sizing and Counting of Airborne Particles in Dust-Controlled Areas Using Instruments Based upon Light Scattering Principles
	Abstract	Methods are given for sampling, particle counting, and data evaluation using light-scattering particle counters in cleanrooms.
D30.6	Document No.	F 91 Source ASTM
	Title	Standard Recommended Practice for Testing for Leaks in the Filters Associated with Laminar Flow Clean Rooms and Clean Work Stations by the Use of a Condensation Nuclei Detector
	Abstract	Provides a method of testing the integrity of HEPA filter installations in laminar flow cleanrooms and clean work stations.
D30.7	Document No.	F 328 Source ASTM
	Title	Standard Practice for Determining Counting and Sizing Accuracy of an Airborne Particle Counter Using Near-Monodisperse Spherical Particulate Materials
	Abstract	Counting and sizing accuracy determination procedures are given for certifying operation of an optical airborne particle counter.
D30.8	Document No.	F 649 Source ASTM
	Title	Standard Practice for Secondary Calibration of Airborne Particle Counter Using Comparison Procedures
	Abstract	Procedures are given for fine-tuning the response of an airborne particle counter to match that of a standard

36

instrument for defining atmospheric
dust, following calibration with
monodisperse latex particles.

D30.9	Document No.	F 661 Source ASTM
	Title	Standard Practice for Particle Count and Size Distribution Measurements in Batch Samples for Filter Evaluation Using an Optical Particle Counter
	Abstract	Procedures are given for sample handling, sample evaluation, and particle count and size analysis in batch samples for use in an optical single particle counter. The method is directed at samples obtained in filter testing, but can be used for any samples.
D30.10	Document No.	IES-RP-CC-002 Source IES
	Title	Laminar Flow Clean Air Devices
	Abstract	Covers definitions, procedures for evaluating performance, and major requirements of laminar flow clean air devices. Sixteen test and performance criteria are considered.
D30.11	Document No.	IES-RP-CC-001 Source IES
	Title	HEPA Filters
	Abstract	Recommends basic provisions for HEPA filters for use in clean air devices and cleanrooms. Five levels of performance and two grades of construction are included.
D30.12	Document No.	IES-RP-CC-006 Source IES
	Title	Testing Clean Rooms
	Abstract	Describes test methods for character- izing the performance of cleanrooms. Performance tests are recommended for three types of cleanrooms at three operational phases.

37

D30.13   Document No.   IES-RP-CC-013   Source IES
          Title   Recommended Practice for Equipment
                          Calibration or Validation Procedures
          Abstract   This Recommended Practice covers
                          definitions and procedures for calibrat-
                          ing instruments used for testing clean-
                          rooms and clean air devices, and for
                          determining intervals of calibration.

D30.14   Document No.   NHB 5340.2   Source MSFC
          Title   NASA Standards for Clean Rooms and Work
                          Stations for the Microbially Controlled
                          Environment
          Abstract   Establishes standard classes of air
                          conditions (both total particles and
                          viable particles) within cleanrooms and
                          clean work stations for the microbially
                          controlled environment.

D30.15   Document No.   IES-CC-009   Source IES
          Title   Compendium of Standards, Practices,
                          Methods and Similar Documents Relating
                          to Contamination Control
          Abstract   Listing of documents.

D30.16   Document No.   MIL-STD-45622   Source NPFC
          Title   Calibration Systems Requirements
          Abstract   Prescribes requirements for
                          establishment and maintenance of a
                          calibration system used to control the
                          accuracy of measuring and test
                          equipment.

D30.17   Document No.   MIL-F-51068   Source NPFC
          Title   Military Specification: Filter,
                          Particulate, High-Efficiency, Fire
                          Resistant
          Abstract   Covers design, construction, and
                          performance of HEPA filters in six sizes
                          and seven types.

D30.18   Document No.   MIL-F-51079   Source NPFC
          Title   Military Specification: Filter Medium,
                          Fire-Resistant, High-Efficiency

38

	Abstract	Provides requirements and test methods for determining compliance for one grade of HEPA filter medium.

D30.19   Document No.   MIL-F-51477   Source NPFC
          Title   Military Specification: Filters, Particulate, High-Efficiency, Fire Resistant, Biological Use
          Abstract   Covers general requirements for particulate filters for use in air cleaning or air filtration systems involving chemical, carcinogenic, radiogenic, or hazardous biological particles.

D30.20   Document No.   NE: F3-41   Source DOE
          Title   In-Place Testing of HEPA Filter Systems by the Single-Particle, Particle-Size Spectrometer Method
          Abstract   Procedures are described for in-place testing of single and tandem HEPA filter installations with DOP challenge and an optical particle counter with sensitivity to 0.1 micrometer.

D30.21   Document No.   NASA SP-5045   Source NTIS
          Title   Contamination Control Principles
          Abstract   Broad overview and guidelines to those designing or planning cleanroom facilities.

D30.22   Document No.   NASA SP-5074   Source NTIS
          Title   Clean Room Technology
          Abstract   Considerable information on history, need, nature, and type of cleanrooms with details of cleanroom environment and operation.

D30.23   Document No.   NASA SP-5076   Source NTIS
          Title   Contamination Control Handbook
          Abstract   Extensive detail on contaminants and their control and cleaning methods.

39

D30.24	Document No.	F 24 Source ASTM
	Title	Measuring and Counting Particulate Contamination on Surfaces
	Abstract	A method for size distribution analysis of particulate contamination, 5 micrometers and larger, either on, or washed from, surfaces of small electronic device components.

D30.25	Document No.	F 51 Source ASTM
	Title	Sizing and Counting Particulate Contaminant In and On Clean Room Garments
	Abstract	A membrane filter/microscope method for determining detachable particulate contaminants, 5 micrometers and larger, on cleanroom garments.

D30.26	Document No.	MIL-HDBK-406 Source NPFC
	Title	Contamination Control Technology – Cleaning Materials for Precision Precleaning and Use in Clean Rooms and Clean Work Stations
	Abstract	Extensive information on selection and use of cleaning materials developed by DOD.

D30.27	Document No.	MIL-HDBK-407 Source NPFC
	Title	Contamination Control Technology – Precision Cleaning Methods and Procedures
	Abstract	Extensive information on cleaning methods used by the military services for gross and precision cleaning of work processed under controlled environment conditions.

40

APPENDIX E

GLOSSARY

E10. Scope. This appendix lists terms used in the other
appendixes, for which further explanation in the context of such
use may benefit the user.

E20. List of terms.

E20.1 Isokinetic. A term describing a condition of sampling, in
which the velocity of gas into the sampling device (at the
opening or face of the inlet) has the same velocity rate and
direction as the ambient atmosphere being sampled.

E20.2 Isotropic particles. Particles with equal, uniform
physical and chemical properties along all axes.

E20.3 Membrane filter. Porous membrane composed of pure and
biologically inert cellulose esters, polyethylene, or other
materials through which the air stream is passed for the purposes
of filtration.

E20.4 Reynolds number. A dimensionless number which is
significant in the design of a model of any system in which the
effect of viscosity is important in controlling the velocities or
the flow pattern of a fluid: equal to the density of a fluid
times its velocity, times a characteristic length, divided by the
fluid viscosity.

Appendix 2

Simplified Cleanroom Certification Procedure

Arizona State University

Cleanroom Monitoring/Certification Procedure

Originator: Edward J. Bawolek   9/13/89   Revision:1/1/90

OBJECTIVE

To describe a procedure which can be employed to certify or routinely monitor a cleanroom or cleanroom area in accordance with Federal Standard 209D.

SCOPE

This method meets the requirements of Fed. Std. 209D, but is not all-inclusive.   Significant deviations from this procedure are possible which still meet the requirements of the standard.   This method is designed to be a convenient subset of the standard.   This procedure does not specify safe operating procedures for cleanroom equipment and personnel.   This procedure does not cover calibration of the optical particle counter.

EQUIPMENT/MATERIALS

1. Optical particle counter with 0.5 um detection capability. (0.3 um capability required for Class 10 and Class 1 measurements.)
2. Data sheet.
3. Statistical Calculator

PROCEDURE

1. Fill out Steps 1 through 6 of the attached worksheet to determine the number of air samples required, the particle size to be measured, and the sample volume.
2. Space the sample locations as uniformly as possible throughout the clean area to be measured. Avoid direct interference with operating equipment. Do not attempt to move equipment to enable sampling a location. When taking a sample, set the sample intake at a height which approximates the working surfaces of the room. The sample tube should be perpendicular to the airflow (this typically means the sample tube should be oriented horizontally at approx. 40" above floor).
3. Set the particle counter to the appropriate particle size and sample volume. Take THREE measurements at each sample location. Record the data on the data sheet.
4. Use the COMMENTS area on the data sheet to record activity in the sample area (at-rest, operational, maintenance, etc.).
5. Follow the mathematical analysis in steps 7 through 13 on the worksheet to determine whether the cleanroom is in statistical control.
6. Report the results to appropriate facilities personnel and file the worksheet and datasheet for future reference.

SAMPLING
----

1. Write the area of the clean work space here: A = |_____| (ft^2)

2. Write the Class number
   of the clean work space here: C = |_____|

3. Check the appropriate box for
   the type of airflow:
   a. Unidirectional | | b. Nonunidirectional | |
      (Laminar)   (Turbulent)

4. Determine the divisor, D, from the following table:

CLASS (Line 2)	Unidirectional Flow (Box 3a Checked)	Nonunidirectional Flow (Box 3b Checked)
100,000	316	316
10,000	100	100
1,000	31.6	31.6
100	25	10
10	25	3.1
1	25	1

   Write the divisor here: D = |_____|

5. Determine the number of sample
   locations, L, by dividing A   L = A / D = |_____|
   (from Line 1) by D (from Line 4):
            (If L is less than 2, write 2 in the box.)
   This is the number of locations you must test in the clean
   room. At each location, THREE air samples will be taken.

6. Determine the particle size to be measured, the air sample volume,
   and the class limit from the following table:

CLASS (Line 2)	Particle Size P (um)	Air Sample Volume V (ft^3)	Class Limit Cl
100,000	0.5	0.1	100,000
10,000	0.5	0.1	10,000
1,000	0.5	0.1	1,000
100	0.5	0.2	100
10	0.3	0.7	30
1	0.3	7.0	3

   6a. Write the particle size, P here: P = |_____| (um)

   6b. Write the air sample volume, V here: V = |_____| (ft^3)

   6c. Write the class limit, Cl here: Cl = |_____|

E. J. Bawolek 9/13/89   Revision: 1/1/90

7. Are any of the numbers in Column 4 LARGER than C1, (from Line 6c)?

                   --   --
           a. NO | | b. YES | |
                   --

   a. If you checked NO, then STOP!! The cleanroom meets the
      statistical control criteria and no further work is necessary!

   b. If you checked YES, proceed with the following statistical analysis
      to determine if room is in control:

8. Take the mean, M, of the numbers in Col 4:   M = | |
   (Use a statistical calculator)

9. Take the Standard Deviation, SD of Col 4:   SD = | |
   (Use a statistical calculator)

10. Calculate the Standard Error by dividing SE = SD / L = | |
    SD (from Line 10) by L (from Line 5):

11. Determine F, the Upper Control Limit Factor from the following
    table:

L	2	3	4	5-6	7-9	10-16	17-29	>29
F	6.3	2.9	2.4	2.1	1.9	1.8	1.7	1.65

                            Write the value for F here: F = | |

12. Determine the UCL by multiplying F (from Line 11) by SE
    (from Line 10) and adding this result to M (from Line 8).

    Write the UCL here: UCL = (SE * F) + M = | |

13. Compare UCL (from Line 12) with C1 (from Line 6c). Check the
    appropriate box:

                       --   ----------------
    a. UCL greater than C1: | | Clean room is | OUT OF CONTROL | .
       (UCL > C1)   --   ----------------
                       --   ------------
    b. UCL less than C1: | | Clean room is | IN CONTROL | .
       (UCL < C1)   --   ------------

  E. J. Bawolek 9/13/89   Revision: 1/1/90

DATA SHEET

```
Date:                              Particle Size:        um
Room or Bay Identification:        Air Sample Volume:    ft^3
-----------------------------------------------------------------
|      |          |          | Col 1    | Col 2    | Col 3    || Col 4* |
|Time  | Location |Comments  | Sample 1 | Sample 2 | Sample 3 || Sample |
|      |          |          | Result   | Result   | Result   || Average|
-----------------------------------------------------------------
|      |          |          |          |          |          ||        |
|      |          |          |          |          |          ||        |
-----------------------------------------------------------------
|      |          |          |          |          |          ||        |
|      |          |          |          |          |          ||        |
-----------------------------------------------------------------
|      |          |          |          |          |          ||        |
|      |          |          |          |          |          ||        |
-----------------------------------------------------------------
|      |          |          |          |          |          ||        |
|      |          |          |          |          |          ||        |
-----------------------------------------------------------------
|      |          |          |          |          |          ||        |
|      |          |          |          |          |          ||        |
-----------------------------------------------------------------
|      |          |          |          |          |          ||        |
|      |          |          |          |          |          ||        |
-----------------------------------------------------------------
|      |          |          |          |          |          ||        |
|      |          |          |          |          |          ||        |
-----------------------------------------------------------------
|      |          |          |          |          |          ||        |
|      |          |          |          |          |          ||        |
-----------------------------------------------------------------
|      |          |          |          |          |          ||        |
|      |          |          |          |          |          ||        |
-----------------------------------------------------------------
|      |          |          |          |          |          ||        |
|      |          |          |          |          |          ||        |
-----------------------------------------------------------------
|      |          |          |          |          |          ||        |
|      |          |          |          |          |          ||        |
-----------------------------------------------------------------
|      |          |          |          |          |          ||        |
|      |          |          |          |          |          ||        |
-----------------------------------------------------------------
|      |          |          |          |          |          ||        |
|      |          |          |          |          |          ||        |
-----------------------------------------------------------------
|      |          |          |          |          |          ||        |
|      |          |          |          |          |          ||        |
-----------------------------------------------------------------
|      |          |          |          |          |          ||        |
|      |          |          |          |          |          ||        |
-----------------------------------------------------------------
                           * Col  4 = (COL 1 + Col 2 + Col 3) / 3
```

Appendix 3

Program and Sample Output

```
10 CLS
20 COLOR 2,0
30 INPUT "AIR DUMPED, (cfm)";F5
40 INPUT "CLEANROOM EXHAUST, (cfm)";F3
50 F1=F3+F5
60 PRINT "MAKE-UP AIR, (cfm)";F1
70 INPUT "MAKE-UP PARTICLE DENSITY, (/ft^3)";P1
80 INPUT "PREFILTER EFFICIENCY, (%)";EF1
90 E1=(100-EF1)/100
100 INPUT "CLEANROOM AIRFLOW, (cfm)"; F2
110 IF F2<F1 THEN GOTO 350
120 PM=100*F1/F2
130 PRINT "% MAKEUP AIR";PM
140 INPUT "HEPA FILTER EFFICIENCY, (%)";EF2
150 E2=(100-EF2)/100
160 INPUT "CLEANROOM FLOOR AREA, (ft^2)";FA
170 INPUT "CLEANROOM HEIGHT, (ft)";HT
180 V=FA*HT
190 INPUT "CLEANROOM PARTICLE GENERATION, (/ft^3-min)";G0
200 P41=G0*V*(1-F3/F2)
210 P42=F1*P1*E1*E2
220 P43=(F2-F3)-(F2-F1)*E2
230 P4=(P41+P42)/P43
240 P2P1=(F2-F1)*P4*E2
250 P2P=(P42+P2P1)/F2
260 PRINT "------------------------------------------------------------
270 PRINT "PARTICLE DENSITY AT CEILING, (/ft^3)"; P2P
280 PRINT "PARTICLE DENSITY AT FLOOR,(/ft^3)";P4
290 PRINT "AVERAGE PARTICLE DENSITY IN CLEANROOM, (/ft^3)";(P2P+P4)/2
300 PRINT "AIRCHANGES (/min)";F2/V
310 PRINT "AVERAGE AIR VELOCITY, (fpm)";F2/FA
320 PRINT "% HEPA REQUIRED";((F2/FA)/90)*100
330 PRINT "------------------------------------------------------------
340 GOTO 30
350 PRINT "CLEANROOM AIRFLOW TOO LOW"
360 GOTO 100
```

```
AIR DUMPED, (cfm)? 0
CLEANROOM EXHAUST, (cfm)? 25000
MAKE-UP AIR, (cfm) 25000
MAKE-UP PARTICLE DENSITY, (/ft^3)? 1000000
PREFILTER EFFICIENCY, (%)? 95
CLEANROOM AIRFLOW, (cfm)? 120000
% MAKEUP AIR 20.83333
HEPA FILTER EFFICIENCY, (%)? 99.97
CLEANROOM FLOOR AREA, (ft^2)? 4000
CLEANROOM HEIGHT, (ft)? 10
CLEANROOM PARTICLE GENERATION, (/ft^3-min)? 500
------------------------------------------------------------
PARTICLE DENSITY AT CEILING, (/ft^3) 3.165404
PARTICLE DENSITY AT FLOOR,(/ft^3) 170.6651
AVERAGE PARTICLE DENSITY IN CLEANROOM, (/ft^3) 86.91524
AIRCHANGES (/min) 3
AVERAGE AIR VELOCITY, (fpm) 30
% HEPA REQUIRED 33.33334
------------------------------------------------------------
AIR DUMPED, (cfm)? 45000
CLEANROOM EXHAUST, (cfm)? 25000
MAKE-UP AIR, (cfm) 70000
MAKE-UP PARTICLE DENSITY, (/ft^3)? 1000000
PREFILTER EFFICIENCY, (%)? 95
CLEANROOM AIRFLOW, (cfm)? 360000
% MAKEUP AIR 19.44445
HEPA FILTER EFFICIENCY, (%)? 99.9997
CLEANROOM FLOOR AREA, (ft^2)? 4000
CLEANROOM HEIGHT, (ft)? 10
CLEANROOM PARTICLE GENERATION, (/ft^3-min)? 100
---------------------------------------------------------------
PARTICLE DENSITY AT CEILING, (/ft^3) 2.895483E-02
PARTICLE DENSITY AT FLOOR,(/ft^3) 11.14223
AVERAGE PARTICLE DENSITY IN CLEANROOM, (/ft^3) 5.585591
AIRCHANGES (/min) 9
AVERAGE AIR VELOCITY, (fpm) 90
% HEPA REQUIRED 100
---------------------------------------------------------------
AIR DUMPED, (cfm)?
```

Index

Heterick Memorial Library
Ohio Northern University

DUE	RETURNED		DUE	RETURNED
1. 12.04.95	DEC 1 0 95	13.		
2. 3|9|96	FEB 5 96	14.		
3. 1-19-97	JAN 2 0 1997	15.		
4. 4-22-9	MAR 1 3 199	16.		
5. 7-21-98	JUL 2 9 1998	17.		
6. 8|24|99	SEP 3 0	18.		
7. 1-10-2000	JAN 2 8 2	19.		
8. 11|7|00	NOV 3	20.		
9.		21.		
10.		22.		
11.		23.		
12.		24.